Fisher-Hunters and Neolithic Pastoralists in East Turkana, Kenya

John Webster Barthelme

Cambridge Monographs in African Archaeology 13

BAR International Series 254
1985

B.A.R.

5, Centremead, Osney Mead, Oxford OX2 0ES, England.

B.A.R.-S254, 1985: 'Fisher-Hunters and Neolithic Pastoralists in East Turkana, Kenya'

ISBN 9780860543251 paperback

ISBN 9781407341408 e-book

DOI https://doi.org/10.30861/9780860543251

A catalogue record for this book is available from the British Library

This book is available at www.barpublishing.com

CAMBRIDGE MONOGRAPHS IN AFRICAN ARCHAEOLOGY

General Editor

John Alexander, M.A., Ph.D., F.S.A.

Dept. Archaeology, Downing Street
Cambridge CB2 3DZ, England

Manuscripts should be submitted to
Dr. Alexander in the first instance.

Titles already published:

1. 'The Niger Delta: Aspects of its Prehistoric Economy and Culture' by Nwanna Nzewunwa. BAR-S75. Price £10.00 post free.

2. 'Prehistoric Investigations in the Region of Jenne, Mali' by Susan Keech McIntosh and Roderick J. McIntosh. BAR-S89. Price £20.00 post free.

3. 'Off-Site Archaeology and Human Adaptation in Eastern Africa: An Analysis of Regional Artefact Density in the Amboseli, Southern Kenya' by Robert Foley. BAR-S97. Price £10.00 post free.

4. 'Later Pleistocene Cultural Adaptations in Sudanese Nubia' by Yousif Mukhtar el Amin. BAR-S114. Price £10.00 post free.

5. 'Settlement Patterns in the Iron Age of Zululand: An Ecological Interpretation' by Martin Hall. BAR-S119. Price £10.00 post free.

6. 'The Neolithic Period in the Sudan, c. 6000 - 2500 B.C.' by Abbas S. Mohammed-Ali. BAR-S139. Price £12.00 post free.

7. 'History and Ethnoarchaeology in Eastern Nigeria: A Study of Igbo-Igala relations with special reference to the Anambra Valley' by Philip Adigwe Oguagha and Alex Ikechukwu Okpoko. BAR-S195. Price £15.00 post free.

8. 'Meroitic Settlement in the Central Sudan: An Analysis of Sites in the Nile Valley and the Western Butana' by Khidir Abdelkarim Ahmed. BAR-S197. Price £14.00 post free.

9. 'Economy and Technology in the Late Stone Age of Southern Natal' by Charles Cable. BAR-S201. Price £13.00 post free.

10. 'Frontiers: Southern African Archaeology Today' ed. M. Hall, G. Avery, D.M. Avery, M.L. Wilson and A.J.B. Humphreys. BAR-S207. Price £18.00 post free.

11. 'Archaeology and History in Southern Nigeria: The ancient linear earthworks of Benin and Ishan' by P.J. Darling. BAR-S215. Price £28.00 post free.

12. 'The Latest Stone Age of Southernmost Africa' by Janette Deacon. BAR-S213. Price £20.00 post free.

13. 'Fisher-Hunters and Neolithic Pastoralists in East Turkana, Kenya' by John Webster Barthelme. BAR-S254. Price £18.00 post free.

TABLE OF CONTENTS

CHAPTER I

INTRODUCTION

During the last decade in East Africa, the tempo of archaeological research into late Pleistocene-Holocene time periods has increased markedly. Recent studies have included the investigation of lacustrine based fishing settlements (Robbins 1974; Phillipson 1977b); documentation and dating of the initial appearance of food production (Bower et al. 1977); assessments of cultural diversity within and between food producing and non-fooding cultural traditions (Barthelme 1981; Clark and Brandt 1984); theoretical modeling of human adaptations to highly diverse environmental and ecological settings (Gifford et al. 1980); and linguistic reconstruction of Neolithic and Iron Age prehistoric societies (Ehret and Posnansky 1982).

The current research studies are an outgrowth of the first systematic investigation of the prehistory of East Africa undertaken by L.S.B. Leakey (1931, 1935). Focusing his research on the Navasha-Nakuru Basins in Central Kenya, Leakey excavated a large number of Paleolithic and Neolithic sites. With the assistance of previous geological field studies (Nilsson 1931), Leakey also identified a succession of former high lake levels near the modern lake Nakuru. Based in part on his own stratigraphic interpretations, as well as previous studies by Gregory (1921) in Kenya and Wayland (1929) in Uganda, Leakey believed East Africa had undergone periodic intervals of dramatically increased rainfall (pluvials) during parts of the Pleistocene and Holocene. The interpretation and correlation of his pluvial-interpluvial climatic sequence have since been abandoned, especially the suggested equivalance to the Quaternary alpine glaciations of Europe (Cooke 1958; Flint 1959). Nevertheless, the pioneering investigations by Leakey and others provided the initial set of paleo-environmental observations that have since confirmed pronounced climatic changes during the late Pleistocene-Holocene of North and East Africa (Butzer et al. 1972).

Leakey also constructed, based in part on Wayland's (1923) field studies in Uganda, the first cultural-historical succession of archaeological traditions in Kenya. Many of the classificatory names employed were borrowed directly from the then accepted chronologies of Europe (e.g. Acheulian, Mousterian and Aurignacian) and South Africa (e.g. Wilton and Stillbay). Through the use of such nomenclature, by implication and through intention, Leakey introduced the concepts of migration and diffusion into the study of prehistory in East Africa.

Today, the earlier large-scale migration-diffusion hypotheses of L.S.B. Leakey (1931, 1935), M.D. Leakey (1945), and Cole (1963), especially for early to mid Holocene fishing settlements and subsequent introduction of domestic animals, are undergoing critical reassessment. Several studies suggest, for example, that early Holocene lacustrine technological and economic patterns may have been derived primarily from pre-existing local cultural traditions (Phillipson 1977b:46; Owen et al. 1982:5). Alternate models have also been proposed to account for the presence of domestic animals in East Africa including, indigeneous domestication of local animal species (Onyango- Abuje 1980), small scale migration of populations (Phillipson 1977b:84; Barthelme 1984), and large scale movements of intrusive populations that assimilated local gatherer-hunter groups (Sutton 1975:15; Ehret 1974:8). As Phillipson (1977b) has stressed, an understanding of the Quaternary prehistory of East Africa requires a new synthesis of information on climatic change, ecological adaptation, prehistoric chronology and linguistic and historical evidence, especially if the important concepts of migration and diffusion are to be adequately understood.

The research reported in this monograph attempts to unify early as well as more contemporary approaches to the later prehistory of East Africa. The study both delineates and synthesizes the rich set of late Pleistocene-Holocene archaeological remains discovered along the northeastern margin of Lake Turkana in northern Kenya. Three principal research objectives are addressed: (1) reconstruction of a stratigraphic and chronologic framework for major Holocene lake level changes; (2) documentation of a late Pleistocene-Holocene cultural-historical succession; and (3) reconstruction and interpretation of Holocene land use patterns (Fig.1).

RESEARCH SETTING

Modern Lake Turkana, one of the oldest and largest African lakes, occupies a predominantly semi-arid basin of internal drainage at the northern end of the Kenyan Rift (Yuretich 1979; Owen et al. 1982). The lake lies essentially between the Kenyan and Ethiopian domal uplifts in the middle portion of the East African Rift Valley System (Baker et al. 1972; King 1978). Main fault zones on the Eastern Rift pass through the southern end of the lake and link up the two major areas of domal uplift (Fitch and Vondra 1976).

Lake Turkana's waters, which are moderately saline and alkaline, are primarily derived from the Ethiopian Highlands, through the Omo River (Butzer 1971). Like most non-outlet lakes, Turkana has been affected by climatically-induced fluctuations in level and hydro- chemistry of variable magnitude, frequency and duration. Vegetation is predominately semi-desert with infrequent riverine gallery forests. Rainfall averages less than 10 inches yearly (Yuretich 1979).

The sedimentary history of the Turkana Basin is complex. During the Miocene, and perhaps as early as the late Mesozoic, coarse sediments accumulated within the Turkana depression (Savage and Williamson 1978). The most extensive series of sediments range in age from Pliocene to Holocene and outcrop along the northeastern sector of the lake basin. They consist of 325 m of fluvial, deltaic and lacustrine units which were deposited disconformably on Upper Miocene and Pliocene volcanics (Vondra and Bowen 1978). Holocene lacustrine sediments lie around the margins of the lake and occur as scattered outcrops over much of the Plio-Pleistocene sediments. To the northeast of the lake near Koobi Fora, Holocene deposits have been informally termed the Galana Boi beds (Vondra et al. 1971). Artifacts,

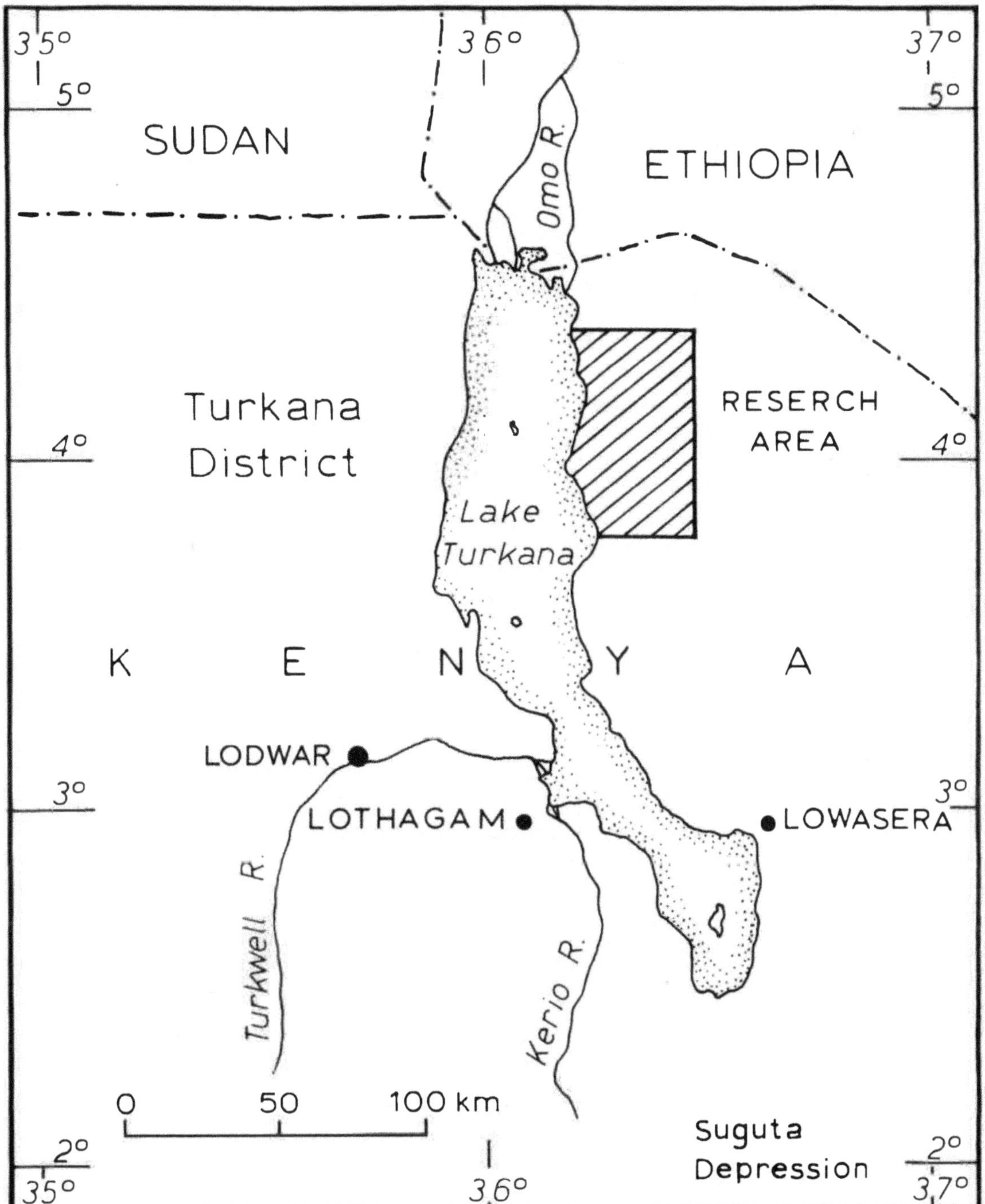

Figure 1. Lake Turkana Basin: study area indicated by hatching.

including microliths, bone harpoons, pottery and abundant faunal refuse are conspicuous features of the Galana Boi deposits.

HOLOCENE LAKE LEVEL FLUCTUATIONS

The African continent has undergone considerable climatic change during the past 20,000 years. As Livingstone (1975) notes, glaciers, lakes, aggrading and degrading rivers, and active dunes have all been more widespread during some part of that time than they are today. An increasing number of radiocarbon dates indicate that, beginning about 12,000 years ago, lake levels in many parts of Africa were higher than they are today (Butzer et al. 1972). Regional studies, including those from North Africa (Faure 1966; Servant, Servant and Delibrias 1969; Maley 1977) and Ethiopia (Grove et al. 1975; Street and Grove 1979), have identified a number of significant lake level fluctuations throughout the Holocene.

Within Eastern Africa, a substantial body of paleo-environmental, stratigraphic and chronologic data also documents major fluctuations in lake levels. Studies of the Holocene sediments in the Magadi Basin (Baker 1958; Butzer et al. 1972), the Naivash-Nakuru Basins (Nilsson 1931, 1940; Leakey 1931; Richardson and Richardson 1972; and Wasbourn-Kamau 1971, 1975), at Lake Victoria (Kendall 1969), at lakes Bogoria and Baringo (Young and Renaut 1979) and in the Saguta Basin (Dodson 1963; Truckle 1976) demonstrate the existence of higher lake strandlines or terraces.

At Lake Turkana, early observations on high lake stands were made by Fuchs (1939) and Arambourg (1943:206). Holocene sediments on the southwestern flank of the lake have been investigated (Walsh and Dodson 1969; Robbins 1972, 1974) and have been informally named the Kabua beds (Whitworth 1965a). On the southeastern side of the lake, Phillipson (1977a) has described Holocene sediments from the archaeological site of Lowasera while preliminary observations of equivalent sediments within the Galania Boi beds to the northeast have been conducted by Bowen and Vondra (1973), Raynolds (1973) and Barthelme (1977). Detailed investigation of the Galana Boi beds has recently been undertaken by Owen (1982 et al.) and Renaut (1980). Further studies of lake level fluctuations were made by Butzer (1971, 1976, 1980) to the north in the Lower Omo Basin. The Holocene deposits there have been formally designated as belonging to Members IVa and IVb of the Kibish Formation (Butzer, Brown and Thurber 1969). Butzer's extensive research has identified three principal high lake stands dated between 10,000 - 7,000 BP, 6,500 - 4,000 BP and roughly 3,250 BP. Periods of water level comparable or lower than the modern lake level were reported at 6,600± 150 BP and after 2,500 BP. Multiple aneriod readings, taken during helicopter traverses, indicated that Lake Turkana reached a maximum transgressive height of 75 meters (Butzer 1980).

The research project's first objective, the reconstruction of a stratigraphic and chronologic framework in order to determine the number and magnitude of Holocene lake level fluctuations, was initiated in order to verify and possibly refine Butzer's findings as well as provide a much needed stratigraphic framework for the abundant archaeological sites. As a result, an intensive geo-archaeological investigation of the Galana Boi beds was carried out. Three separate but complimentary studies were undertaken. First, 26 geological and archaeological trenches were dug along a continuous 4 km exposure of Holocene sediments near Koobi Fora. All lithological units were measured with surveying instruments and levelled correlated sections were drawn up. Analysis of the sections led to the identification of three major transgressive-regressive cycles. Second, a surveying aneroid was used to establish the elevation of each lake transgression as well as the elevation of numerous archaeological sites. Third, a series of 28 radiocarbon determinations helped erect a chronological framework for the three transgressive-regressive cycles and dated the occupation of many of the archaeological settlements.

CULTURAL-HISTORICAL SUCCESSION

The second major research objective focused on the reconstruction for the northeastern sector of the Turkana Basin of a late Pleistocene-Holocene cultural-historical succession. Site survey transects were utilized to locate archaeological sites on various topographic features both within and outside the sedimentary basin. It proved possible to identify a number of successive cultural traditions including those of late Pleistocene-Holocene hunter-gatherers, an early to middle Holocene lacustrine based fishing phase, mid-Holocene Pastoral Neolithic settlements and occupation of the area by recent to sub-recent pastoralist groups.

Fishing Settlements

Particular attention was centered on the numerous fishing sites. Associated with the well documented rise in many lake and river elevations during parts of the Holocene in Africa are widespread traces of archaeological settlements whose economic basis depended on the rich fishing resources these waters at one time offered (Sutton 1977). These settlements are characterized by microlithic stone tool assemblages, barbed bone harpoon heads, and food refuse consisting chiefly of fish and non-domestic animal remains. Decorated and/or undecorated pottery is usually present. The sites are located on or near the shores of former lakes and rivers. The majority of the fishing settlements date to the early and middle part of the Holocene, between 9500-4000 BP (Phillipson 1977b:45-49).

The first detailed description of this primarily lacustrine economic adaptation was made by Arkell (1949) from the site of Early Khartoum in the Sudan. Barbed bone harpoons and a characteristic type of pottery decorated with wavy lines were recovered from excavations. Since the excavation of Early Khartoum, numerous fishing settlements have been discovered in the Sahara and Sahel regions of North Africa (Monod and Mauny 1957; Huard and Massip 1964; Camps-Fabrer 1966; Petit-Maire et al. 1983; Clark 1971; Smith 1974) and along the flanks of the White Nile in the Sudan (Adamson, Clark and Williams 1974) and

Egypt (Wendt 1966; Wendorf, Said and Schild 1970). A single site, located in Zaire, on the northwestern shore of Lake Edward, has been reported from Central Africa (de Heinzelin 1957, 1962). Within East Africa, from the Nakuru Basin, a single sherd of possible wavy line pottery and a barbed bone harpoon were recovered associated with early Holocene beach deposits at Gamble's Cave (L.S.B. Leakey 1931: Plate XII; Oakley 1961).

By far, the most extensive study of fishing sites in East Africa has been made along the Holocene shorelines of Lake Turkana. Artifacts have been recovered from surface collections and excavations located along the southwest portion of the lake (Whitworth 1965b; Robbins 1972, 1974, 1979), to the southeast (Phillipson 1977a), and to the far north, within the Lower Omo Basin (Arambourg et al. 1943; Butzer, Brown and Thurber 1969; F.H. Brown 1975; and Butzer 1980). Previous excavations had mainly concentrated on single, large, multi-component sites (Robbins 1974; Phillipson 1977a). A regional approach, focusing on the temporal-spacial reconstruction of technological, economic and settlement patterns from several sites had not been attempted. Therefore, I decided to initiate a study that centered on the sampling and excavation of a large number of Holocene fishing settlements. Such a research design would, it was hoped, permit a diachronic interpretation of the data within a regional perspective. Differences and similarities in the artifact assemblages were analyzed for each site, between sites, and through time. The information also had, of course, broader implications and comparisons were made with other fishing sites inside and outside of the Turkana Basin.

Pastoral Neolithic[1]

The beginning of food production in Africa has long been a source of investigation, speculation and, not infrequently, of spirited controversy (Monod 1975; Harlan, de Wet and Stemler 1976; Clark and Brandt 1984). A key question has focused on the center of origin, whether indigenous or introduced, of cultivated crops and domestic animals. With the exception of wheat and barley, believed to have diffused from southwestern Asia (Trigger 1968), an indigenous origin for most African crops seems likely (Vavilov 1926; Harlan 1971). Recent work in Upper Egypt, along the Nile Valley, has shown, however, that barley too may have been domesticated on the African continent (Wendorf et al. 1979). A similar debate concerns whether cattle and ovicaprids were first domesticated in northern Africa (Mori 1964) or were originally derived from southwestern Asia (Epstein 1969).

The earliest evidence for animal domestication in Africa comes from Egypt and the central Sahara (Wendorf et al. 1976; Clark 1976) and dates to between the 8th and 7th millennium BP. In the Sudan, early domestic ovicaprids have been identified at Esh Shaheinab (Arkell

---

[1] I have adopted the use of this classificatory term as defined by Bower et al. 1977:119, footnote 2.

1953: 15-18) and recent investigations at Kadero 30 kms to the south-east have indicated the presence of domestic stock (primarily cattle) at 5500±70 BP (Krzyzaniak 1975, 1978, 1982).

Within East Africa, extensive studies of early pastoralist cultures have been conducted in the Naivasha-Nakuru Basins (M.D. Leakey 1945; M.D. Leakey and L.S.B. Leakey 1950; J. Brown 1966; Bower et al. 1977; Bower and Nelson 1978; Onyango-Abuje 1977, 1980) while other areas of western (Chapman 1967), central (Odner 1972; Sutton 1973; Siirianinen 1977), and southern Kenya (Gramly 1976) and northern Tanzania (Bower 1973, 1977), have also been investigated. A large suite of radiocarbon dates place domestic animals in central Kenya and northern Tanzania between the first millenniums BC and AD (Abrose 1984).

While isolated stone bowls, suggesting the presence of former pastoralist traditions, had been collected in the Koobi Fora area, the discovery of three large Pastoral Neolithic habitation sites came as an unanticipated and welcome surprise. Evidence for domestic stock in the Turkana Basin and surrounding areas had previously been limited to Robbin's work (1972, 1979) and excavations at North Horr, 150 kms east of the lake (Phillipson 1977b; Stiles 1982). The importance of the East Turkana sites as key segments in the regional cultural-historical succession and as possible geographical and chronological links with early Pastoral Neolithic settlements to the north and south required that they be thoroughly investigated. The sites were excavated therefore with several objectives in mind including the reconstruction of their paleo-geographic setting, determination of the site's principal technological and economic features and the radiocarbon dating of each occupation. Comparative studies were conducted between each of the three pastoralist occupations as well as between Pastoral Neolithic sites in the Sudan and in central Kenya. (Fig.2)

LAND USE PATTERNS

The study of land use patterns has emerged as an important objective in the reconstruction of prehistoric lifeways (Vita-Finzi and Higgs 1970; Jochim 1976; Bower et al. 1977). As Isaac (1981) notes, the determination of land use patterns can be viewed as inferences that are drawn about the spatial configuration of daily life and about aspects of the use of the landscape by past human societies.

The third major objective addressed in this monograph focused on the reconstruction of land use patterns for both the Holocene fishing and Pastoral Neolithic settlements. The study concentrated primarily on the identification of site distribution patterns. Six survey transects were drawn up in order to systematically sample a series of different paleo-topographic features. Data were collected on both clusters (sites) and scatters (localities) of artifacts. The site distribution survey helped elucidate, as a first step, regional economic, settlement, and ecological relationships for each of the cultural traditions under study. Particular hypotheses were generated that can be tested in future field studies.

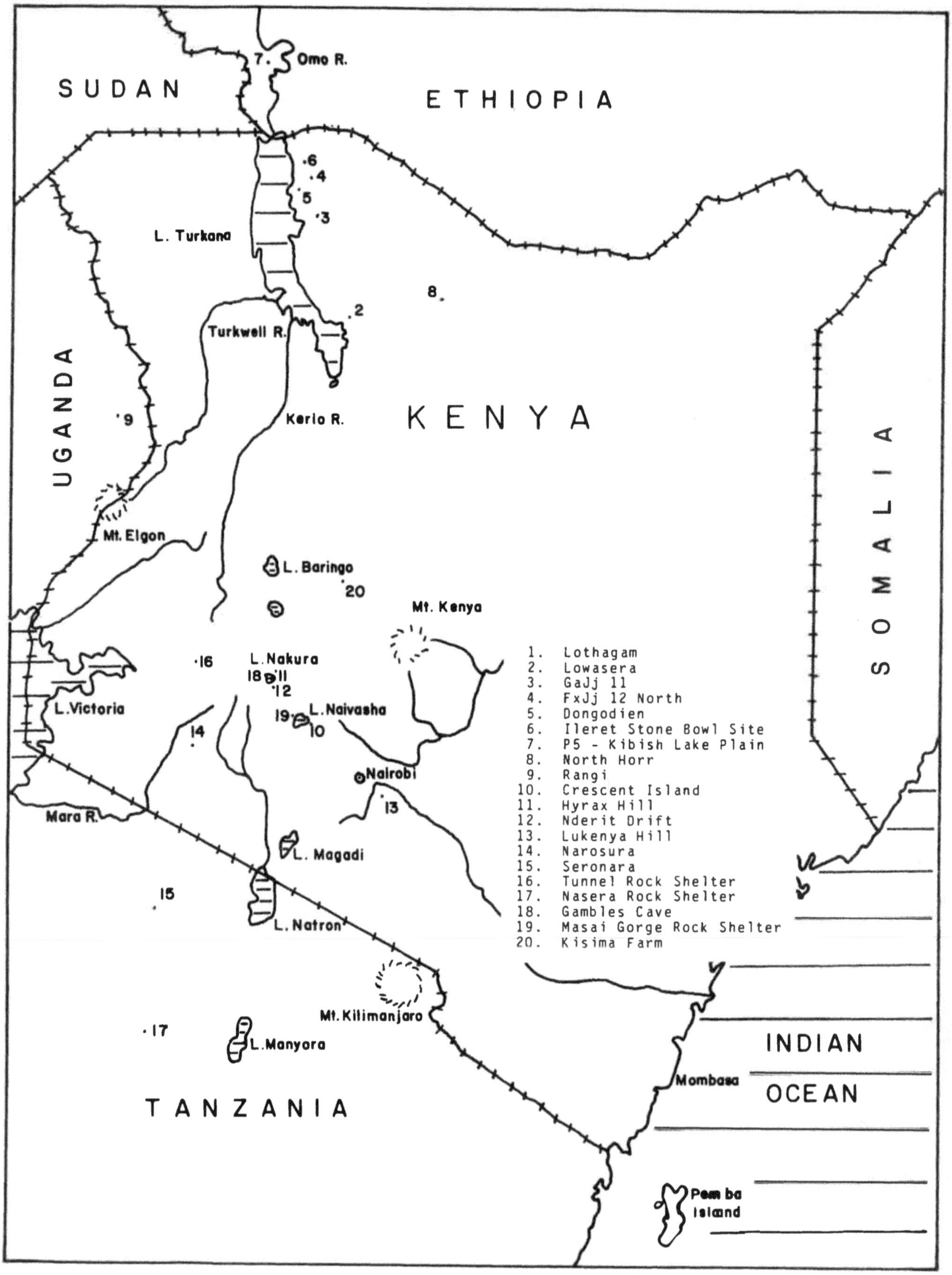

Figure 2. Distribution of selected East African late Pleistocene-Holocene archaeological sites.

FIELDWORK

The area to the northeast of Lake Turkana is topographically rugged, extremely hot and has few established tracks. Most of my area of research, over 2200 square km, was archaeologically unexplored. The initial phase of the research project, field survey, began during the summer of 1975. The survey's objectives included determination of the general character and distribution of Holocene sediments as well as the identification of promising sites for future excavation. With the assistance of three members from the Koobi Fora Research Project's archaeological field crew, a six week exploratory survey of the Galana Boi beds was started. Previously identified Holocene exposures in Areas 102, 103 and 117 (Raynolds 1973; Bowen 1974; Isaac pers. comm. and my own observations) were surveyed first. Thereafter, fly camp safaris prospected the Holocene exposures to the north near Ileret, to the south, inland from Allia Bay, and into the eastern volcanic highlands. During the survey, exposures of Galana Boi deposits, archaeological sites and localities were recorded on aerial photographs (scale 1:23,000). The base camp at Koobi Fora served as the project's main supply center.

After returning to Nairobi for additional supplies and crew members, the second phase of fieldwork was initiated in October. Assisted by a crew of six workmen and a cook, excavation of selected sites commenced. Nearly 2½ months were spent excavating sites in Areas 10, 102 and at Nderati Wells. Because of logistical considerations, including the transportation of food, water and petrol, fly camp based excavations tended to be smaller and of shorter duration (3-10 days) than excavations located within a 6-10 km radius of the Koobi Fora base camp.

The third phase of fieldwork was conducted between April and June 1976. With the help of four workmen, an American volunteer and a cook, excavation of sites resumed. Both the site survey transects and the extensive reconstruction of the Holocene stratigraphy in Area 102 were completed during this phase of fieldwork. The entire collection of surface and excavated materials was deposited with the National Museum in Kenya.

Excavation Procedures

When a site was chosen for excavation, it was assigned a collection area number (Fig. 3) and SASES (Nelson 1971) coordinates (Fig. 4). Artifacts on the surface were then selectively plotted and collected or systematically sampled. A particular area of the site was chosen for excavation and a surface grid was laid. Overburden, when present, usually consisted of reworked unconsolidated colluvial-alluvial sands. This material was occasionally removed with pick and shovels and was always sieved. During excavation, natural stratigraphic boundaries were followed. Whenever a distinct sedimentary unit of considerable thickness was encountered, artificial spits were utilized. These spits ranged in thickness from 20 cm to 5 cm, with the 5 cm spits used extensively during excavation of artifact clusters, features and for distributional mapping. Sediment from the first excavated

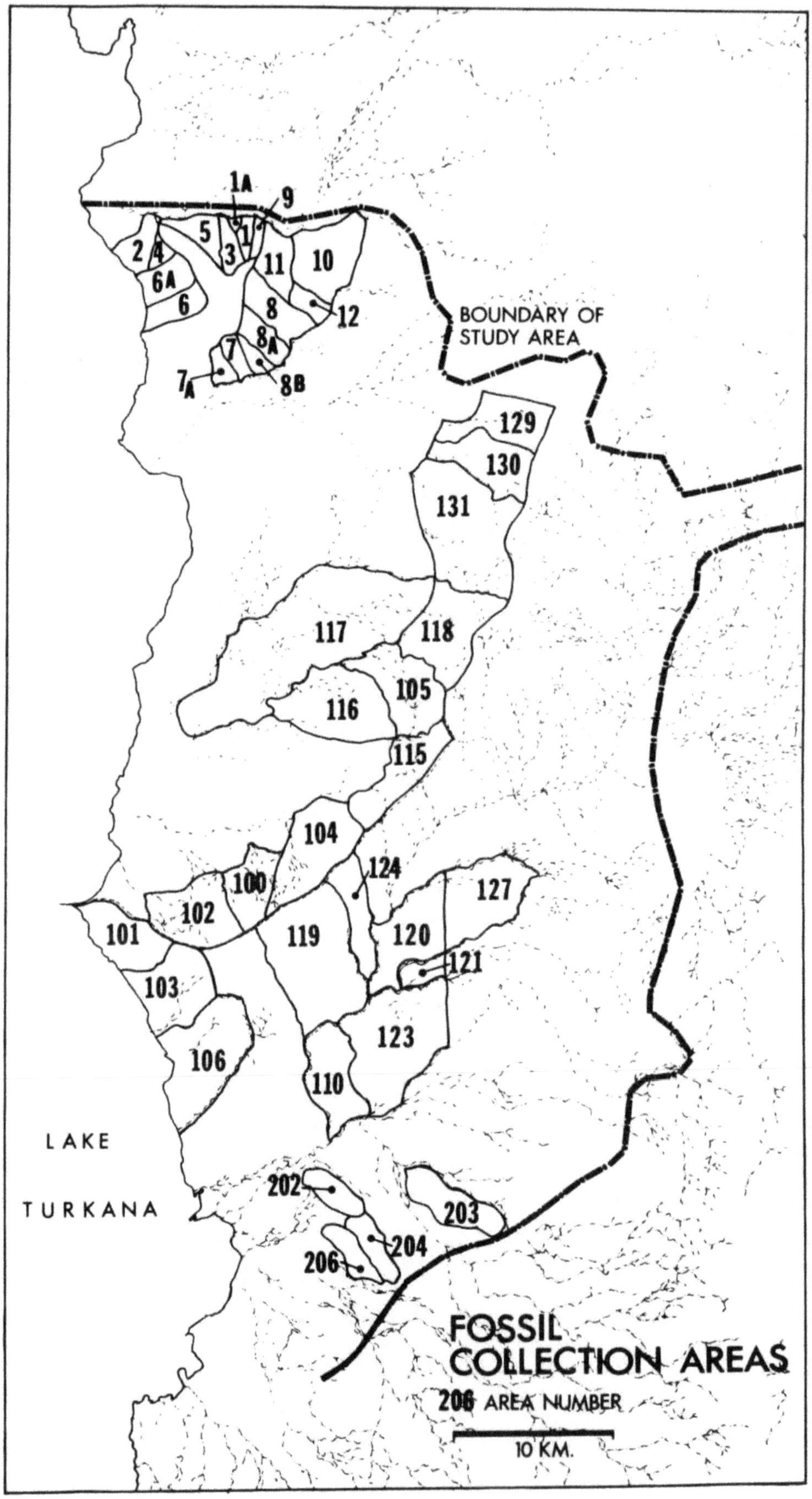

Figure 3. Location of major paleontological collection areas at East Lake Turkana.

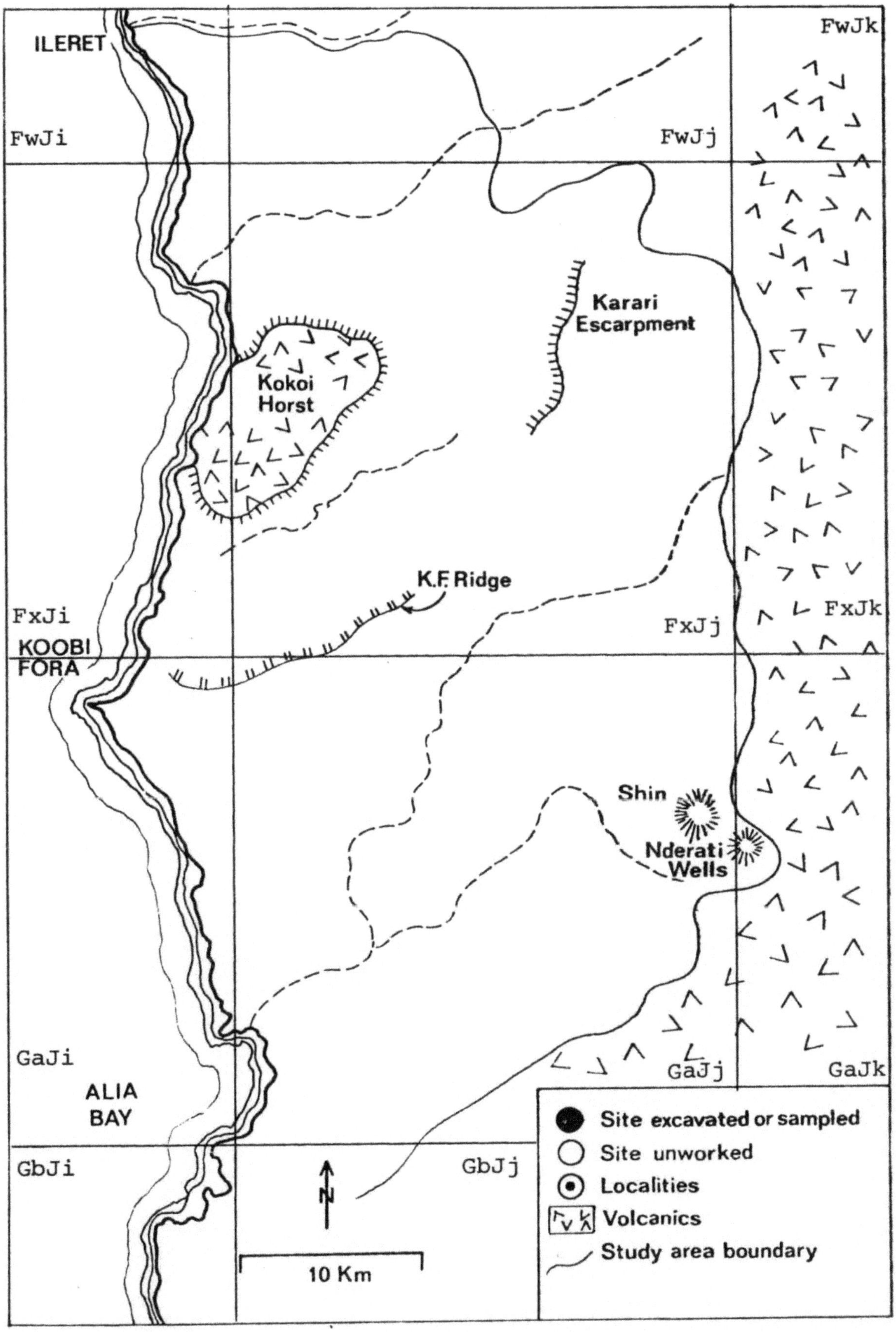

Figure 4. East Turkana study area. Location of grid squares used to catalogue sites (GaJi, GaJj, etc.) are indicated.

site, GaJj 1, was sieved through an 8 mm wire mesh. This gauge proved too large, however, to recover ostrich eggshell beads, small bone fragments, and fish teeth. Thereafter, all excavated sediment was passed through a 4 mm mesh sieve.

Sampling

The systematic location of archaelogical sites and the collection of representative samples of artifacts from each site comprised an important part of the project's research design. Large scale excavations of single sites, similar to those at Lothagam (Robbins 1974) and Lowasera (Phillipson 1977a) were not undertaken. Rather, field investigation centered on the sampling of surface artifacts and the excavation of test trenches from a number of sites with similar technological and economic features. Sampling was viewed as an important technique for examining part of the available artifactual material in order to derive statements about the whole body of material (Redman 1974).

Two different sampling strategies were employed. First, with the aid of survey transects, archaeological sites were located in relation to specific paleogeographic features. The second approach centered on the intensive sampling of surface artifacts from 13 sites. A major objective of this strategy was the systematic sampling of at least 5% of a site's surface area and the recovery of a representative collection of cultural materials.

After a site had been selected for sampling, the site's dimensions were detemined by surveying instruments. Depending on a series of variables including topographic relief, size of the occupation area, artifact density, and time available, an appropriate sampling strategy (e.g., simple random, stratified randon, linear transect, systematic checkerboard, etc.) was chosen (Redman 1974). A grid system was then laid out with surveying instruments or compass and workmen were assigned to collect each sampling unit. All units consisted of 5 m squares and, unless noted in a specific site report, all artifacts within each unit were collected. Cultural remains were tabulated by sampling unit and, in several cases, the bulk of the stone artifacts were returned to the site. The great majority of the cultural materials however, were analyzed later.

SITE GROUPS

When sampled or excavated sites demonstrated markedly similar patterns of technology, economy and settlement arrangement, they were combined to form site groups. Such groupings permitted a more thorough discussion of the differences and similarities between cultural phases at East Turkana. As well, it enabled comparative discussion of sites and cultural traditions outside of the Basin. Four site groups were identified: (1) fishing settlements; (2) Pastoral Neolithic sites; (3) other Later Stone Age sites with pottery and (4) aceramic Later Stone Age sites without barbed bone harpoons. A fifth series of sites, consisting of relatively recent pastoral encampments and burial cairns, was identified but not investigated thoroughly.

Chapter II

GEOLOGY, STRATIGRAPHY AND CHRONOLOGY OF THE GALANA BOI BEDS

GEOLOGY

The Galana Boi beds at Lake Turkana consist of predominantly lacustrine and marginal lacustrine diatomaceous silts and fine to coarse sands (Owen et al. 1982). The deposits rest uncomformably and occur as a discontinuous mantle over the widespread Plio-Pleistocene sediments. Where exposed the beds are usually less than 10 m thick but they reach an aggregate thickness of 32 m in Area 103 (Raynolds 1973). The beds lie up to 80 m above the modern lake and, except locally near the Kokoi horst where they have been uplifted, they are unaffected by tectonics. Over the last 10,000 years, Lake Turkana's flucuating lake elevations have contributed to the formation of lenticular and interdigitating sedimentary units which commonly exhibit vertical and lateral facies variations over short distances. Thick beds of lacustrine silts have infilled paleovalleys while littoral sandy units crop out as discontinuous bodies that are parallel to former Holocene shorelines. Though the Galana Boi sediments are usually horizontally bedded, primary depositional dips of up to 15° have occurred where the deposits are draped over topographic highs.

Recent investigations by Owen (1981; 1982 et al.) have helped define and interpret the intricate lithofacies of the Galana Boi beds. His work has shown that nearshore, moderate to high energy environments such as beach bars and sandy littoral shelves, are represented by fine to coarse grained heterogeneous sands that are typically well sorted, show local heavy mineral laminae and usually form linear bodies less than 2 m in thickness. Molluscs, as well as fish bone, are common in such nearshore sediments and include *Mutela emini*, *Etheria elliptica* and *Pila ovata*. Delta plain and distal alluvial environments are associated with various sublithofacies that are moderately to poorly sorted. Stratification of these sediments is generally poor and the units are usually less than 2 m in thickness. Molluscs, calcium carbonate concretions, root-marked paleosols, fish and other vertebrates are locally present. Low energy lacustrine environments are represented by clay, diatomaceous silt and fine sand lithofacies. These sediments are usually well laminated and exhibit a maximum thickness of approximately 20 m. Molluscs are abundant and are dominated by *Melanoides tuberculata*, *Corbicula africana*, *Cleopatra pirotrii* and *Bellamya unicolor*. Fishbone, sponge spicules, ostracods and pollen are also well represented.

Owen's research has also identified three important biogenic lithofacies. In high energy environments, *Etheria elliptica* may form reefs up to 4 m thick and, based on my archaeological work in Area 127, up to 8 kms long. Within the second biogenic lithofacies, *Melanoides tuberculata* and *Corbicula africana* were identified as the predominate molluscan species in silty sand units less than 10 cm thick. Shell wear indicated that both _in situ_ and reworked assemblages were present. Thirdly, the presence of former lagoons and quiet water embayments was recorded from several outcrops of impure diatomite.

STRATIGRAPHY AND CHRONOLOGY

The reconstruction of Holocene lake level fluctuations required precise information on the stratigraphy, altimetry and chronology of the Galana Boi beds. Major research efforts were undertaken to establish the sequence of rock (Barthelme 1981) and diatom (Owen et al. 1982) stratigraphic units in Areas 102 and 103. As well, a large suite of radiocarbon dates and altimeter measurements was complied from selected geological and archaeological deposits.

Geological Transect

A thick succession of Galana Boi sediments occurred in Area 102 banked against the northern flank of the Koobi Fora Ridge. The deposits consisted of lacustrine, fluvial and colluvial units and revealed a complex pattern of lake level flucuations. In order to reconstruct in greater detail the number and magnitude of lake level changes, three weeks were spent digging twenty-six geological trenches along a 4 km section of the Holocene exposures. Each lithological contact was mapped and leveled with surveying instruments. A leveled topographic profile based on a fixed datum was drawn up and, where possible, individual lithologies were associated. Correlated stratigraphic sections and diagrammatic relationships of the geological transect are presented in Figures 5 and 6.

The lithological units in trenches A-F and trench 18 revealed a substantial sequence of alternating layers of consolidated and unconsolidated diatomaceous silts. Horizontal bands of calcium carbonate concretions with limonite staining occurred within the massive unconsolidated silts. Fine layers of horizontally bedded diatomite were present within the silty horizons and indicated periods of low energy, stable lacustrine environments of deposition. The sequence of diatomaceous silts thinned to the east, where it interfingered with coarse beach sands of the 75-80 m lake transgression and the fishing settlement of GaJj 1. Three layers of coarse gravelly sands and reworked colluvial-fluvial sediments overlay the silt units. Sediments in trenches 10-11 and 16-17 revealed several units of fine and coarse gravelly beach sands. These units were poorly exposed to the east of trench 17. The Holocene deposits in trenches 1-9 and 12-15 showed a complex succession of sands, silty clays, fine silts and coarse sands that were underlain by Plio-Pleistocene indurated sandstones. The units displayed an alternating series of coarse and fine grained deposits that coarsened upward in the section. Two Pastoral Neolithic sites occurred within layers of fine to medium coarse littoral sands.

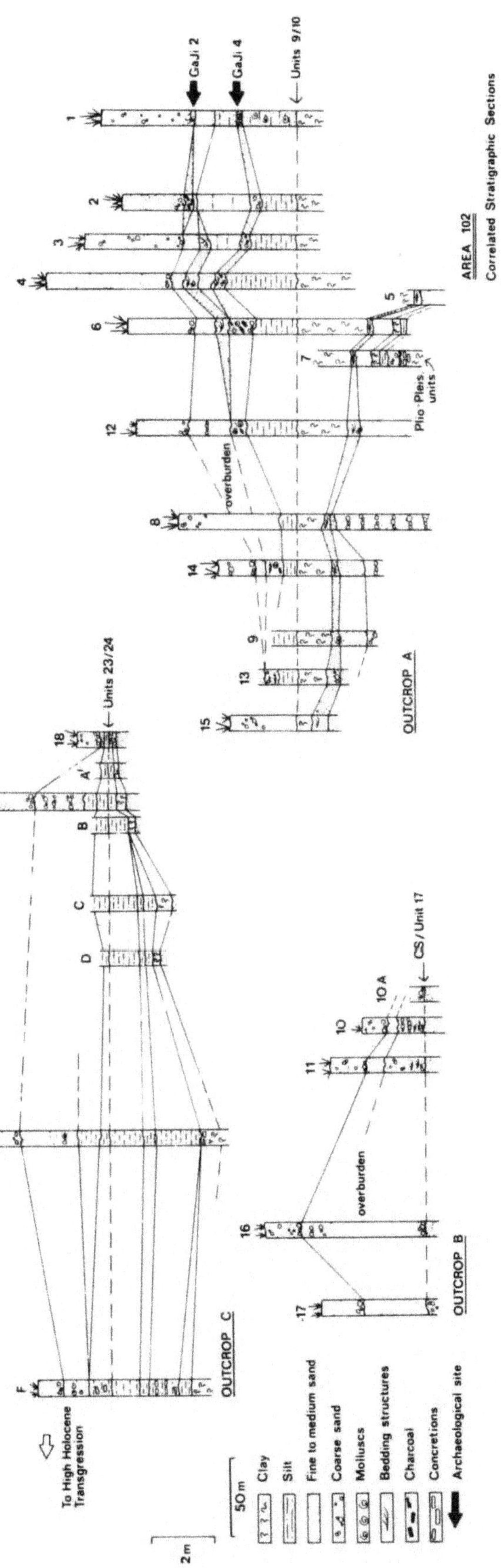

Figure 5. Geological transect in Area 102: Correlated Sections.

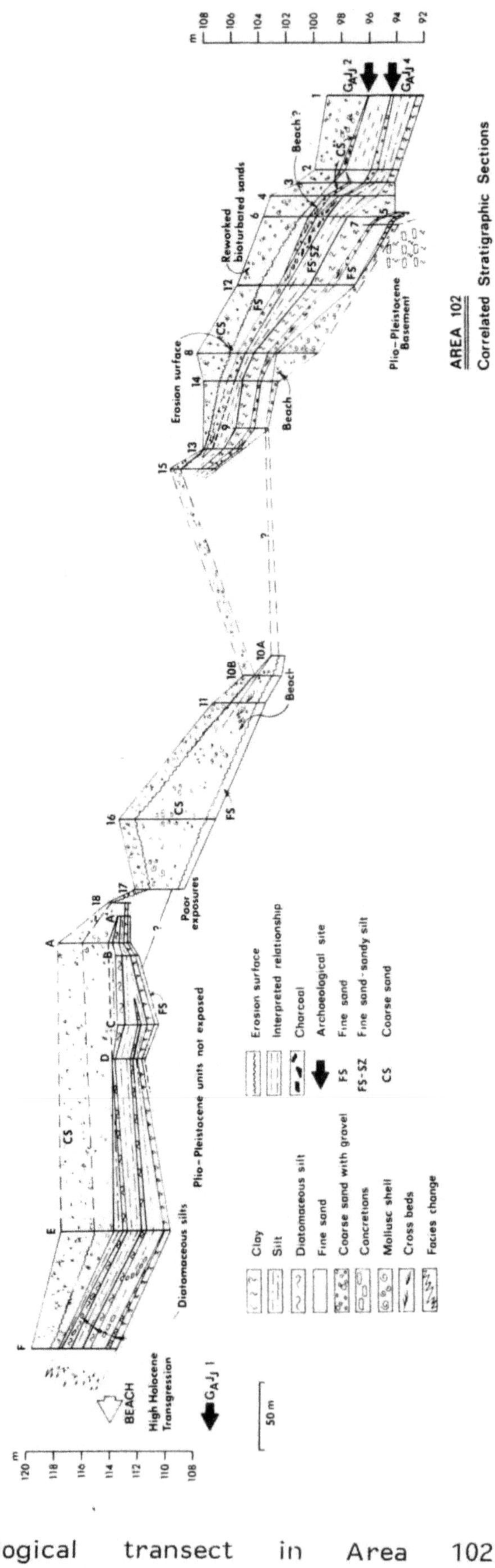

Figure 6. Geological transect in Area 102: Interpretive Reconstruction.

Altimeter Measurements

With the exception of the Survey of Kenya's Series Y503 Edition 1-SK topographical maps (contour lines at 1000 meter intervals) and preliminary map making efforts by members of the Koobi Fora Research Project, the area to the northeast of Lake Turkana lacked detailed topographic maps with closely defined contour intervals. The absence of such maps made it difficult to determine the former flucuating heights of the Holocene lake and to compare the results with other lake level measurements in the Basin (Butzer et al. 1972; Butzer 1980; Phillipson 1977a). Fortunately, a surveying aneriod was kindly loaned by the British Institute in Eastern Africa and used in the field from September 1975 to June 1976. The instrument provided a crucial series of elevation measurements on the heights of several transgressive lake phases as well as elevations of selected archaeological sites.

Twenty-nine altimeter readings are represented in Table 1 and Figure 7. When altimeter measurements were taken within a morning's driving distance from Koobi Fora, the instrument was set at 375 m (modern elevation of Lake Turkana; Butzer 1971) beside the shoreline of the present lake betweeen 6:30-7:00 a.m. This reading was subsequently subtracted from a measurement recorded on the exposures. For example,

Galana Boi outcrop	450 M
Modern shoreline	375 M
Difference in elevation	75 M

In order to check single readings, additional measurements were usually taken the following morning. When altimeter readings were recorded from Galana Boi outcrops at a greater distance from the lake and from up-country work camps, a different set of procedures was followed. The altimeter was set at 375 m in the early morning at Koobi Fora and the elevation difference was not read until the same time on the following morning. In order to check the first reading, during the subsequent morning, the aneriod was re-set at 375 m and the elevation difference was once again recorded. The two separate readings were invariable within 1-3 meters of each other.

Radiocarbon Dates

Table 2 presents 29 radiocarbon determinations from the late Pleistocene-Holocene deposits at East Turkana. Twenty-four dates are associated with the archaeological excavations and Galạna Boi stratigraphy reported in this monograph. For consistency with previous research, all shell dates have been recalibrated 400 years younger (see Butzer et al. 1969). The set of dates received from Geochron and Radiocarbon Ltd. laboratories has been corrected for C-13 fractionation (Table 3). Bone apatite dates were run on fish (F), mammal (M), and human (H) bone and are designated accordingly. No determinations have been recalibrated following the dendrochronologic time scale (Damon et al. 1974).

Table 1

ALTIMETER MEASUREMENTS

Sites	Area	Readings Above Level of Lake Turkana: in Meters September 1975-June 1976
Fishing		
GaJi 3	102	55-56
GaJj 1	102	81, 78-80, 76-77
GaJj 2	100	78
GaJj 11	127	75, 73-75, 73-75
GaJj 12	127	75, 74-76
GaJj 16	---	79, 77-78
FxJj 12	116	95
FxJj 12 North	117	100
FxJj 108	North of Area 117	140, 135
Pastoral Neolithic		
GaJi 2	102	45-46, 45, 43-45
GaJi 4	102	46-47, 44-45
Other		
FxJj 102	129	175
GaJj 9	103	61
Areas		
Six Mile Airstrip		86-87, 86
Wave Cut Beach Terraces:	119	
Upper Terrace		82, 80, 79, 77-78
Middle Terrace		72, 72
Lower Terrace		55

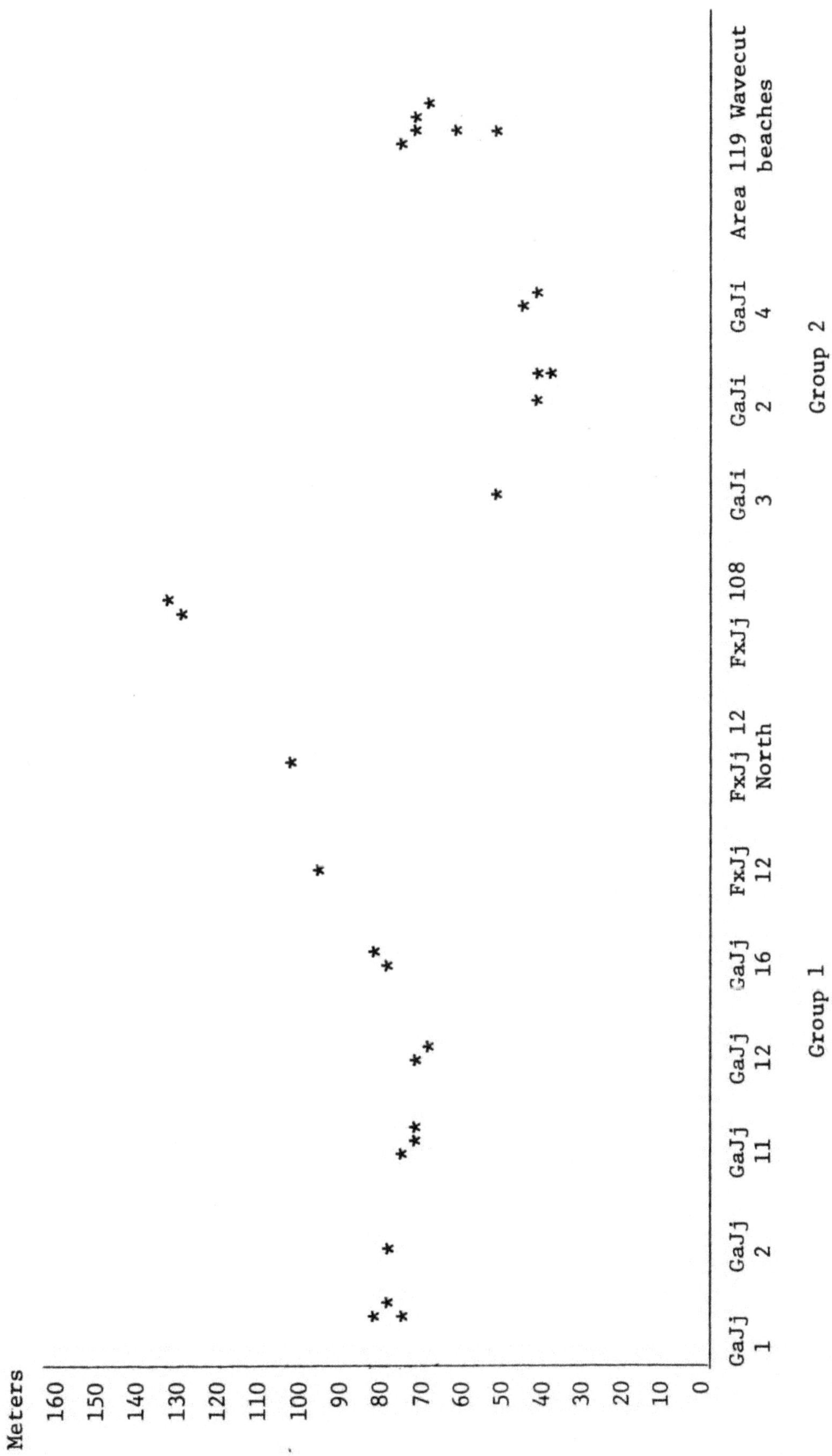

Figure 7. Plot of altimeter measurements for selected archaeological sites. Measurements record elevation above water level of Lake Turkana from September 1975 to June 1976.

Research investigations into the geology, stratigraphy and chronology of the Galana Boi beds have made it possible to identify three major high lake level transgressive-regressive phases. It should be kept in mind that such phases do not represent continuously stable lake levels, but rather indicate periods of varying duration under which Lake Turkana's water level flucuated between +50 and +80 m (Owen et al. 1982).

During the early Holocene the first transgressive-regressive phase is well represented by the thick sequence of diatomaceous silts in Area 102. The silts thicken to the south and east where they are laterally equivalent to the +75 to +80 m beach deposits at GaJj 1. In Area 103, the silty units from the base of the section have been dated to 9880±670 BP (Raynolds 1973). Other radiocarbon determinations on molluscs from littoral sands south of the Kokoi highlands and near Ileret, have yielded dates of 9540±260 and 9360±135 BP. In Area 127, dates of 8710±130 and 8520±130 BP on fish bone and mollusc shell from +73 to +75 beach sands have also been obtained. The radiocarbon dates from phase 1 at Lake Turkana indicate that early Holocene high lake levels are in close agreement with other East African lakes (Butzer et al. 1972).

Butzer's (1980) research in the Omo has indicated that, following the early Holocene high transgression, Lake Turkana may have fallen to modern levels about 6600±150 BP. While studies of the diatom stratigraphy indicate lake levels had receeded at this time, data are inconclusive whether Lake Turkana experienced a drop of such magnitude (Owen et al. 1982). This regression was succeeded by flucuating higher lake levels of phase 2.

The second transgressive-regressive cycle is well documented from Areas 102 and 103. Dates of 5060±245 (+70 m) and 4540±230 (+67) BP have been recorded from littoral sands. Figure 5 reveals a succession of coarse cross-stratified sands from the archaeological sites of GaJi 2 and Dongodien that have been dated to 4100±125 (+47 m) and 3960±60 (+45 m) BP. While the early middle Holocene phase 2 period was represented by flucuating lake levels of up +70 m, subsequent stages were characterized by lower transgressive heights at between +55 and +50 m.

The third phase of expanded lake levels is recorded by 6 m of white diatomaceous silts located five kilometers southeast of Koobi Fora. The base of this sequence lies roughly +35-40 m and is interbedded with a lense of white volcanic tuff. Over the last two millennia, a widespread and poorly defined series of beach ridges, regressive littoral and colluvial sands indicate that Lake Turkana continues to experience flucuating lake levels. While phase 3 has yet to be radiometrically dated, it appears to be of similar age to late Holocene deposits in the Omo region which date to 3250 BP (Butzer et al. 1969).

TABLE 2

RADIOCARBON DATES

Site	Lithology/ environment of dated unit	Material dated	Laboratory	Laboratory date no.	Date (yr BP)
Fishing settlements					
GaJi 3	Beach sands	Bone apatite (M)	Geochron	Gx-5475-A	4,560±185
GaJj 11	*Etheria* reef	*Etheria* shell	Geochron	Gx-5477-A	Modern
GaJj 11	*Etheria* reef	*Etheria* shell	Helsinki	Hel-1275	Modern
GaJj 11	Sand bar	Shell	Helsinki	Hel-1276	8,520±130
GaJj 11	Sand bar	*Etheria* shell	Helsinki	Hel-1277	8,710±130
GaJj 11	Sand bar	Bone apatite (F)	Geochron	Gx-5476-A	7,855±160
FxJj 12	Beach sands	Shell	Geochron	Gx-5479	9,260±235
FxJj 12	Beach sands	Shell	Radiocarbon Ltd.	Rl-954	
FxJj 12	Beach sands	Bone apatite (M)	Geochron	Gx-5481-A	8,395±270
FxJj 12	Beach sands	Bone apatite (M & F)	Geochron	Gx-5480-A	8,355±235
FxJj 12-N	Beach sands	Bone apatite (H)	Geochron	Gx-4733-A	3,215±155
Pastoral Neolithic					
GaJi 2	Beach sands	Charcoal	Un. Penn.	P-2609	3,970±60
GaJi 2	Beach sands	Charcoal	Un. Sydney	SUA-634	4,160±110
GaJi 2	Beach sands	Shell	Un. Sydney	SUA-635	8,915±140
GaJi 4	Beach sands	Charcoal	Un. Sydney	SUA-637	3,945±135
GaJi 4	Beach sands	Humic acid from SUA-637	Un. Sydney	SUA-637-B	4,100±125
GaJi 4	Beach sands	Charcoal	Un. Penn.	P-2610	3,960±60
GaJi 4	Beach sands	Bone apatite (M)	Geochron	Gx-4642-I-A	3,405±130
GaJi 4	Beach sands	Bone apatite (F)	Geochron	Gx-4642-II-A	4,580±170
GaJi 4	Beach sands	Shell	Un. Sydney	SUA-638	10,320±150
FwJj 5	Fluvial sands	Bone apatite (M)	Geochron	Gx-4643-A	4,000+140
Other sites					
Nderati	Sandy silts	Shell	Geochron	Gx-5478	13,440±640
GaJj 9	Burial site 1	Bone apatite (H)	Geochron	Gx-4641-A	Modern
GaJj 9	Burial site 2	Bone apatite (H)	Geochron	Gx-6400-A	3,125±210
Previously obtained dates					
Ileret[1]	Sands	Shell	---	---	9,360±135
Area 103[2]	Coquina	Shell	---	---	9,880±670
Area 103[2]	Sandy silts	Shell	---	---	5,060±245
Area 103[2]	Silts	Charcoal	---	---	4,390±235
Area 102[3]	Beach sands	Shell	Un. Birm.	Birm-540	4,540±230

[1] Vondra and Bowen 1976

[2] Raynolds 1973

[3] Bishop 1976

TABLE 3

RADIOCARBON DATES: CORRECTED AND UNCORRECTED AGES

Site	Material Dated	Date (yr BP)	Uncorrected Date (yr BP)	C^{13} Value
<u>Fishing Settlements</u>				
GaJi 3	Bone apatite (M)	4,560 ± 185	4,160 ± 185	+0.5
GaJj 11	*Etheria* shell	modern		-0.1
GaJj 11	Bone apatite (F)	7,855 ± 160	7,460 ± 160	+0.3
FxJj 12	Shell	9,540 ± 260	9,140 ± 260	-0.1
FxJj 12	Shell	9,260 ± 235	8,950 ± 235	-5.0
FxJj 12	Bone apatite (M)	8,395 ± 270	8,030 ± 270	-1.9
FxJj 12	Bone apatite (M & F)	8,355 ± 235	8,000 ± 235	-2.4
FxJj 12 - N	Bone apatite (H)	3,245 ± 155	2,945 ± 155	-6.0
<u>Pastoral Neolithic</u>				
GaJi 4	Bone apatite (M)	3,405 + 130	3,075 ± 130	-3.8
GaJi 4	Bone apatite (F)	4,580 ± 170	4,230 ± 170	-2.6
FwJj 5	Bone apatite (M)	4,000 ± 170	3,660 ± 170	-3.5
<u>Other Sites</u>				
Nderati	Shell	13,440 ± 640	13,100 ± 640	-3.1
GaJj 9	Bone apatite	Modern		-10.3
GaJj 9	Bone apatite	3,125 ± 210	2,880 ± 210	-9.3

CHAPTER III

FISHING SETTLEMENTS

Chapter III presents a collection of seven site reports from the early to middle Holocene fishing settlements at East Turkana (Fig. 8). Each report includes sections on stratigraphic context, paleogeographic setting, sampling and excavation methodologies, as well as the description of stone, bone and ceramic artifacts. Preliminary identifications of mammal remains have been completed for two of the sites. The remaining mammal bones, as well as fish and reptile faunal material, are currently undergoing analysis. Radiocarbon dates have been received from four sites. Summary interpretations are included at the conclusion of each site report.

INTRODUCTION

The site of GaJj 1 was located along the Koobi Fora Ridge in Area 102 (Fig. 8). Numerous stone artifacts, fish and mammal remains, bone harpoons and fragments of pottery were discovered eroding out of a series of sand and gravel exposures. Cultural material was dispersed along approximately 400 m of the outcrop and substantially concentrated for nearly 150 m. Artifacts exposed by erosion were sampled systematically and two areas of the site were excavated. Three altimeter readings indicated the site lay 75-80 m above the June 1976 lake level.

STRATIGRAPHY

The Galana Boi deposits at GaJj 1 lay on unexposed units of the Koobi Fora Formation and had a maximum thickness of just under three meters. To the west, the Holocene sediments became finer grained and eventually reached a thickness of 32-35 meters. The principal sedimentary exposure was oriented roughly north-south (Fig. 9) and was composed of a fluctuating sequence of fine sands, silty clays and coarse gravels. These sediments suggest the archaeological site was situated close to a beach shoreline which was intermittently submerged by oscillations of the Holocene lake. Mollusc remains were not common and occurred only within the northwestern portion of the site. Recent eolian activity has mantled the Galana Boi units with a thin coat of poorly sorted coarse sand. Frequently, this sand will accumulate around Commiphora bushes and form small stabilized dunes.

Two areas of the site, approximately 150 m apart, were selected for test excavation. Both areas contained significant concentrations of stone artifacts, bone harpoons, pottery and associated faunal material.

Excavation I

The stratigraphic section in Fig. 10 was composed of a sequence of fine to coarse sands and gravel horizons that overlayed Upper Member Plio-Pleistocene silty clays and course sands. A single lens of silty clay interrupted the predominately sandy series. Artifacts were dispersed throughout the Holocene beds especially within several gravel lenses. Within these lenses, the artifacts were often heavily water abraided and frequently broken. The lowermost Galana Boi horizon (lithology G) contained rolled stone artifacts and a very weathered and broken uniserially barbed bone harpoon. Details of the excavation section are described below.

Lithology	Description
A	unconsolidated fine to coarse light orange sand; several stringers of coarse gravels with sands; silt coated grains near top; abundant modern root casts; abrupt wavy boundary with underlying bed; artifacts scattered and at times concentrated; 25-40 cms.

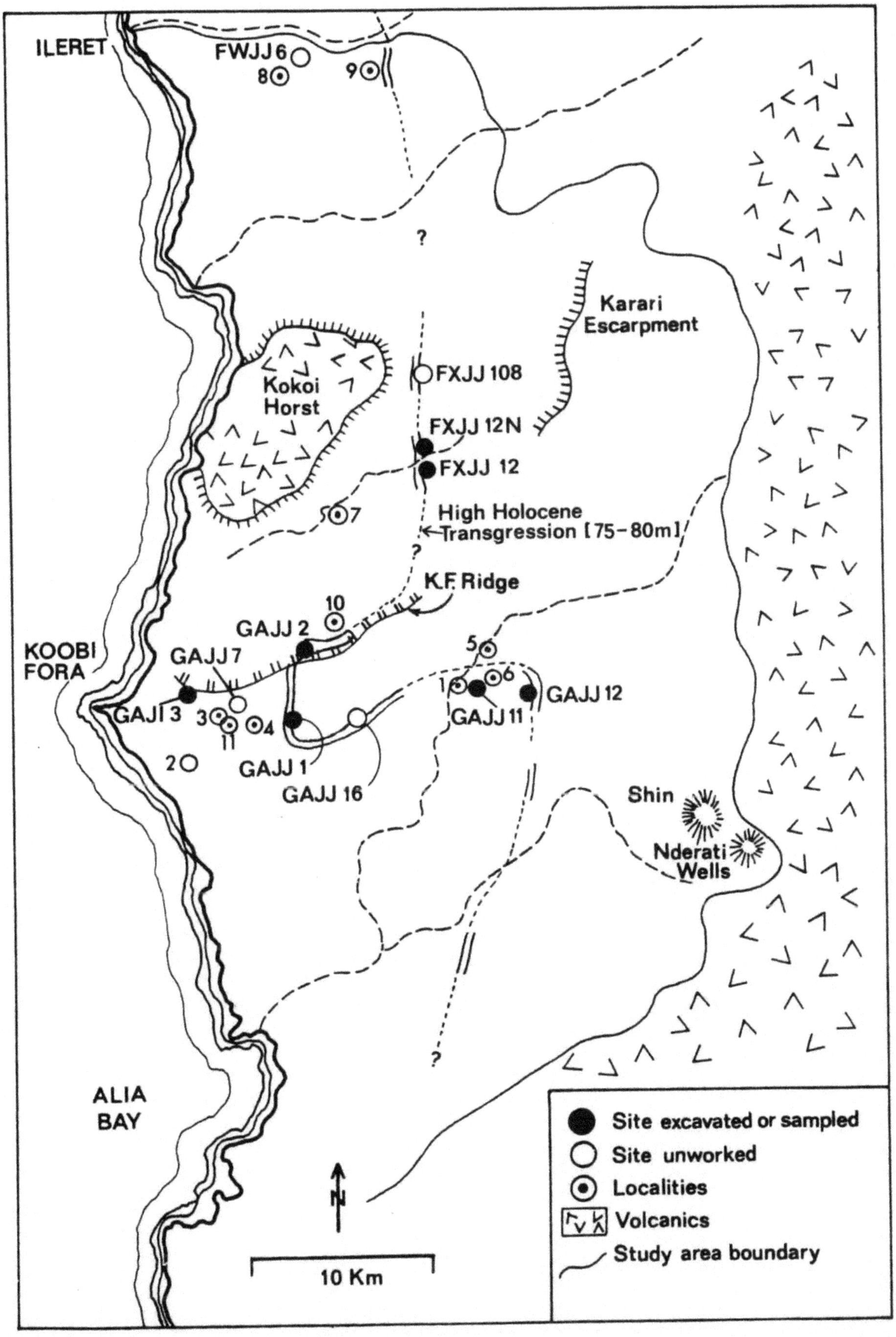

Figure 8. Distribution of early to middle Holocene fishing sites, localities and reconstructed shoreline of the high Holocene lake transgression.

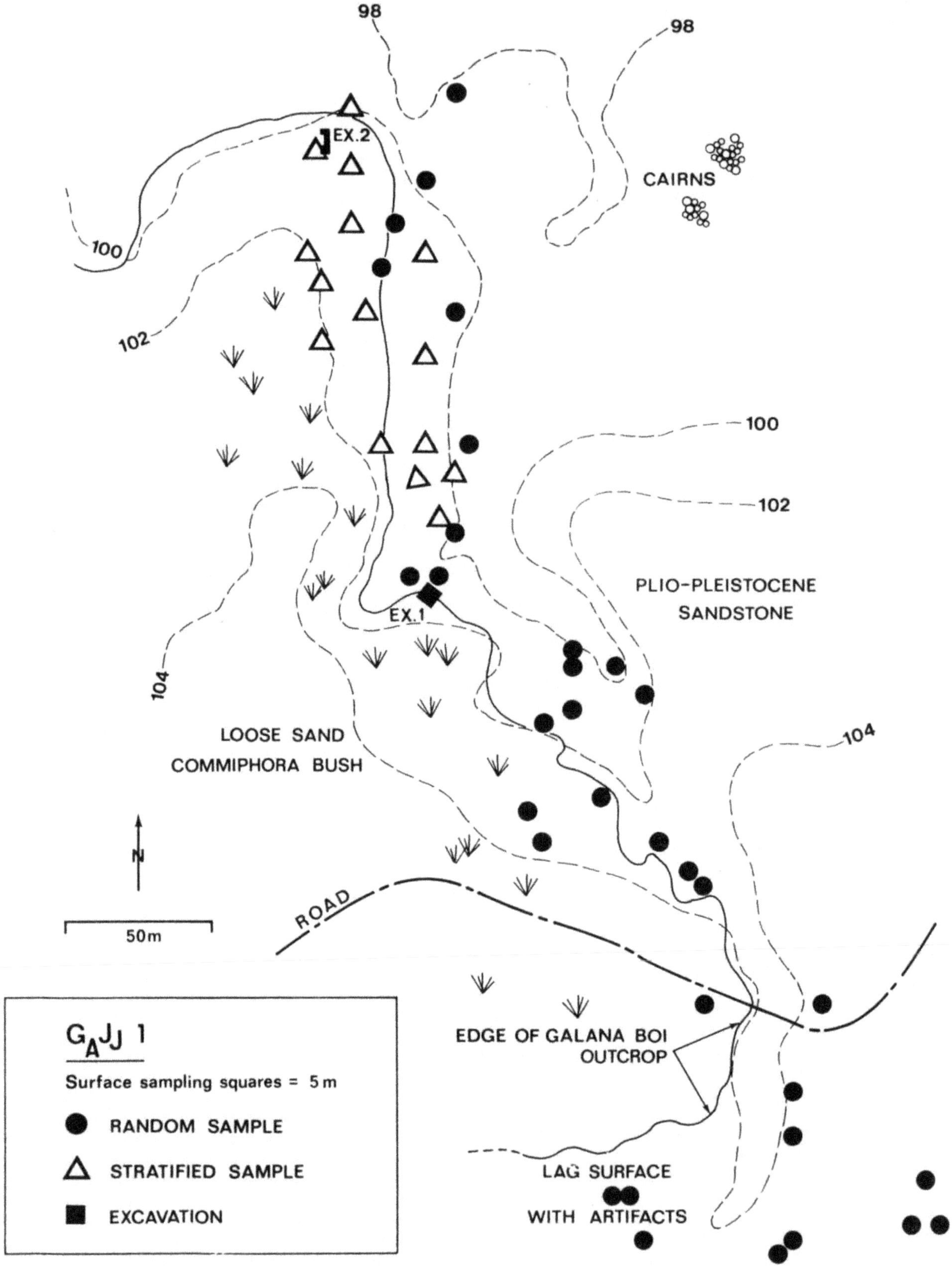

Figure 9. Contour plan of GaJj 1. Random and stratified sampling units are indicated.

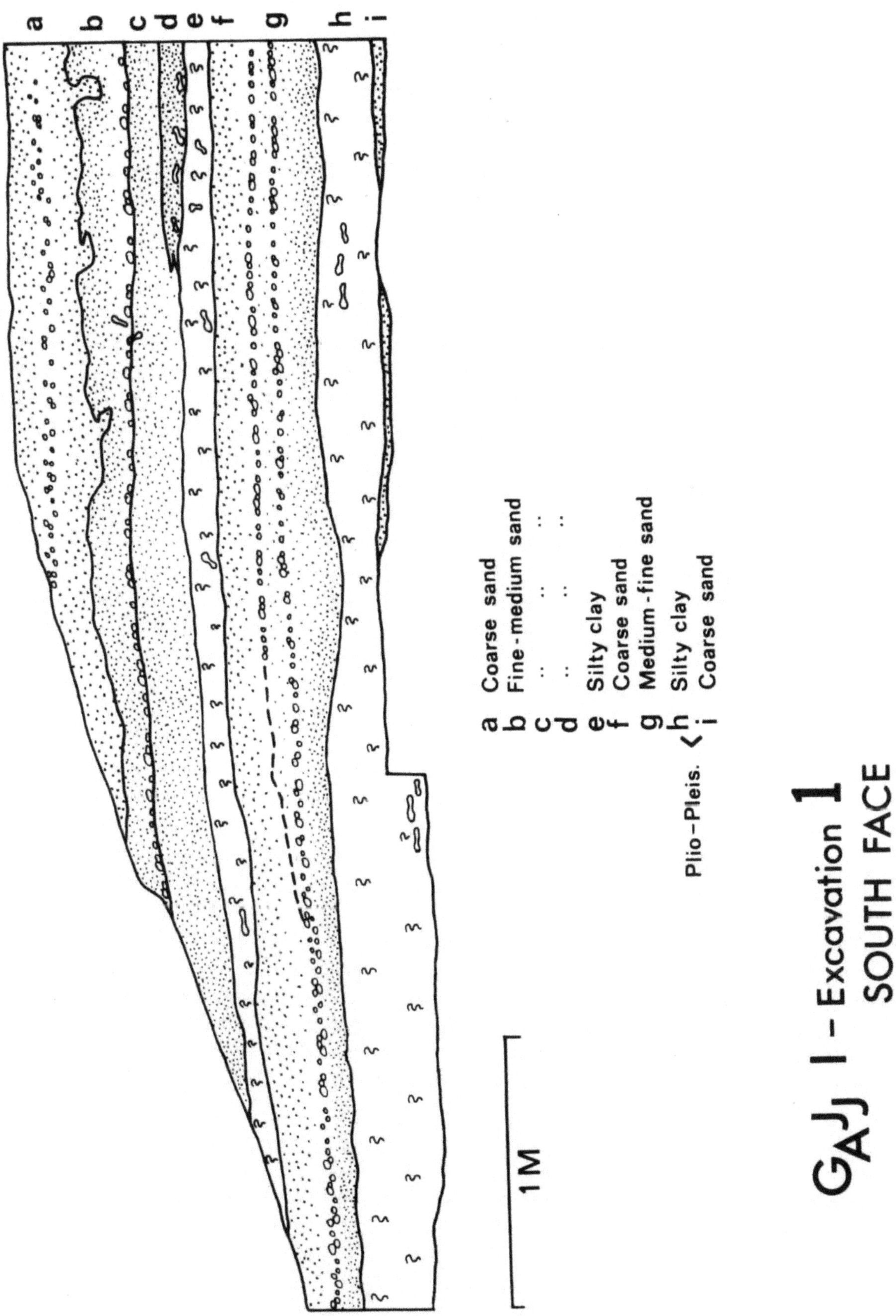

Figure 10. GaJj 1. Section of the south face of Excavation 1.

B	indurated fine to medium orange sand; considerable moisture content; heavy mineral laminae and $CaCO_3$ nodules throughout; gravel lens at base; scattered artifacts; 15-20 cms.
C	semi-consolidated fine to medium orange sand; abundant heavy mineral laminae; few artifacts; 15-25 cms.
D	consolidated fine to medium light gray sand; abundant $CaCO_3$ staining and nodules at base; artifacts numerous; 8-10 cms.
E	consolidated light brown silty clay; $CaCO_3$ nodules and sand grains in vertical cracks at contact with D; few artifacts; probably derived from overlying sand lens; 10-15 cms.
F	consolidated fine to medium coarse light gray sand; $CaCO_3$ nodules; few artifacts; 10-15 cms.
G	semi-consolidated fine to medium orange sand; two lenses of sandy gravels with associated artifacts; 20-25 cms.
H	consolidated brown silty clay; slickensides; thin lenses of fine sand interbedded with clay; conspicuous oxidation stains; abrupt wavy boundary with overlying bed; Plio-Pleistocene unit; no artifacts; 20-40 cms.
I	very consolidated medium to coarse grained gray sand; large $CaCO_3$ concretions; manganese staining at base; Plio-Pleistocene unit; base not exposed; no artifacts; 5 cms.

Excavation 2

The second excavation was located within a concentration of dispersed artifacts and broken mollusc shells including the remains of Melanoides, Cobicula, Mutela and Pila ovata. The section diagram (Fig. 11) revealed a series of fine to coarse sands and gravels which lay on Upper Member sands and silty sands. Similar to the section from Excavation 1, these sediments were interpreted as littoral deposits. The majority of artifacts were recovered from lithology A either immediately below the surface or to a maximum depth of 50 cms. A second set of heavily water abraided artifacts were collected from gravel lenses within lithology C. Two distinct horizons of silty clays occurred near the section's base. Cumulative sediment thickness was just over 3 meters. The excavation section is described below.

Lithology	Description
A	unconsolidated poorly sorted light orange sand; lenses of gravels and broken shell; silt covered sand grains;

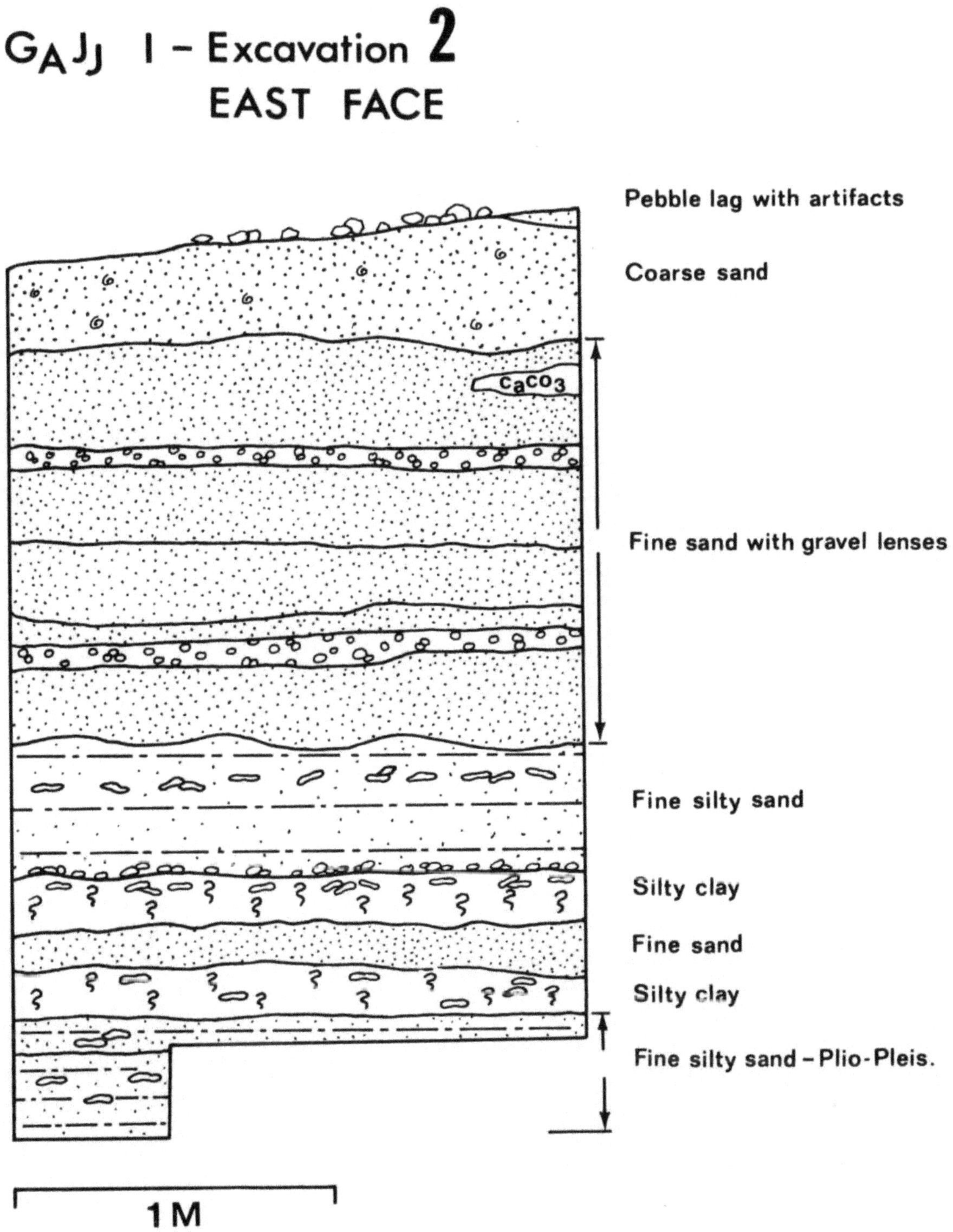

Figure 11. GaJj 1. Section of the east face of Excavation 2.

artifacts dispersed throughout horizon and at times numerous; 25-55 cms.

B a fluctuating series of sand horizons ranging from fine sands with heavy mineral laminae to coarse sandy gravels with rolled artifacts; 75-85 cms.

C alternating sequence of fine sands and silty sands; gravel lens at base; few artifacts; 35-40 cms.

D consolidated light brown silty clay; $CaCO_3$ nodules and sand filled cracks at top; oxidation stains throughout bed; few artifacts; 10-15 cms.

E semi-consolidated light gray fine sand with heavy mineral laminae at base; scattered artifacts; 10-15 cms.

F second bed of consolidated light brown clay; scattered $CaCO_3$ concretions; silt fraction higher than in lithology D; few artifacts; 10-20 cms.

G consolidated light gray fine sand and silty sand; $CaCO_3$ nodules abundant; oxidation stains; Plio-Pleistocene bed; base of unit not exposed; 25 cms.

<u>Summary: Excavations 1 and 2</u>

The sequence of fine to coarse sands and sandy gravels present in both excavations is thought to represent deposition along the fluctuating margin of a beach. The Galana Boi beds lay upon silty clays and sands of the Upper Member of the Koobi Fora Formation. Nearly all horizons were well cemented with calcium carbonate and irregular $CaCO_3$ concretions were a regular feature of the sand lenses. Evidence for periodic lake shoreline oscillations was indicated by silty clay horizons in both excavations. These units may represent back beach lagoons. For each excavation, only the basal Plio-Pleistocene and uppermost Holocene sand units are thought to be satisfactorily correlated (Fig. 12). However, one or more of the silty clay horizons may also be correlated between sections. A number of the sand and gravel horizons probably display facies interrelationships but their complex interconnections have not been worked out.

Archaeological horizons were present throughout each section including the bottommost horizons at each excavation. They are believed to represent a series of occupation episodes that span, as yet, an unknown duration of time. The great majority of artifacts were water rolled and most probably in secondary context. The stone artifacts often exhibited a cream colored surface patina, believed to be the result of either prolonged surface exposure or reworking by water action.

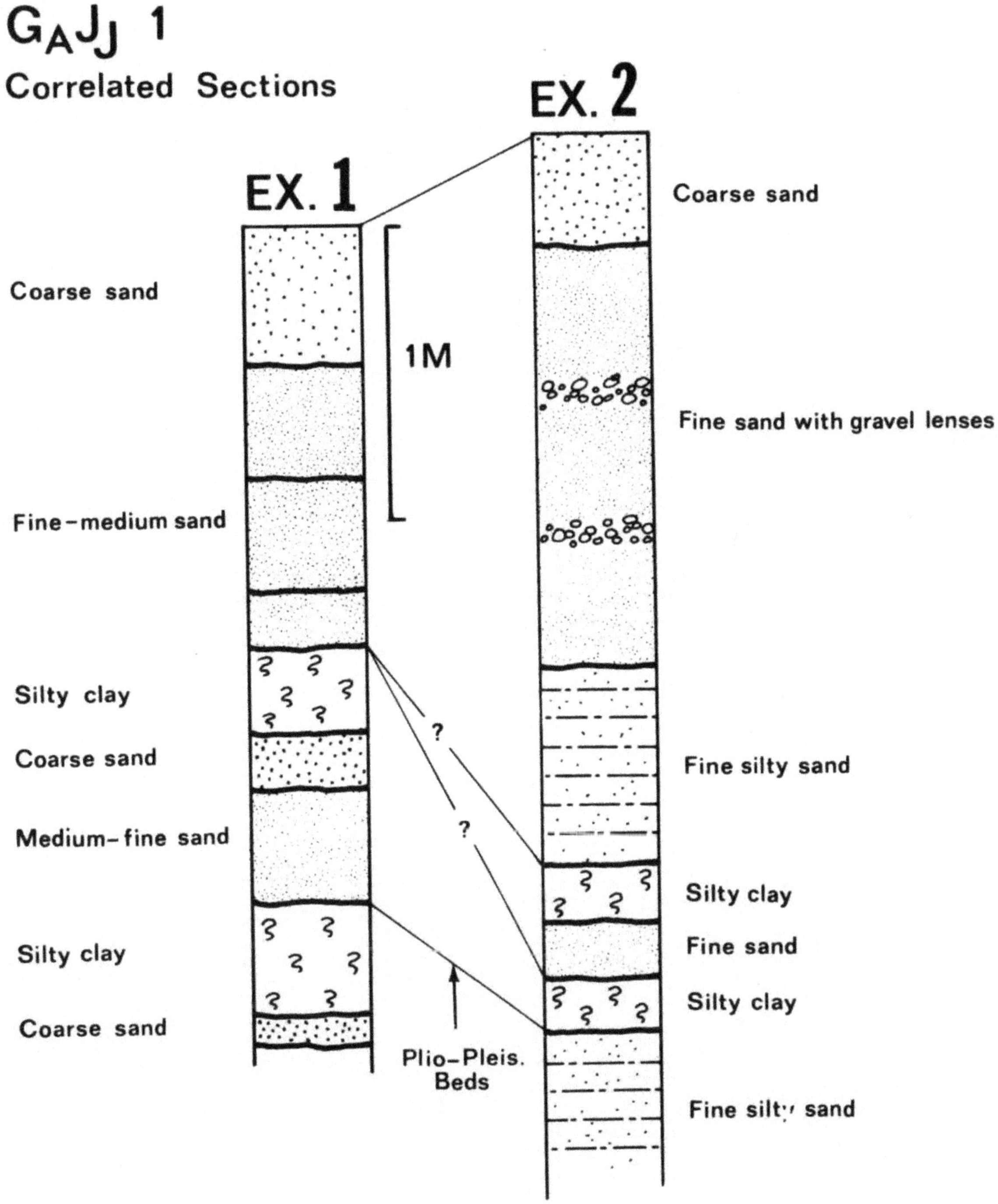

Figure 12. GaJj 1. Correlated sections from Excavations 1 and 2.

SAMPLING

Both simple random and stratified random sampling methods were used at GaJj 1. The procedures employed for simple random sampling were as follows: site dimensions were first determined by use of surveying instruments and found to be 400 m E/W by 65 m N/S. The site was gridded into eleven blocks of 100 five meter square sampling units. Utilizing a table of random numbers (Redman 1974), three numbers were selected for each collecting block. The five meter collecting squares were then laid out with 30 m tapes and a compass. Hence:

potential sampling units:	1100
units sampled:	33
3.0% of the population sampled	

It became apparent that the simple random technique did not adequately collect the area of greatest artifact density. Further sampling of this area was desirable and consequently a stratified sampling method was developed. An area of substantial artifact concentration was determined by surveying instruments and ploted on graph paper. Three blocks of 80 five meter sampling units were laid out. Using a random numbers table, 5 sampling units were selected for each 80 m block (Fig. 9). Hence:

potential sampling units:	240
units sampled:	15
6.3% of the population sampled	

Combining both sampling techniques yielded:

total potential sample units:	1340
total units sampled:	48
3.6% of the population sampled	

GaJj 1 represented the first lake shoreline site to be sampled systematically, and in retrospect, several of the collecting procedures were not clearly defined. Given the great wealth of fish bones and the potential difficulty of distinguishing naturally occurring faunal refuse from discarded food remains, instructions were given to collect only fish teeth, pectoral spines and large vertebrae, elements particularly valuable for taxonomic identification. Similarly, for mammal and reptile remains, workmen were instructed to collect only tooth fragments and bones with articular surfaces. Nonidentifiable bone fragments were left within the sampling units. This selective collection procedure proved to be unsatisfactory. The faunal inventory counts for GaJj 1 are therefore underrepresented. For all subsequent sites that were sampled, the entire surface assemblage of faunal elements was collected.

CULTURAL REMAINS: SURFACE

Tables 4 and 5 represent separate inventories by sampling unit for both random and stratified surface collections. In the following summary discussion, artifacts from both collections are treated as a composite group. Cultural remains recovered during the 1975 field survey season, surface scrapings from each excavation area, and individually surface plotted artifacts are tabulated as well. A comprehensive inventory of all surface material is presented in Table 6.

Stone Artifacts (Tables 7-9)

Nearly 3000 stone artifacts were collected from the surface at GaJj 1. The lithic assemblage was primarily derived from small single platform, casual and broken chert cores. Microlithic implements accounted for the largest shaped tool category followed by outils écaillés, core/choppers and various scrapers. Several technical burins and awls were also present. Crescents constituted over 40% of the microlithic tools. A few micro-percoirs were included in the assemblage. The great majority of the debitage consisted of small flakes and flake blades manufactured from small chert pebbles. Lava chippage accounted for only 3% of the lithic debris. Obsidian was present but only eight fragments (less than 1%) were recovered (Figs. 13 and 14).

Three fragments of groundstone were identified. Two examples were manufactured from indurated sandstone while the third artifact was shaped from an unidentified metamorphic stone. Each specimen displayed a shallow smoothed and polished interior depression. No traces of red ochre staining were visible although small haematite fragments were recovered from lithologies A and B in Excavation 1. Direct evidence for seed grinding or vegetable processing, if originally present, was not identified.

The lithic artifacts displayed a wide range of weathering states. Many pieces were substantially water abraided and often exhibited a surface patina. Other artifacts had sharp, minimally weathered, dorsal scar ridges and flake margins. Some archaeological horizons were apparently buried with little secondary disturbance while other horizons, particularly wave sorted gravel lenses, were subject to considerable reworking. Surface exposure over time also contributed to weathering of the stone artifacts.

Evidence for possible hearth stones came from seven large fragments of indurated bioclastic sandstone. The fragments were differentially colored black, gray and dull red. Such stones were a recurrent feature of the shoreline fishing and hunting settlements. They may well have been used in cooking activities or for assisting the firing of clay pots.

Bone Artifacts

Forty-nine bone artifacts were collected, 14 of which were recovered from the sampling units. The remainder were either individually

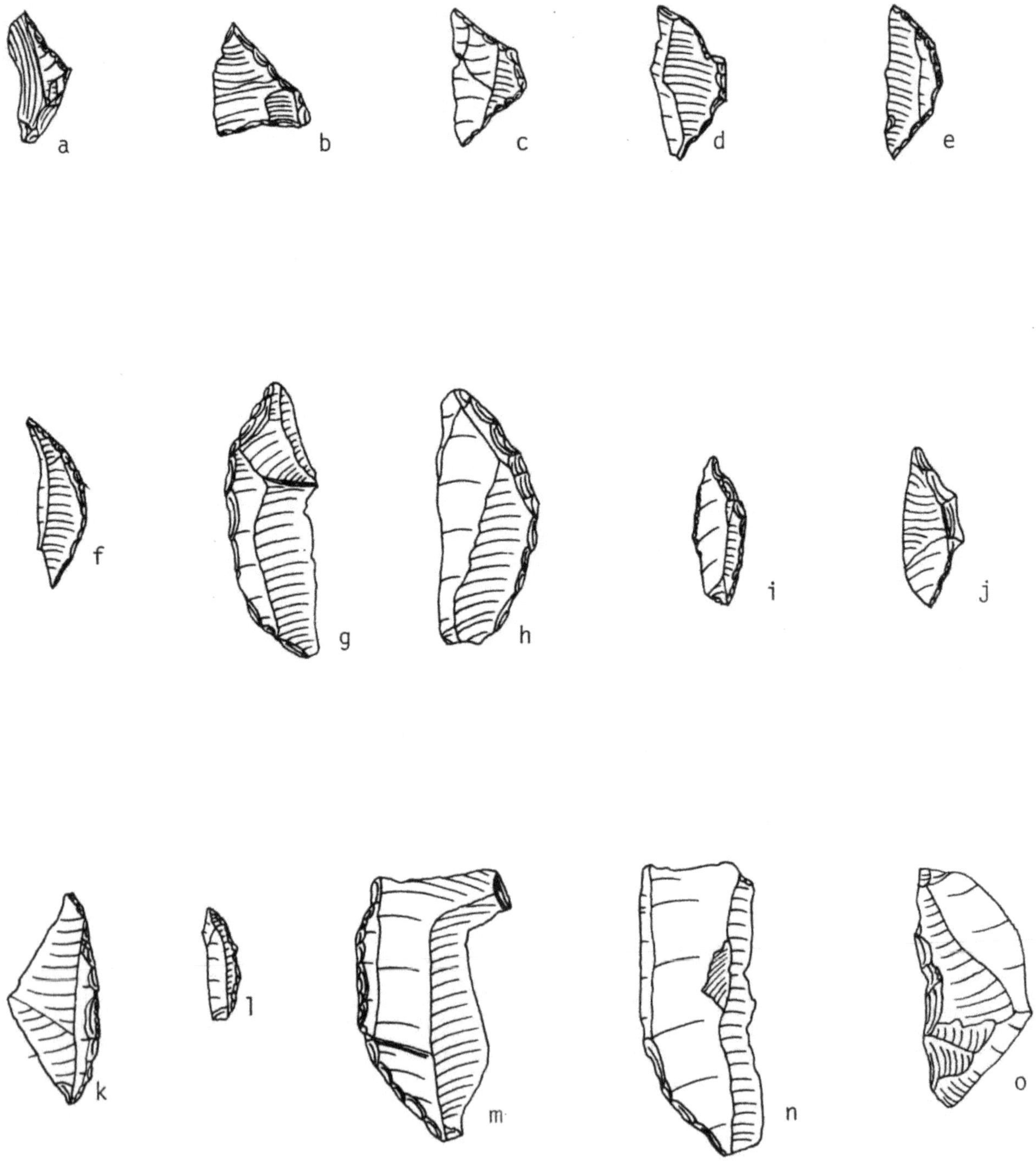

Figure 13. GaJj 1. Backed tools: Surface.
(a-b) triangles; (c-d) trapezoids; (e-j) crescents; (k-n) curved backed flakes; (o) straight backed flake. Natural size.

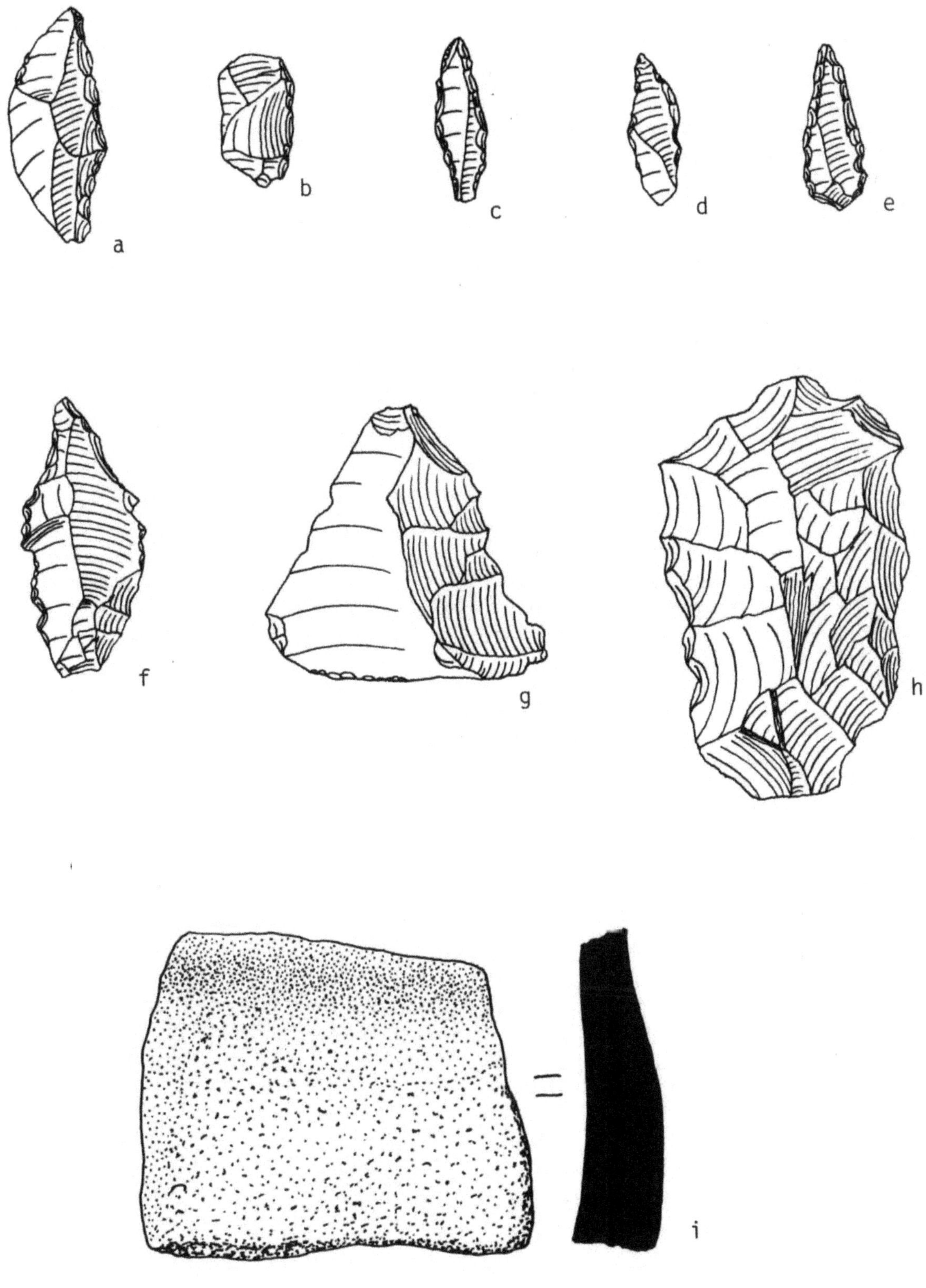

Figure 14. GaJj 1. Stone artifacts and pottery: Surface.
(a-b) straight backed flakes; (c) double backed flake; (d-e) micro-percoirs; (f) awl; (g) side scraper; (h) double side and end scraper; (i) undecorated body sherd. Natural size.

plotted on the site contour map or removed during the 1975 field survey. One example was complete while the rest of the artifacts were broken and in various weathering stages. Most of the bone tools consisted of uniserially and biserially barbed harpoon heads (Fig. 15). Several barbless forms were also identified. The uniserial harpoons exhibited both basal notching and circular grooved incisions.

Pottery

Fifty-two undecorated body sherds were collected from the surface. The sherds tended to be friable, poorly fired, and tempered with a coarse sandy grit. Because of the absence of rim sherds, size and shape dimensions of the original vessels were not determinable. However, the profiles of two body sherds indicated the presence of comparatively thick walled vessels (Fig. 14, i).

Beads

A small collection of beads including five ostrich eggshell examples, a bone bead and a specimen manufactured from microcline feldspar were recovered. Similar feldspar crystals may be collected from basement outcroppings to the northeast or from river cobbles within the larger dry stream beds draining the basement rocks.

Fauna

A preliminary analysis of the mammal bones indicated few elements could be identified to either the family or generic level. Those taxa identified presently inhabit the East Turkana area (Table 15).

EXCAVATION METHODS

Excavation 1

An area of substantial artifact density located adjacent to the outcrop's edge was selected for the first excavation A 5 m square was gridded and subdivided into units of one, two, and one by two meter squares. Units A and B were removed over the entire 5 m block. Beginning with lithology C, a decision was made to limit further excavations to two right angle trenches, each 1 m wide and 3-5 m in length. Once the association of the surface and excavated artifacts was clearly established, primary attention was focused on determining the total thickness of the Galana Boi deposits and their stratigraphic relationship to the underlying Plio-Pleistocene units. Consequently, efforts shifted from artifact collection to a combination of contextual and stratigraphic reconstruction.

Excavation 2

A second area of high artifact density was selected for excavation in the northeastern sector of the site. Initially, a 3 m square was gridded and subdivided into four one and one-half meter units. Further expansion of the excavation grid was subsequently undertaken. Lithology A was completely removed in each square. Like Excavation

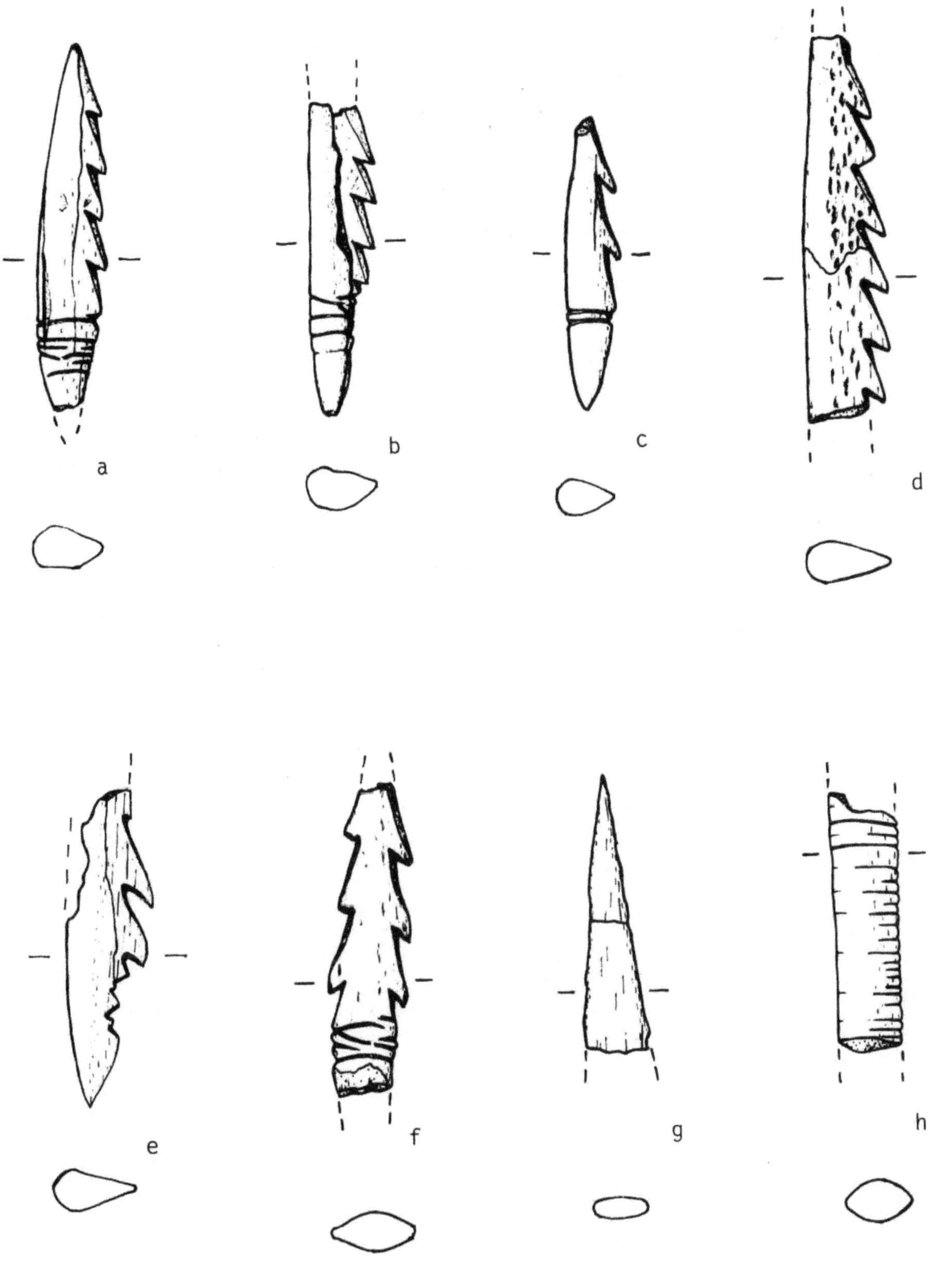

Figure 15. GaJj 1. Bone artifacts: Surface.
(a-e) uniserially barbed harpoons; (f) biserially barbed; (g) barbless; (h) barbless with series of circular incisions. Natural size.

1, a decision was made to limit the areal dimensions of the second excavation and to concentrate on determining the depth of the Galana Boi deposits and exposing the underlying Plio-Pleistocene beds.

CULTURAL REMAINS: EXCAVATION 1

Artifactual remains were recovered from each stratigraphic level. The greater part of the artifacts were derived from wave sorted coarse gravels and medium to coarse sands. No archaeological features were recognized. Table 10 presents an inventory of the finds by lithology and split elevation. The following description summarizes the artifacts and bone refuse for the entire sample.

Stone Artifacts (Tables 11-13; Figs. 16-17)

Microlithic implements constituted over 90% of the shaped stone tools followed by core/choppers and casually trimmed pieces. Scrapers and *outils écaillés* were not identified. Débitage consisted primarily of small flakes and flake blades struck from single platform and casual chert cores. A small but persistent collection of blades and single and double platform blade cores was also present. Two core/choppers were recovered *in situ*. They are morphologically similar to specimens collected from the Plio-Pleistocene beds in Area 105 (Isaac 1973).

Bone Artifacts

Five harpoon segments were recovered from Excavation 1, including two basal fragments from the lowest excavated Galana Boi unit. Three of the bases had incised circular lines while the fourth example had a notched base (Fig. 17 c-e). All specimens were broken and showed evidence of substantial water abrasion.

Pottery

A total of six undecorated body sherds were identified. The excavated sherds, like their surface counterparts, were poorly fired and tempered with quartz and small rock fragments. Their external surface characteristically exhibited a light pink to orange oxidation color, indicative of intense and incomplete firing (Shepard 1956). It was not possible to reconstruct vessel shape or size from the small sample.

Fauna

The excavated faunal material was fragmentary and showed considerable evidence of water abrasion. Fish bone outnumbered mammal bone fragments by almost six to one. A preliminary list of the identified mammalian taxa is presented in Table 15.

CULTURAL REMAINS: EXCAVATION 2

Artifact remains were collected from all six excavation levels. The majority of finds came from coarse gravels within levels A and B. Archaeological features were not identified. Table 10 provides an

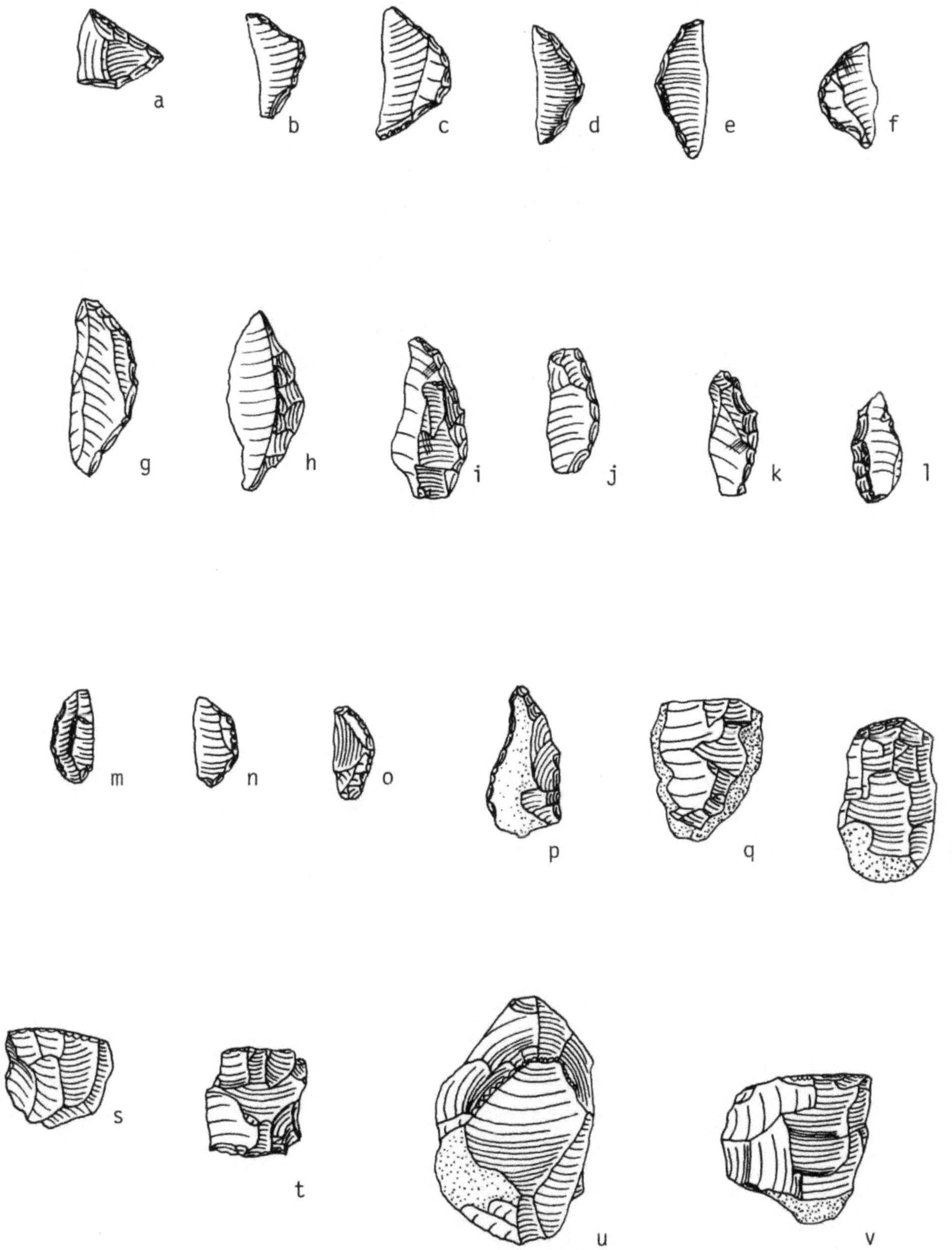

Figure 16. GaJj 1. Stone artifacts: Excavation 1.
(a) triangle; (b) trapezoid; (c-h) crescents; (i-o) curved backed flakes; (p) double backed flake; (q-s) single platform cores; (t) bipolar core; (u-v) double platform cores. Natural size.

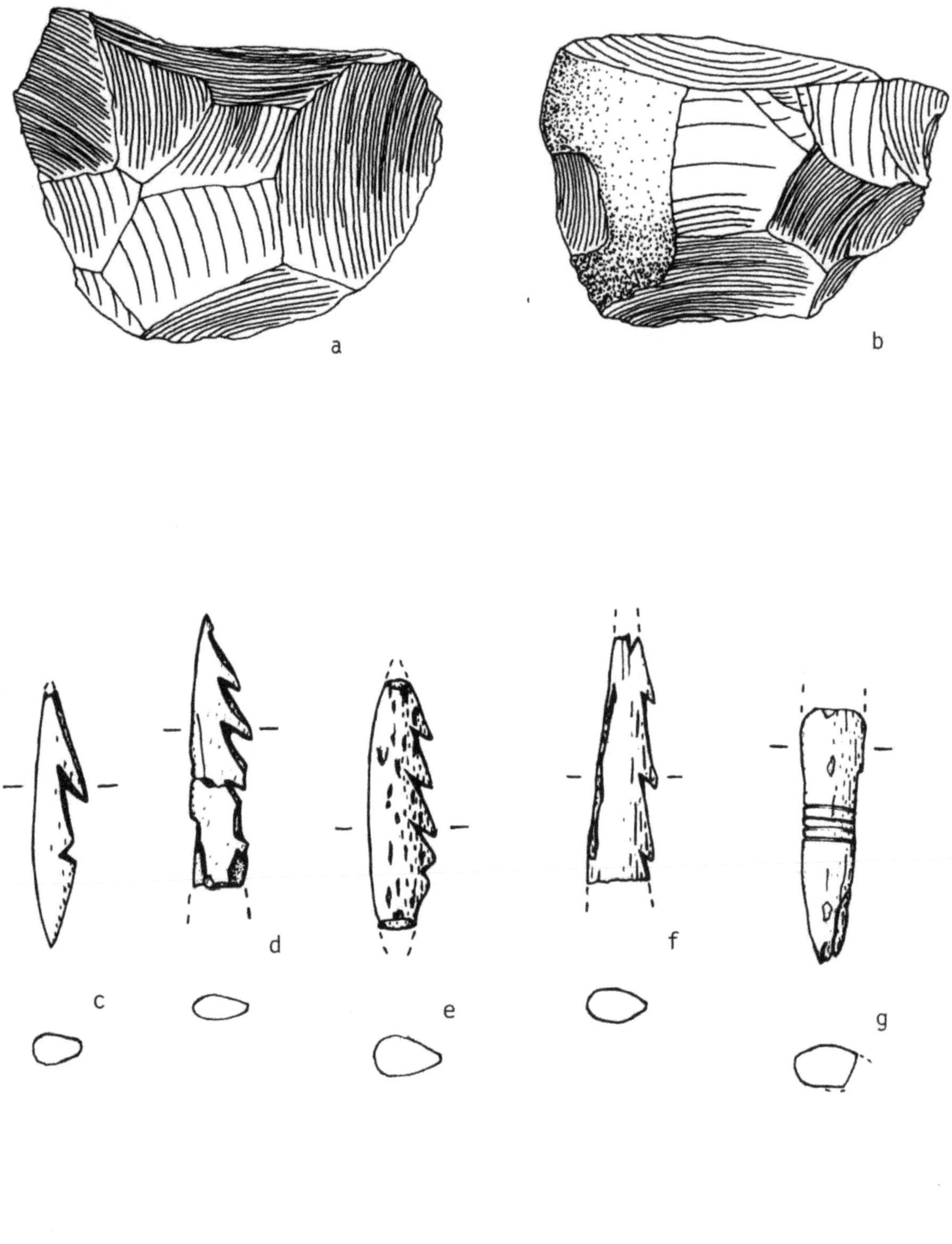

Figure 17. GaJj 1. Core tools and bone artifacts: Excavation.
(a) unifacial core/chopper; (b) polyhedron; (c-e) uniserial harpoons, Excavation 1; (f-g) uniserial harpoons, Excavation 2. Natural size.

inventory of finds by lithology and split evaluation. Cultural materials were similar throughout the section and have been combined for purposes of general discussion.

Stone Artifacts (Tables 11-13)

Microlithic elements accounted for over 90% of the shaped stone tools. Unlike the sample from Excavation 1, outils écaillés were present but core/choppers and scrapers were absent. Except for the recovery of a single obsidian flake fragment, debitage patterns were similar to those of Excavation 1.

Bone Artifacts

Two barbed bone harpoon fragments were identified, including a basal fragment from the lowest Galana Boi stratigraphic horizon (Fig. 17, g). Both fragments were very water worn.

Pottery

A total of nine undercorated body sherds were collected. They were comparable in color, paste and temper to those from Excavation 1.

Beads

Four ostrich eggshell beads were identified. They were similar in size and shape to several surface beds collected near the excavation.

Faunal Remains

The faunal material was fragmentary and heavily water rolled. While broken Pila ovata shells were collected from the surface, direct evidence for their consumption or possible use as fishing bait such as that reported from Shabona along the Nile River (J.D. Clark 1973) was not identified.

PRELIMINARY INTERPRETATION

A field strategy of extensive sampling and limited excavation yielded large quantities of stone artifacts, barbed bone harpoons, pottery and faunal remains that are thought to have accumulated along the 75-80 m Holocene beach shoreline. Cultural material was dispersed along a surface outcrop for at least 400 m and was present in at least 7 distinct excavated horizons. A significant proportion of the artifacts were heavily rolled and broken. This observation strongly suggests that many of the original occupation surfaces are in secondary context. The original cultural material was often concentrated by repeated wave action within lenses of coarse gravelly sands. Archaeological features, such as distinct activity areas and hearths, if originally present, have not survived. Unfortunately, the amount of bone from a single excavation spit proved insufficient to submit for radiocarbon determination. The stratigraphic evidence demonstrates, in any case, a considerable percentage of the surface and excavated cultural remains should be viewed as a composite collection. The total assemblage is thought to represent many settlement episodes of discarded artifactual and bone refuse from a broadly similar technological and economic tradition.

Table 4
GaJj 1

INVENTORY OF SURFACE FINDS BY SIMPLE RANDOM SAMPLING

Unit	1	2	3	4	5	6	7	8	9	10	11	12	13	14	15	16	17	18	19
Stone:																			
Microliths	-	-	-	-	-	-	1	6	3	-	1	-	1	1	-	-	-	-	-
Scrapers	-	-	-	-	-	-	-	-	-	-	1	-	-	2	-	-	-	-	-
Outil écaillés	-	-	-	-	-	-	-	1	2	-	-	2	-	-	-	-	1	-	-
Burins	-	-	-	-	-	-	-	-	-	-	-	-	-	-	-	-	-	-	-
Awls	-	-	-	-	-	-	-	-	-	-	-	-	-	-	1	-	-	-	-
Core/choppers	-	-	-	-	-	-	-	-	-	-	-	-	2	-	-	-	-	1	-
Cas. trimm.	-	-	-	-	-	-	1	-	-	-	1	-	1	-	-	-	-	-	-
Cas. retouch	-	-	-	-	-	-	-	-	-	-	-	-	-	-	-	-	-	-	-
Whole flakes	2	-	-	1	-	-	3	5	-	-	1	2	6	-	8	2	1	2	-
Broken flakes	2	1	-	3	1	-	3	16	6	-	7	5	8	12	10	2	3	2	-
Angular frags.	-	2	-	7	3	-	7	16	11	1	8	7	10	2	4	2	3	3	-
Cores	-	-	-	-	-	2	3	3	1	-	1	2	3	3	3	2	-	-	-
Pottery:																			
Undecorated	-	-	-	-	-	-	-	-	-	-	-	-	-	-	-	-	-	-	-
Bone harpoons:																			
Uniserial	-	-	-	-	-	-	-	-	-	-	-	-	-	-	-	-	-	-	-
Bones w/art. sur.	-	-	-	-	-	-	-	-	-	-	-	-	-	-	-	-	-	-	-
Raw counts:																			
Stone	4	3	-	11	4	2	18	47	23	1	20	18	32	23	26	8	8	8	-

Table 4 Continued

GaJj 1

INVENTORY OF SURFACE FINDS BY SIMPLE RANDOM SAMPLING

Unit	20	21	22	23	24	25	26	27	28	29	30	31	32	33	Total (N)
Stone:															
Microliths	-	3	-	-	-	-	-	-	2	17	-	2	-	10	47
Scrapers	-	-	-	-	-	-	1	-	-	-	-	-	-	-	4
Outil écaillés	-	-	1	-	2	-	-	1	-	-	-	-	-	4	14
Burins	-	-	-	-	-	-	-	-	-	1	-	-	-	-	1
Awls	-	-	-	-	-	-	-	-	-	-	-	-	-	-	1
Core/choppers	-	-	-	-	-	-	-	-	-	2	-	-	-	-	5
Cas. trimm.	-	-	-	-	-	-	-	1	-	1	-	-	-	-	5
Cas. retouch	-	-	-	-	-	-	-	-	-	-	-	-	-	-	-
Whole flakes	4	5	4	2	11	-	1	5	3	40	-	9	3	23	143
Broken flakes	4	3	6	3	16	1	2	15	6	62	5	15	13	58	292
Angular frags.	2	7	4	1	24	1	5	10	5	55	3	15	13	51	282
Cores	2	2	2	1	12	-	-	12	8	24	-	7	7	23	123
Pottery:															
Undecorated	-	-	-	-	-	-	-	3	-	2	-	2	-	-	7
Bone harpoons:															
Uniserial	-	-	-	-	-	-	-	-	-	1	-	-	1	-	2
Bones w/art. sur.	-	-	-	-	-	-	-	-	-	-	-	1	-	-	1
Raw counts:															
Stone	12	20	17	7	65	2	9	44	24	200	8	48	36	169	917

Table 5

GaJj 1

INVENTORY OF SURFACE FINDS BY STRATIFIED RANDOM SAMPLING

Unit	34	35	36	37	38	39	40	41	42	43	44	45	46	47	48	Total (N)
Stone:																
Microliths	4	3	4	–	12	1	7	9	13	3	7	21	41	5	8	138
Scrapers	–	1	–	–	–	–	–	–	–	–	–	1	1	–	–	3
Outils écaillés	–	1	–	–	–	–	–	–	–	–	–	–	–	–	–	1
Burins	–	–	–	–	–	–	2	–	–	–	–	–	–	–	–	2
Awls	–	–	–	–	–	–	–	–	–	–	–	–	–	1	–	1
Core/choppers	1	2	–	1	–	2	–	–	–	1	–	–	–	1	–	8
Cas. retouch	1	1	–	–	–	–	–	–	–	–	–	–	1	–	–	3
Cas. trimm.	–	1	–	–	–	–	–	–	–	–	1	–	–	1	–	3
Whole flakes	23	7	18	–	23	–	25	57	27	16	14	26	89	16	11	352
Broken flakes	17	11	30	3	32	3	33	83	42	11	29	35	143	24	14	510
Angular frags.	10	18	14	–	42	3	36	92	31	14	14	29	165	23	9	500
Cores	14	14	24	–	16	3	12	15	29	13	8	20	39	13	4	233
Grindstones	–	–	–	–	–	–	–	–	–	–	–	–	–	1	–	1
Hammerstones	–	–	–	–	–	–	–	–	–	3	–	–	–	–	–	3
Bone harpoons:																
Uniserial	2	–	–	–	–	–	–	–	5	1	1	–	–	3	–	12
Pottery:																
Undecorated	1	–	–	–	2	–	1	3	8	3	–	5	2	–	–	25
Beads:																
Ostrich eggshell	–	–	–	–	–	–	–	1	1	–	–	–	–	–	–	2
Raw counts: stone	70	59	90	4	125	12	115	256	142	58	73	132	479	84	46	1754

Table 6
GaJj 1

INVENTORY OF MISCELLANEOUS COLLECTIONS AND FINAL SURFACE TOTALS

	Sampling Units Ex. 1	Sampling Units Ex. 2	Plotted	Collection 1974-75	Simple Random (N)	Stratified Random (N)	Other (N)	Assemblage Total (N)
Stone:								
Microliths	2	12	-	9	47	138	23	208
Scrapers	-	-	5	-	4	5	4	13
Outils écaillés	-	1	-	1	14	1	2	17
Burins	-	-	-	-	1	2	-	3
Awls	-	1	-	2	1	1	3	5
Core/choppers	-	-	1	-	5	8	1	14
Cas. retouch	-	-	-	-	-	3	-	3
Cas. trimm.	-	-	-	-	5	3	-	8
Whole flakes	5	38	-	2	143	352	45	540
Broken flakes	11	55	-	2	292	510	67	870
Angular frags.	6	68	-	9	282	500	83	865
Cores	2	27	1	6	123	233	36	392
Groundstone	-	-	1	1	-	1	2	3
Hammerstones	-	-	-	-	-	3	-	3
Bone harpoons:								
Uniserial	-	1	26	6	2	12	33	47
Biserial	-	-	1	-	-	-	1	1
Barbless	-	-	1	-	-	-	1	1
Pottery:								
Undecorated	-	1	9	9	7	25	19	51
Bone frags. (NID):								
Mammal	-	-	-	-	-	-	-	-
Fish	-	-	-	-	-	-	-	-
Turtle	-	-	-	-	-	-	-	-
Bones w/art. sur.	-	-	3	-	1	-	3	4
Teeth:								
Mammal	-	-	8	-	-	-	8	8
Fish	-	-	-	8	-	-	8	8
Beads:								
Ostrich eggshell	-	1	2	-	-	2	3	5
Bone	-	-	1	-	-	-	1	1
Stone	-	-	1	-	-	-	1	1

Table 7
GaJj 1: Surface

FREQUENCY OF TRIMMED PIECES

	Chert	Lava	Quartz	Total (N)	Total (%)
Microliths:				(208)	(77.0)
Triangle	3	-	-	3	1.1
Trapezoid	3	-	-	3	1.1
Crescent	51	-	-	51	18.9
Curved backed flake	51	-	-	51	18.9
Straight backed flake	12	-	-	12	4.4
Micro-percoir	3	-	-	3	1.1
Truncations	2	-	-	2	0.7
Miscellaneous backed	5	-	-	5	1.9
Fragmentary	78	-	-	78	28.9
Scrapers:				(13)	(5.5)
Side	1	-	-	1	0.5
Double side	2	-	-	2	0.7
Side/end	1	-	-	1	0.5
Denticulate	1	-	-	1	0.5
Double side/end	1	-	-	1	0.5
Notch	2	-	-	2	0.7
Steep	1	2	-	3	1.1
Nosed	1	-	-	1	0.5
Core scraper	1	-	-	1	0.5
Outils écaillés	12	-	4	16	5.9
on a crescent	1	-	-	1	0.5
Burins	3	-	-	3	1.1
Awls	4	1	-	5	1.9
Core/choppers	1	13	-	14	5.2
Cas. trimm.	6	2	-	8	3.0
Cas. retouch	3	-	-	3	1.1
Raw material (N)	249	18	4	271	100.0

Table 8

GaJj 1: Surface

FREQUENCY OF CORE TYPES

	Chert	Lava	Quartz	Obsidian	Total (N)	Total (%)
Single platform	64	6	-	-	70	17.9
Double: alternate	42	-	-	-	42	10.9
other	42	2	-	-	44	11.1
Multi	44	-	-	-	44	11.1
Disc	-	1	-	-	1	0.3
Casual	119	-	1	-	120	30.6
Fragmentary	71	-	-	-	71	18.1
Raw material (N)	382	9	1	-	392	100.0
(%)	97.4	2.3	0.3	-		

Table 9

GaJj 1: Surface

FREQUENCY OF MAJOR ARTIFACT CLASSES

	Chert	Lava	Quartz	Obsidian	Total (N)	Total (%)
Trimmed pieces	249	18	4	-	271	9.2
Cores	382	9	1	-	392	13.4
UTFF	2110	114	43	8	2275	77.4
Raw material (N)	2741	141	48	8	2938	100.0
(%)	93.3	4.8	1.6	0.3		

Table 10

GaJj 1

INVENTORY OF FINDS FROM EXCAVATIONS 1 AND 2

Excavation Unit	Excavation 1								Excavation 2							Mixed Levels	Total (N)
	A	B	C	D	E	F	G	(N)	A	B	C	D	E	F	(N)		
Stone:																	
Microliths	8	8	7	15	2	1	-	41	6	6	4	4	1	-	21	-	62
Scrapers	-	-	-	-	-	-	-	-	-	-	-	-	-	-	-	-	-
Outils écaillés	-	-	-	-	-	-	-	-	1	1	-	-	-	-	2	-	2
Burins	-	-	-	1	-	-	-	1	-	-	-	-	-	-	-	-	1
Awls	-	-	-	-	-	-	-	-	-	-	-	-	-	-	-	-	-
Core/choppers	-	1	1	-	-	-	-	2	-	-	-	-	-	-	-	-	2
Cas. trimm.	-	-	-	2	-	-	-	2	1	-	-	-	-	-	1	-	3
Cas. retouch	-	-	-	-	-	-	-	-	-	-	-	-	-	-	-	-	-
Whole flakes	12	5	27	32	1	8	7	92	25	42	9	8	-	-	84	17	193
Broken flakes	35	30	25	55	5	9	10	169	31	52	15	9	1	-	108	24	301
Angular frags.	15	20	36	46	10	6	4	137	29	62	16	8	6	-	121	26	284
Cores	3	3	10	25	1	-	-	42	8	13	2	1	3	2	27	1	70
Groundstone	-	-	-	-	1	-	-	1	-	-	-	-	-	-	-	-	1
Hammerstones	-	-	-	2	-	-	-	2	-	-	-	-	-	-	-	-	2
Bone harpoons:																	
Uniserial	-	1	1	-	-	1	2	5	-	1	-	-	-	1	2	1	8
Pottery:																	
Undecorated	-	-	-	2	1	3	-	6	2	2	2	2	-	1	9	1	16
Bone frags. (NID):																	
Mammal	21	-	28	51	10	8	17	135	9	34	4	17	-	-	64	24	223
Fish	107	74	152	144	62	26	44	609	9	21	13	31	-	-	74	107	790
Turtle	-	-	2	4	-	-	-	6	-	-	-	-	-	-	-	-	6
Bones w/art. sur.	-	-	-	-	1	-	1	2	-	-	-	-	-	-	-	-	2
Teeth:																	
Mammal	3	1	1	4	-	-	-	9	-	1	-	-	-	-	1	2	12
Fish	1	-	1	1	2	-	2	7	1	-	1	-	-	-	2	3	12
Beads:																	
Ostrich eggshell	-	-	-	-	-	-	-	-	-	1	3	-	-	-	4	-	4

Table 11

GaJj 1: Excavated

FREQUENCY OF TRIMMED PIECES

	Excavation 1		Excavation 2		Total	
	Chert	Lava	Chert	Quartz	(N)	(%)
Microliths:					(62)	(89.0)
Triangle	1	-	-	-	1	1.4
Trapezoid	1	-	-	-	1	1.4
Crescent	10	-	4	-	14	20.0
Curved Backed flake	13	-	5	-	18	25.7
Truncations	1	-	1	-	2	2.9
Miscellaneous backed	-	-	1	-	1	1.4
Fragmentary	15	-	10	-	25	35.8
Outils écaillés	-	-	-	1	1	1.4
On a micro-percoir	-	-	1	-	1	1.4
Burin	1	-	-	-	1	1.4
Core/chopper	-	2	-	-	2	2.9
Cas. trimm.	2	-	1	-	3	4.3
Raw material (N)	44	2	23	1	70	100.0

Table 12

GaJj 1: Excavated

FREQUENCY OF CORE TYPES

	Excavation 1	Excavation 2	Total	
	Chert	Chert	(N)	(%)
Single platform	3	4	7	10.1
Double: alternate	1	4	5	7.2
other	5	2	7	10.1
Multi	8	4	12	17.6
Disc	2	-	2	2.9
Casual	9	8	17	24.6
Fragmentary	14	5	19	27.5
Raw material (N)	42	27	69	100.0

Table 13

GaJj 1: Excavation

FREQUENCY OF MAJOR ARTIFACT CLASSES

	Excavation 1			Excavation 2				Mixed		Total	
	Chert	Lava	Quartz	Chert	Lava	Quartz	Obsidian	Chert	Lava	(N)	(%)
Trimmed Pieces	44	2	–	23	–	1	–	–	–	70	7.7
Cores	42	–	–	27	–	–	–	1	–	67	7.3
UTFF	379	11	8	305	2	5	1	63	4	778	85.0
Raw material (N)	465	13	8	355	2	6	1	64	4	915	100.0

Table 14

GaJj 1

SURFACE AND EXCAVATED TOTALS COMBINED

	Surface	Excavated	Assemblage Total (N)
Stone:			
Microliths	208	62	270
Scrapers	13	-	13
Outils écaillés	17	2	19
Burins	3	1	4
Awls	5	-	5
Core/choppers	14	2	16
Cas. retouch	3	-	3
Cas. trimm.	8	3	11
Whole flakes	540	193	733
Broken flakes	870	301	1171
Angular flakes	865	284	1149
Cores	392	70	462
Groundstone	3	1	4
Hammerstones	3	2	5
Bone harpoons:			
Uniserial	47	8	55
Biserial	1	-	1
Barbless	1	-	1
Pottery:			
Undecorated	51	16	67
Bone frags. (NID):			
Mammal	-	223	223
Fish	-	790	790
Turtle	-	6	6
Bones w/art. sur.	4	2	6
Teeth:			
Mammal	8	12	20
Fish	8	12	20
Beads:			
Ostrich eggshell	5	4	9
Bone	1	-	1
Stone	1	-	1

GaJj 1 Table 15

MAMMALIAN IDENTIFICATION

SURFACE

HIPPOPOTAMIDAE

Hippopotamus (*Hippopotamus* sp.)

BOVIDAE

Grant's Gazelle[1] (*Gazella granti*)

Oryx[1] (*Oryx beisa*)

EXCAVATION

SUIDAE

Warthog (*Phacochoerus* sp.)

RHINOCEROTIDAE

Rhinocerus? (sp. indet)

BOVIDAE

Grant's Gazelle[2] (*Gazella granti*)

---

[1]Tentative identification based on size of post cranial elements

[2]Tentative identification based on size of dental elements

INTRODUCTION

The site of GaJj 2 was located in Area 100 on the northern flank of the Koobi Fora Ridge (Fig. 8). Abundant lithic artifacts, bone harpoons, pottery and fish and mammal remains were discovered exposed on the eroding outcrops of a thick complex of coarse sands and fine gravels. Fragments of a human skeleton were also identified from a sandy horizon lower in the section. The artifacts exposed by erosion were sampled at two locations within the site. No excavations were undertaken. Two altimeter readings indicated the site lay 76-78 m above the June 1976 level of Lake Turkana.

STRATIGRAPHY

The Galana Boi deposits in Area 100 formed a relatively thick sequence of sandy units that lay unconformably on Upper Member beds of the Koobi Fora Formation. The sands graded from relatively fine at the base of the section to coarser grained at the top. Finer grained sediments, diatomaceous silts and clays, were noticeably absent. Molluscs were present in several of the sandy horizons. A well defined coquina of Corbicula, Melanoides and Mutela shells occurred at the base of sections B and C. Dispersed fragmentary shells were present in the artifact bearing horizon.

Outcrops of Holocene deposits were located around the edges of a plateau-like erosional remnant, roughly 400 m E/W by 700 m N/S (Fig. 18). The deposits once formed a continuous mantle along the flanks of the Koobi Fora Ridge. They are presently undergoing extensive erosion by modern gullies from the Bogi Karuma Valley. Well exposed sections were present in the eastern, western, and southern sectors of the outcrop. Geological sections were drawn at three principal exposures. Each section is briefly described below.

Eastern Exposure (Fig. 19, Section A)

The Galana Boi deposits overlay Upper Member dark brown clays and coarse sands. The initial Holocene incursion was marked by a lens of well sorted fine sand with pockets of gravels and broken Etheria shells near the base. The lower levels were virtually devoid of cultural material but numerous artifacts were observed eroding from the uppermost sandy layer. Artifacts exposed as a consequence of erosion were sampled.

Western Exposure (Fig. 19, Section B)

Upper Member sediments were not exposed. The Galana Boi units consisted of a series of fine to coarse sands with a 40 cm lens of Cleopatra shells near the base of the section. Artifacts exposed by erosion were sampled from the topmost coarse sands.

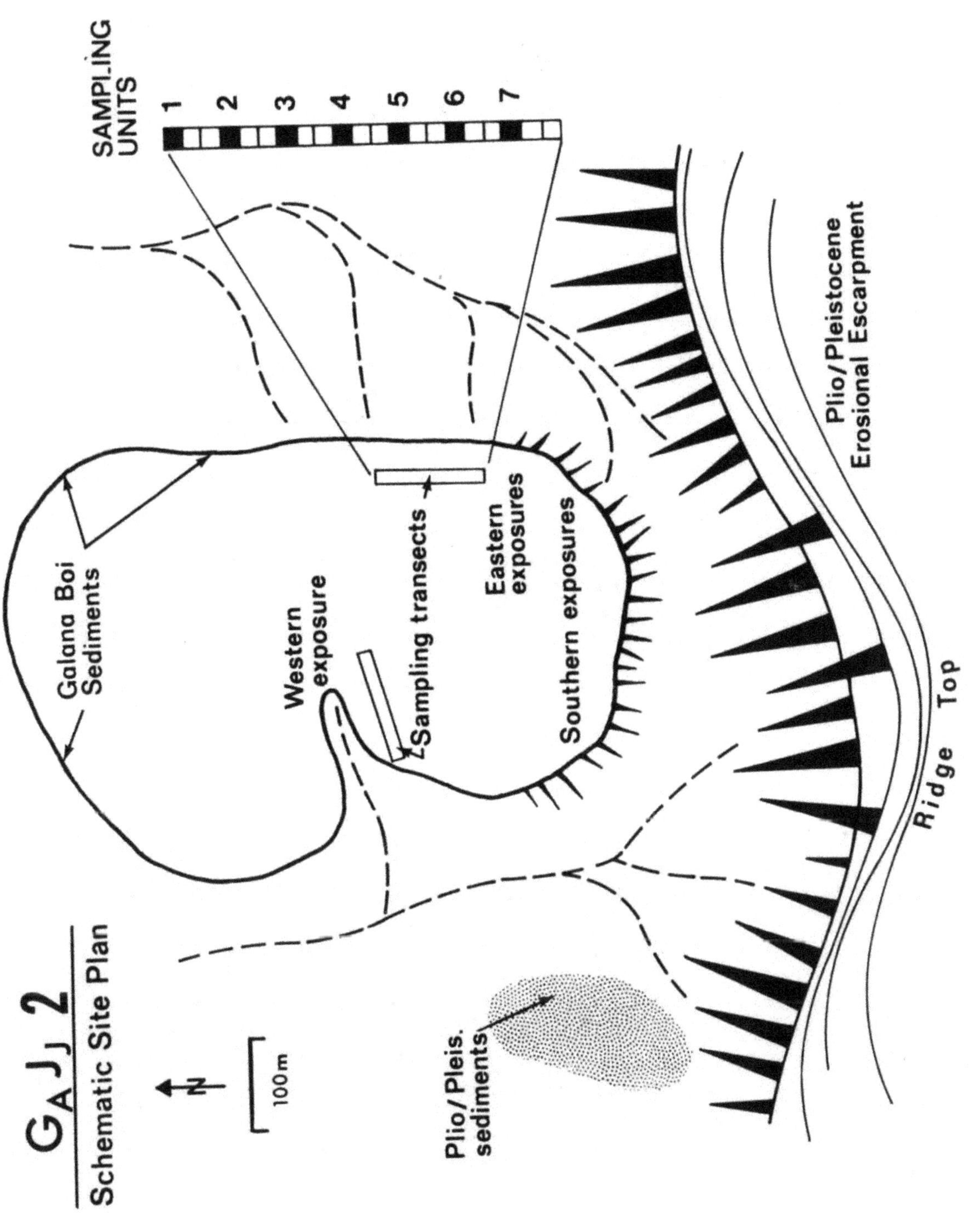

Figure 18. GaJj 2. Schematic site plan. Sampling transects are indicated.

Southern Exposure (Fig. 19, Section C)

Upper Member sediments were not present in this section. The Galana Boi units consisted predominately of fine grained, semi-consolidated silty sands. Fragments of a human skeleton, including a broken mandible and parts of a cranium, were eroding from one lens of silty sand just below the artifact bearing horizon. The skeletal fragments were encrusted with thick calcium carbonate cement and surrounded by a concentration of irregularly shaped $CaCO_3$ concretions. A careful search along the outcrop failed to yield associated artifacts. This exposure was not sampled.

Summary

Towards the top of each section there occurred a layer of coarse sands and sandy fine gravels which is thought to represent deposition along a fluctuating beach shoreline. The eroded outcrops of this layer were littered with artifacts and faunal remains and is believed to be the main archaeological horizon at the site. Additional cultural material may be eroding from underlying horizons but a diligent search failed to locate any artifacts in situ. It should be noted however, that the fragmentary human skeleton was found at just such a lower level. Because of the discontinuous nature of the outcrop and widespread facies differences, correlations between the three sections proved difficult. It does appear likely, nevertheless, that the zone of coarse gravelly sands and artifacts can be correlated. The sediments at the base of sections B and C also appeared to correlate (Fig. 19).

SAMPLING

Artifacts exposed by erosion were collected from the eastern and western exposures of the site. The unconsolidated sandy sediments at each exposure have been extensively eroded which has resulted in a network of thin often discontinuous ridges and steep sided gullies. Such ridges were usually several hundred meters long and only seven to eight meters wide. At both of the large exposures, one of the narrow ridgetops was sampled by the linear transect method (Fig. 18). Each transect was laid out parallel to the alignment of a ridgetop (NS alignment for the eastern exposures; SW for the western exposures). This sampling technique maximized coverage of both the ridgetop surfaces as well as sections of the steep sided gullies. For the sampled area, 21 collecting units were laid out, each unit consisting of a 5 m square.

CULTURAL MATERIAL

Table 16 provides an inventory of the surface finds by sampling unit. Artifactual material collected during the 1974 and 1975 field seasons is also tabulated.

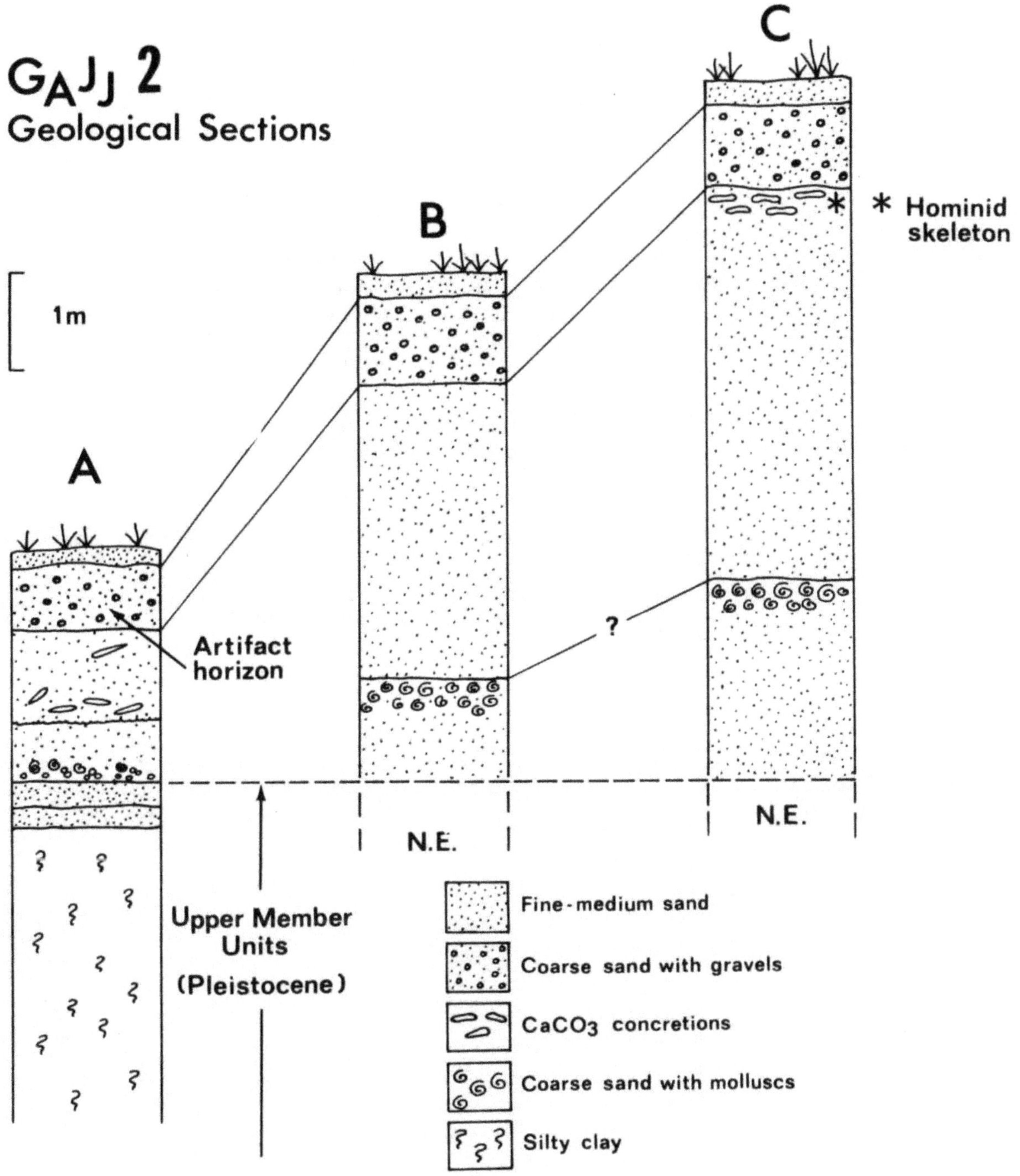

Figure 19. GaJj 2. Correlated geological sections.

Stone Artifacts (Tables 17-19; Fig. 20)

Small nodules of chert were the predominant raw material at GaJj 2. Obsidian was not recovered. Microlithic tools constituted the largest group of trimmed stone artifacts followed by various scrapers and core/choppers. Outils écaillés were present. Crescents represented nearly 50% of the microlithic forms. The debitage consisted primarily of small chert flakes and flake blades. Cobbles of lava were selected for the removal of larger flakes.

Bone Artifacts (Fig.21, a-f).

Of the twenty-five bone harpoons collected from the surface, seven were recovered from the sampling squares. The remaining eighteen specimens were collected outside the sampling units, scattered along the outcrop's edge. Most of the harpoons possessed a single row of barbs but a broken biserial and two barbless examples were also recovered. One of the barbless artifacts displayed a series of grooved circular incisions (Fig. 21, f). Both notching and circular grooved incisions were present at the base of the uniserially barbed harpoons.

Pottery (Figs. 21, g-l; 22)

Both decorated and undecorated sherds were collected from GaJj 2. Nearly 80% of the sample was recovered from the eastern exposure, with 40% of the total sherds coming from unit six. This concentration of material in one sampled square suggests the presence of a single vessel. Three of the associated sherds were later joined together (Fig. 22, a). Unfortunately, neither rim profile nor shape of the pot could be determined.

In general, the pottery remains were poorly fired and possessed a coarse quartz temper. Several decorated sherds had faint rows of impressions that were similar to decorated sherds from FxJj 12 North (Fig. 21, j, k; 22, b). A single decorated rim was collected as well as two undecorated examples with tapering lips. Wavy-line pottery was not identified.

Fauna

The faunal collection consisted of a well mineralized assortment of fish and mammal remains. The terrestrial mammal fauna was diverse and included a considerable size range. Warthog, equids, small to medium-large bovids, hippopotamus, rhinoceros and elephant have been identified. The majority of the identifications were made from dental remains. A list of identified species is presented in Table 20.

Human Remains

During the 1976 field season fragments of a human skeleton were discovered along the eroding margins of the southern exposure. The human remains were identified by body part but were not removed.

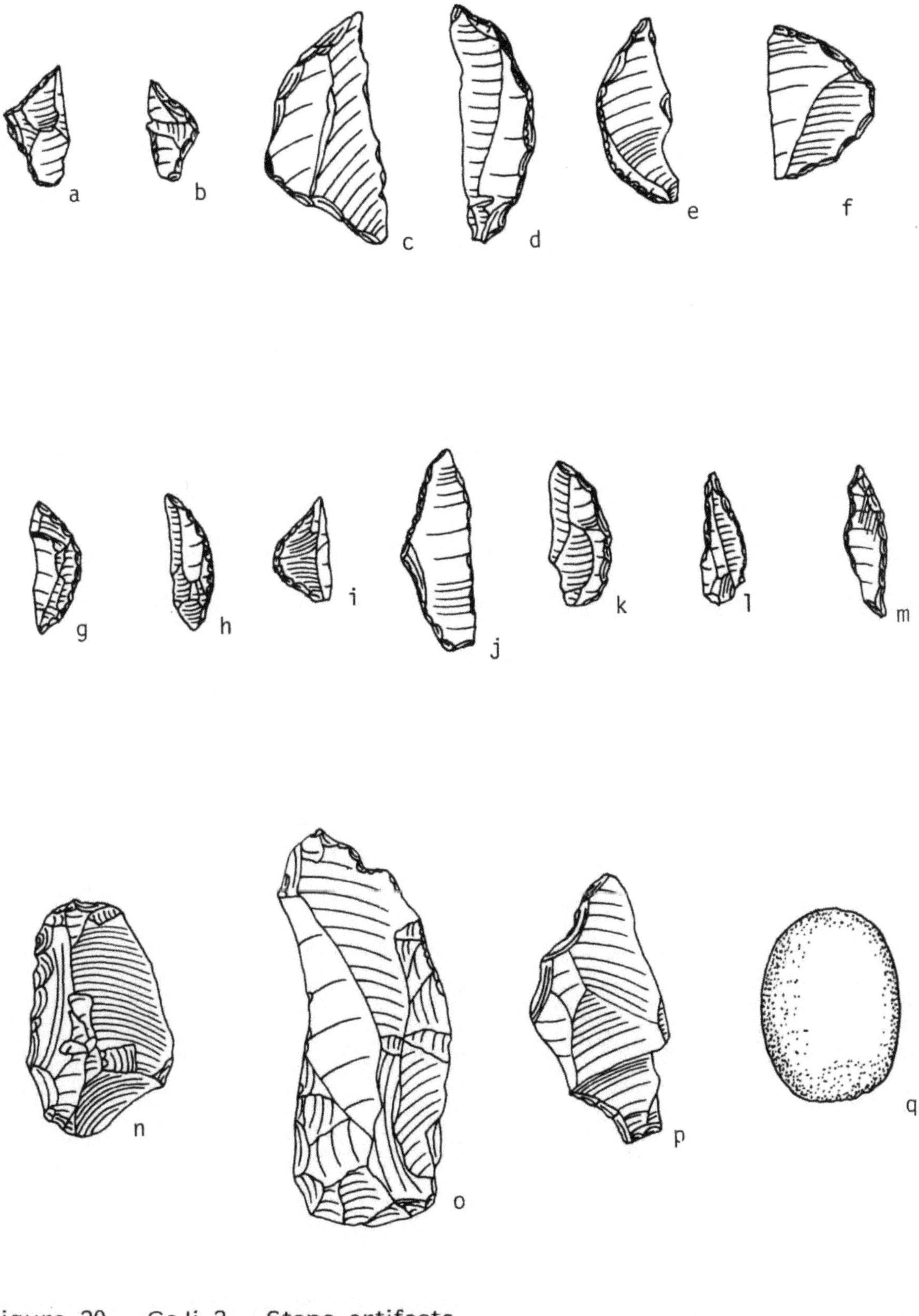

Figure 20. GaJj 2. Stone artifacts.
(a) triangle; (b) trapezoid; (c-i) crescents; (j-k) curved backed flakes; (l-m) micro-percoirs; (n) side scraper; (o) casual retouch; (p) core trimming flake; (q) hammerstone. Natural size.

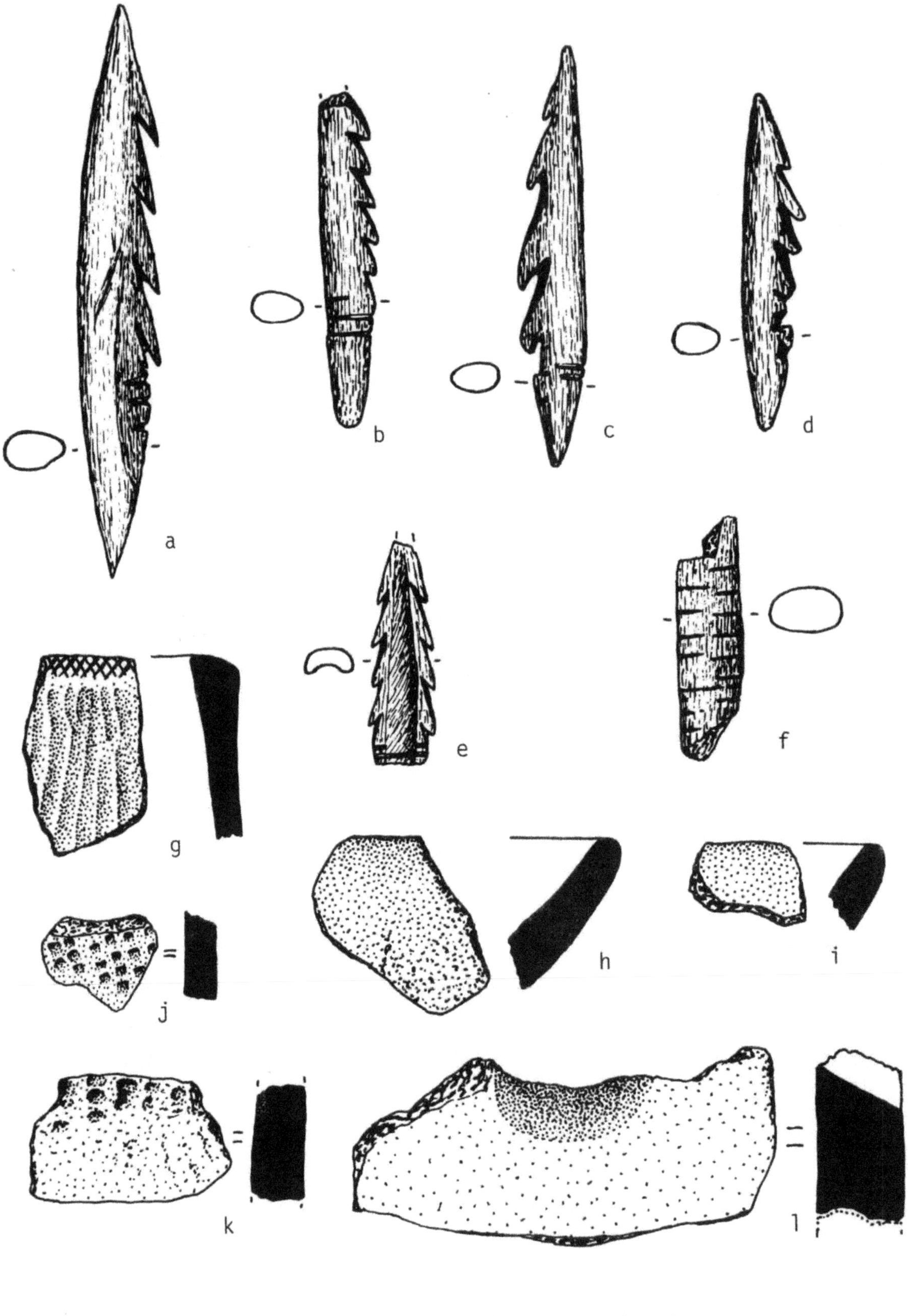

Figure 21. GaJj 2. Bone artifacts and pottery sherds.
(a-d) uniserial harpoons; (e) biserial harpoon; (f) barbless with circular incisions; (g) decorated rim; (h-i) undecorated rims; (j-l) decorated body. Natural size.

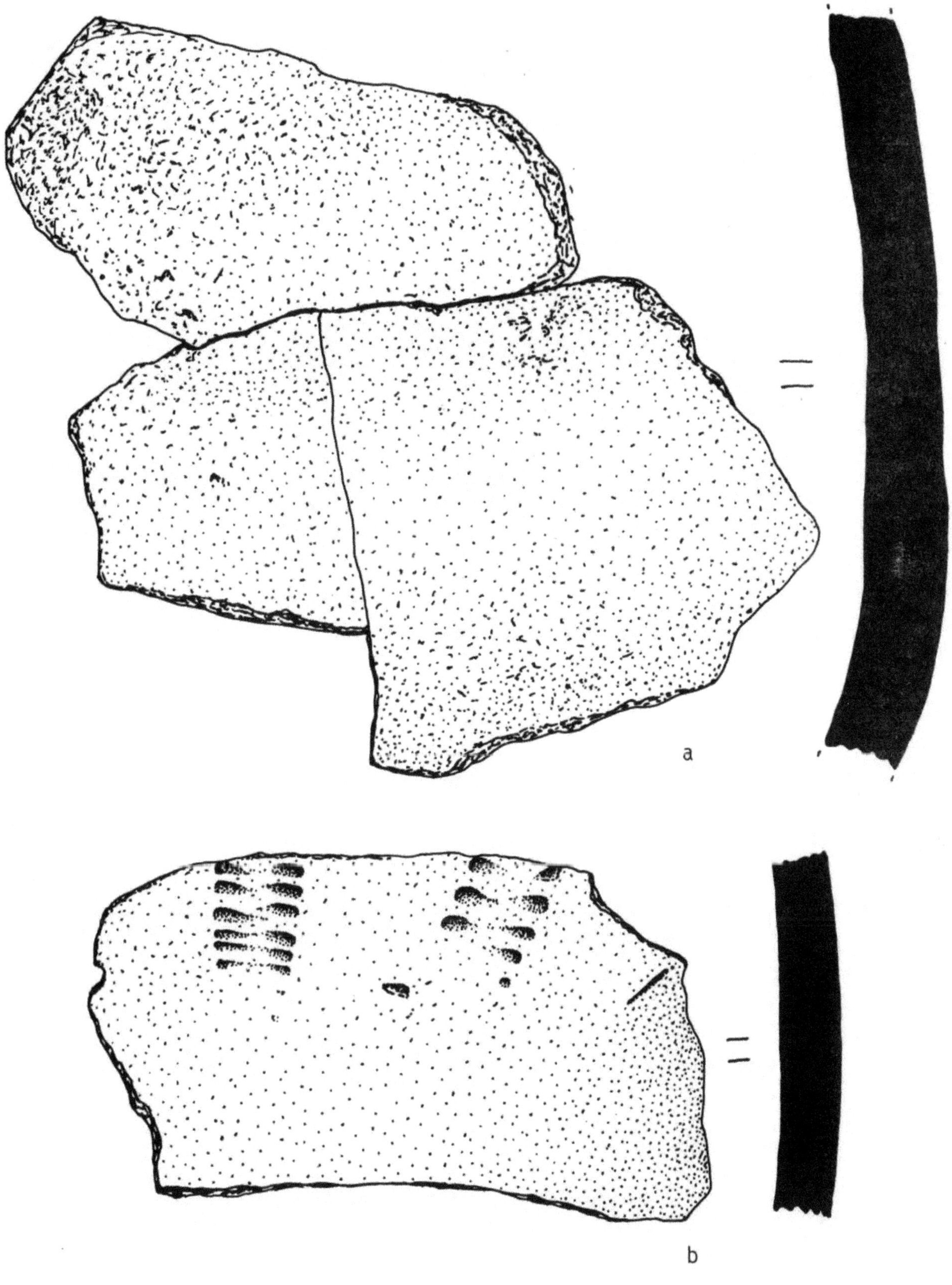

Figure 22. GaJj 2. Pottery sherds.
(a) undecorated body; (b) decorated body. Natural size.

The fragmentary skeleton was subsequently collected by staff of the National Museum during the summer of 1979. Most of the skeleton was represented but was unfortunately poorly preserved. Many of the long bone articular ends were present but usually lacked associated shafts. Both innominates were identified and indicated the individual was male. Three slightly worn molar teeth were preserved in the broken mandible. They suggest the individual was possibly in his mid twenties at time of death.

PRELIMINARY INTERPRETATION

GaJj 2 clearly shows many similarities with other near-shore lacustrine settlements located along the 75-80 m high lake transgression. Stone and bone artifacts, pottery and bone refuse apparently accumulated along a Holocene beach while wave action was still reworking the sands. Presumably, a series of recurrent occupations is represented but, given the circumstances of study, the presence of individual artifact clusters from a single occupation could not be distinguished. In all likelihood, settlement features, if originally present, such as hearths or windbreak foundations, have not been preserved. The archeological materials are thus believed to represent an aggregation of artifacts and food refuse discarded by human groups of similar technological and economic traditions, during a sequence of re-occupations of unknown number and duration.

Table 16

GaJj 2

INVENTORY OF SURFACE FINDS BY SAMPLING UNIT

Unit	1	2	3	4	5	6	7	8	9	10	11	12	13	14	1974-75	Total (N)
Stone:																
Microliths	1	8	1	4	5	9	5	1	1	4	3	5	2	2	6	54
Scrapers	-	-	1	-	1	-	2	-	-	1	-	1	-	-	1	7
Outils écaillés	-	-	-	-	-	-	-	-	-	-	-	-	-	-	-	-
Bec	-	-	-	-	-	1	-	-	-	-	-	-	-	-	-	1
Core/chopper	-	-	-	-	-	-	-	-	-	-	2	-	-	-	2	4
Pick	-	-	-	-	-	-	-	-	-	-	-	-	-	-	1	1
Cas. trimm.	-	-	-	-	-	-	1	-	-	1	1	-	-	-	-	3
Cas. retouch	-	-	-	-	-	-	-	-	-	-	-	1	-	-	-	1
Whole flakes	19	31	9	6	11	15	21	9	-	13	13	10	10	12	-	179
Broken flakes	24	20	14	12	22	19	36	16	2	16	21	24	13	14	-	253
Angular Frags.	14	19	11	6	13	18	28	7	2	13	10	12	18	10	-	181
Cores	30	28	9	3	10	8	26	7	1	4	12	11	6	5	1	160
Hammerstones	-	-	-	-	-	1	-	1	-	-	-	-	-	-	-	2
Bone harpoons:																
Uniserial	1	-	2	-	-	-	1	1	-	1	-	-	-	1	15	22
Biserial	-	-	-	-	-	-	-	-	-	-	-	-	-	-	1	1
Barbless	-	-	-	-	-	-	-	-	-	-	-	-	-	-	2	2
Pottery:																
Decorated	-	-	-	-	-	3	-	-	-	-	-	-	-	-	3	6
Undecorated	2	2	1	5	-	19	5	3	-	-	-	-	-	1	8	46
Bone frags. (NID):																
Mammal	24	15	23	7	1	2	22	21	5	18	19	43	45	30	-	275
Fish	6	3	10	-	1	-	65	4	1	3	3	3	27	18	-	144
Turtle	-	1	-	-	-	-	6	-	-	-	-	-	-	1	-	8
Bones w/art. sur.	-	-	1	-	1	1	8	-	-	-	3	3	4	2	-	23
Teeth:																
Mammal	1	4	4	-	2	1	6	6	1	-	4	8	6	4	-	47
Fish	-	1	-	-	-	-	-	-	-	-	-	-	-	-	-	1
Ostrich eggshell	-	-	-	-	1	-	-	-	-	-	-	-	-	1	-	2
Unmodified stone	12	4	8	9	8	10	6	7	-	-	1	-	2	7	-	74
Raw counts:																
Stone	89	106	45	29	61	71	119	41	6	52	62	64	49	43	11	848
Bone/teeth	31	24	38	7	5	4	107	31	7	21	29	57	82	55	-	498

Table 17

GaJj 2 Surface

FREQUENCY OF TRIMMED PIECES

	Chert	Lava	Total (N)	Total (%)
Microliths:			(54)	(76.1)
Triangle	1	-	1	1.4
Trapezoid	1	-	1	1.4
Crescent	24	-	24	33.8
Curved backed flake	7	-	7	9.9
Micro-percoir	2	-	2	2.8
Truncations	2	-	2	2.8
Miscellaneous backed	6	-	6	8.5
Fragmentary	11	-	11	15.5
Scrapers:			(7)	(9.8)
Side	3	1	4	5.6
Partial side	1	1	2	2.8
Notch	1	-	1	1.4
Bec	-	1	1	1.4
Core/chopper	-	4	4	5.6
Pick	-	1	1	1.4
Cas. trimm.	2	1	3	4.3
Cas. retouch	1	-	1	1.4
Raw material (N)	62	9	71	100.0

Table 18

GaJj 2

FREQUENCY OF CORE TYPES

	Chert	Lava	Quartz	Obsidian	Total (N)	Total (%)
Single platform	23	4	-	-	27	16.9
Double: alternate	3	-	-	-	3	1.9
other	13	-	-	-	13	8.1
Multi	27	-	-	-	27	16.9
Disc	7	-	-	-	7	4.3
Casual	58	2	-	-	60	37.5
Fragmentary	21	1	1	-	23	14.4
Raw material (N)	152	7	1	-	160	100.0
(%)	95.0	4.4	0.6	-		

Table 19

GaJj 2

FREQUENCY OF MAJOR ARTIFACT CLASSES

	Chert	Lava	Quartz	Obsidian	Total (N)	Total (%)
Trimmed pieces	62	9	-	-	71	8.4
Cores	152	7	1	-	160	19.0
UTFF	526	76	11	-	613	72.6
Raw material (N)	740	92	12	-	844	100.0
(%)	87.7	10.9	1.4	-		

Table 20

GaJj 2

MAMMALIAN IDENTIFICATION

HIPPOPOTAMIDAE

Hippopotamus (*Hippopotamus amphibius*)

SUIDAE

Warthog (*Phacochoerus aethiopicus*)

EQUIDAE

Zebra (*Equus* sp.)

RHINOCEROTIDAE

Rhinocerus (sp. ident)

ELEPHANTIDAE

Elephant (*Loxodonta africana*)

BOVIDAE

Dik-dik (*Madoqua* sp.)

Topi? (*Damaliscus karrigum*)

Grant's Gazelle[2] (*Gazella granti*)

Oryx[1] (*Oryx beisa*)

---

[1]Tentative identification based on size of postcranial elements.

[2]Tentative identification based on size of dental elements.

INTRODUCTION

At the site of GaJj 11, abundant artifacts, bone refuse and mollusc shells were observed on the surface of a low ridge of coarse sands and gravels. Over 140 bone artifacts, including uniserial, biserial and triserial harpoons, were recovered from the surface of the ridge. The site was extensively sampled and a small excavation was undertaken. A well defined outcrop of Etheria elliptica lay 100 m to the northeast on the same alignment as the sand and gravel ridge. Several interpretations are offered to account for the stratigraphic and depositional interrelationships between the archeological site and the series of Etheria outcrops. Three altimeter readings indicated the site lay 73-75 m above the June 1976 level of Lake Turkana.

STRATIGRAPHY

The archeological site was located within an area of relatively low topographic relief to the northeast of Allia Bay (Fig. 8). Modern river courses in the area have exposed sediments of the Kubi Algi and Koobi Fora Formations. Galana Boi deposits were distributed discontinuously over the landscape and were usually perched atop these older units. The main archaeological horizon consisted of a low ridge of coarse sands and sandy gravels which overlay finer grained silts and silty clays. These Holocene beds rested in turn on a Plio-Pleistocene bioclastic cemented sandstone.

Figure 23 presents the profile of an 18 m geological trench excavated at right angles to the long axis of the ridge. The basal Holocene units consisted of alternating lenses of fine diatomaceous silts and silty clays. These fine grained horizons were abruptly truncated by a very coarse body of sandy gravels that contained fragmentary shells of the freshwater oyster Etheria elliptica, large broken fragments of Mutela emini, water worn cobbles, bone fragments and stone artifacts. This coarse sandy gravel unit underwent a fining upward sequence that was interrupted several times by additional lenses of sands and gravels with weak forset bedding structures. The forsets had a south to southwest orientation. Near the top of the section Melanoides and Corbicula shells were present in substantial numbers. The final 30 cms of the section consisted of a medium fine to coarse gravelly sand with associated uniserial bone harpoons, stone artifacts and small fragments of charcoal.

To the northeast of the site lay a thick reef-like body of conjoined Etheria shells (Fig. 24). The cemented shells rested on a thin lens of coarse Holocene sand and gravels which is turn underlay Plio-Pleistocene bioclastic sandstones. Four additional outcroppings of the Etheria reef lay 2-4 km to the north of the archaeological site while a sixth exposure lay 3 km to the south (Fig. 25). The cemented shells were anchored to a Holocene sandy substrate, had a north-south orientation and averaged 70-90 cm in thickness (Plate 1). The alignment (essentially parallel to the 75-80 m shoreline) of the Etheria outcrops suggests the cemented oyster shells originally grew on a single or discontinuous series of offshore sandbars.

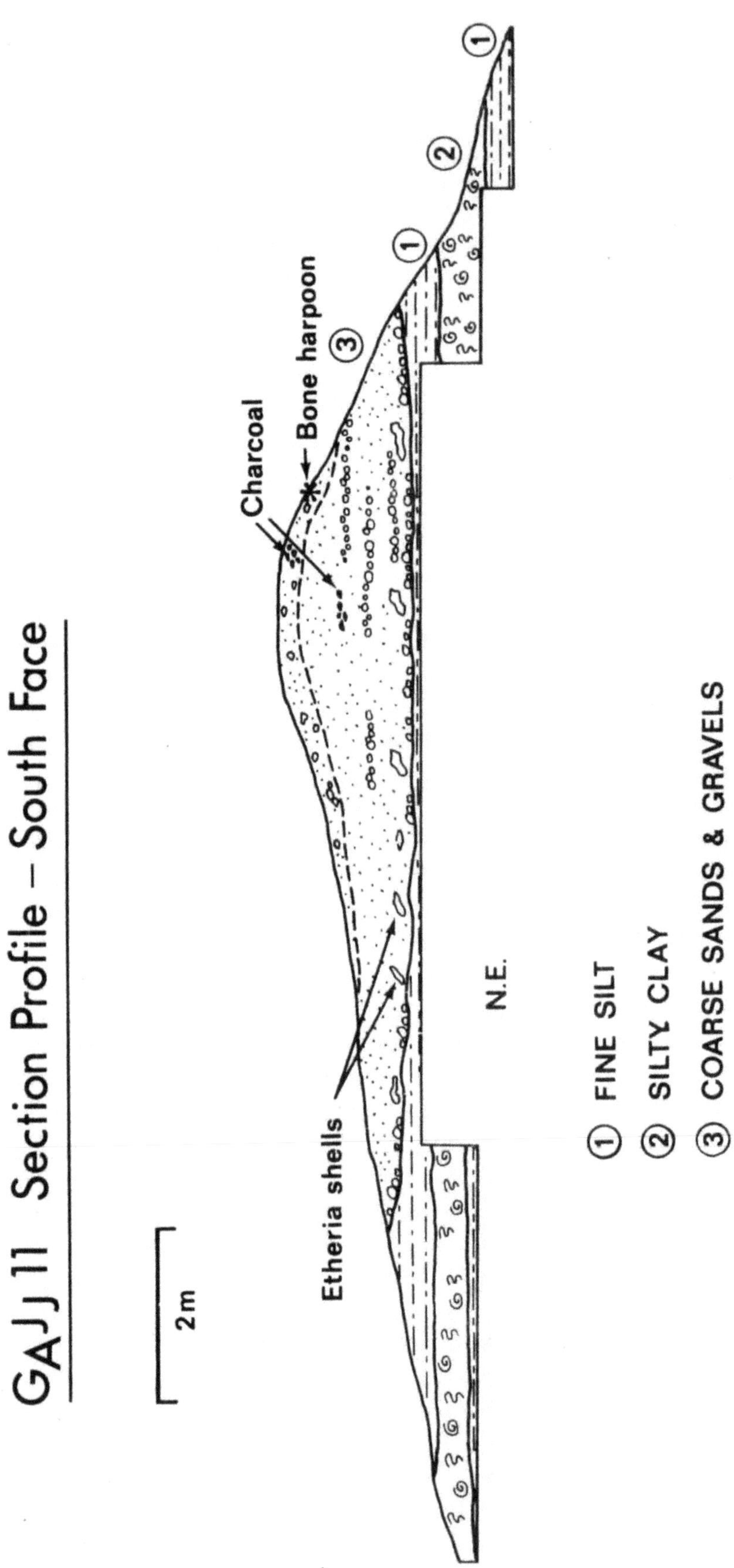

Figure 23. GaJj 11. Principal geological section. South Face.

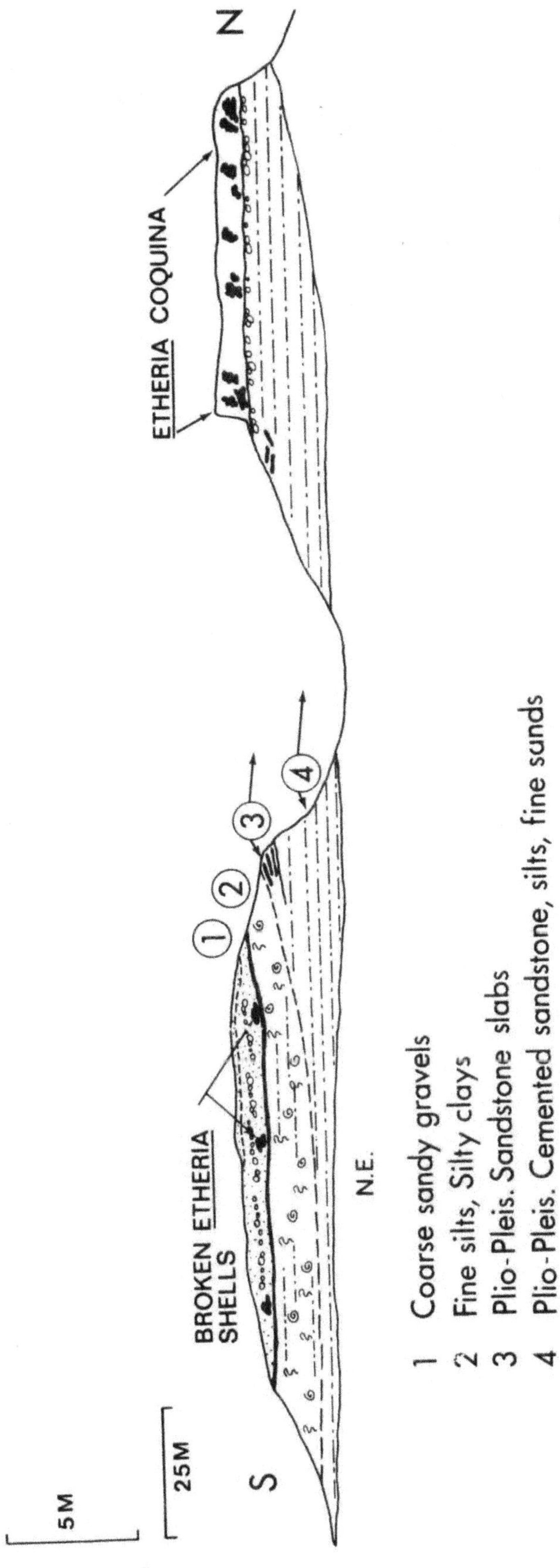

Figure 24. GaJj 11. Topographic profile of the archaeological site and nearby Etheria outcrop.

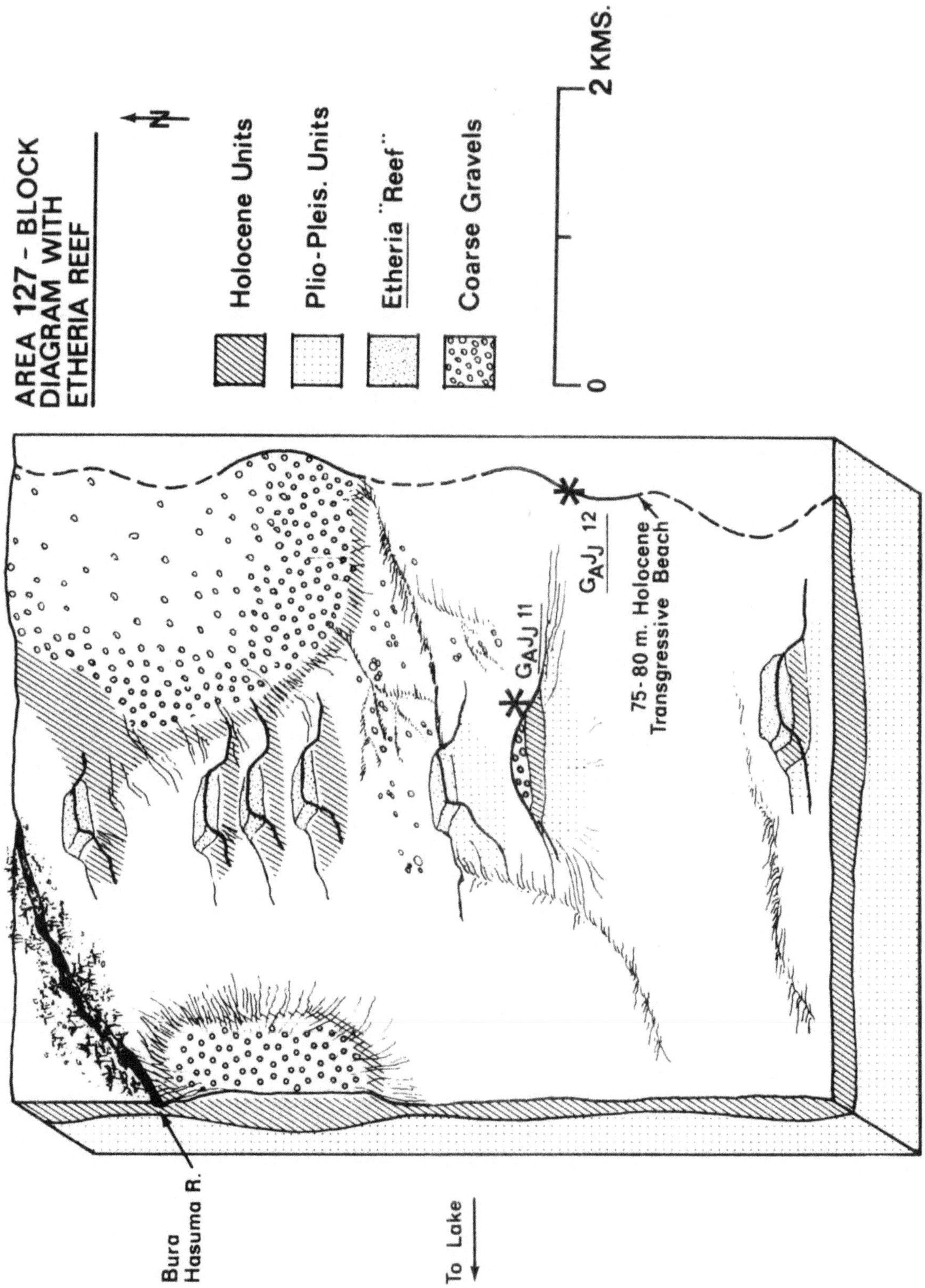

Figure 25. GaJj 11. Schematic block diagram illustrating the location of the archaeological site to the N-S alignment of the Etheria outcrops.

Plate 1. Area 127: Remnant outcrops of Etheria reef.

Etheria elliptica inhabits relatively shallow water, high energy environments such as along the margins of major river channels (Mandahl-Barth 1954). The oyster may be found in the Omo River today, apparently anchoring itself onto compacted bed shoals (Butzer 1971:102). It appears that Etheria also grew along the margins of Holocene offshore sand and gravel bars that were exposed to repeated wave activity. The oyster outcrops contained substantial amounts of broken shell on their western side and decreased in thickness from east to west. These observations suggest the Etheria banks were growing and building out from their westward flank in the direction of wave agitation.

With the exception of the Etheria outcrops and the 75-80 m transgressive beach, the Galana Boi deposits in Area 127 were covered by a mantle of coarse sandy gravels. Their origin may be associated with the first lake recession and subsequent stream incision, erosion and deposition. Alternatively, the gravels may have been deposited by channel activity during phases of the second or third major lake regression. Reworked cultural materials, including bone harpoons, were a consistent feature of these gravels.

A first interpretation of the depositional nature of the archaeological site and its relationship to the Etheria reef is reconstructed as follows. Between 10,000 and 9,500 years ago the Holocene lake reached a transgressive height of 75-80 m. Beach deposits were laid down in Area 127 and a nearby fishing settlement was established along the lake's shoreline at GaJj 12. At some point, between 10,000 and 8,500 years ago, an offshore sandbar was deposited approximately one kilometer west of the high Holocene shoreline. It remains unclear as to whether the bar was laid down as a continuous or discontinuous series of sand and gravel ridges. The sandbar subsequently served as an occupation site (GaJj 11), and, to judge by the considerable quantities of barbed bone harpoons and fish bone refuse, served primarily as a fishing encampment.

During the high transgressive phase, fresh water oysters began to grow on the sandy substratum of the offshore bar. Broken Etheria shells associated with the archaeological deposits suggest the occupation at GaJj 11 was at least partially contemporaneous with the formation of the Etheria reef. However, the reef remains undated and further radiocarbon determinations are needed to confirm this suggestion. Subsequently, small scale transgressive oscillations submerged the offshore bar and permitted patches of Etheria shells to form a reef of nearly one meter in thickness, over 6 kms in length and with a pronounced linear alignment. All of the Etheria outcrops except the example nearest GaJj 11 were anchored to sandy deposits underlain by a thick sequence of Holocene silty clays and fine grained silts (Fig. 26). The Etheria outcrop immediately to the north of the site rested on an elevated remnant of Plio-Pleistocene sandstone and a thin lens of Holocene gravels. These gravels are thought to be laterally equivalent to the sand and gravel deposits at GaJj 11 and to represent an eroded surface of the original offshore bar. The Holocene lake subsequently underwent a major regression approximately 7000-6500 BP (Owens et.

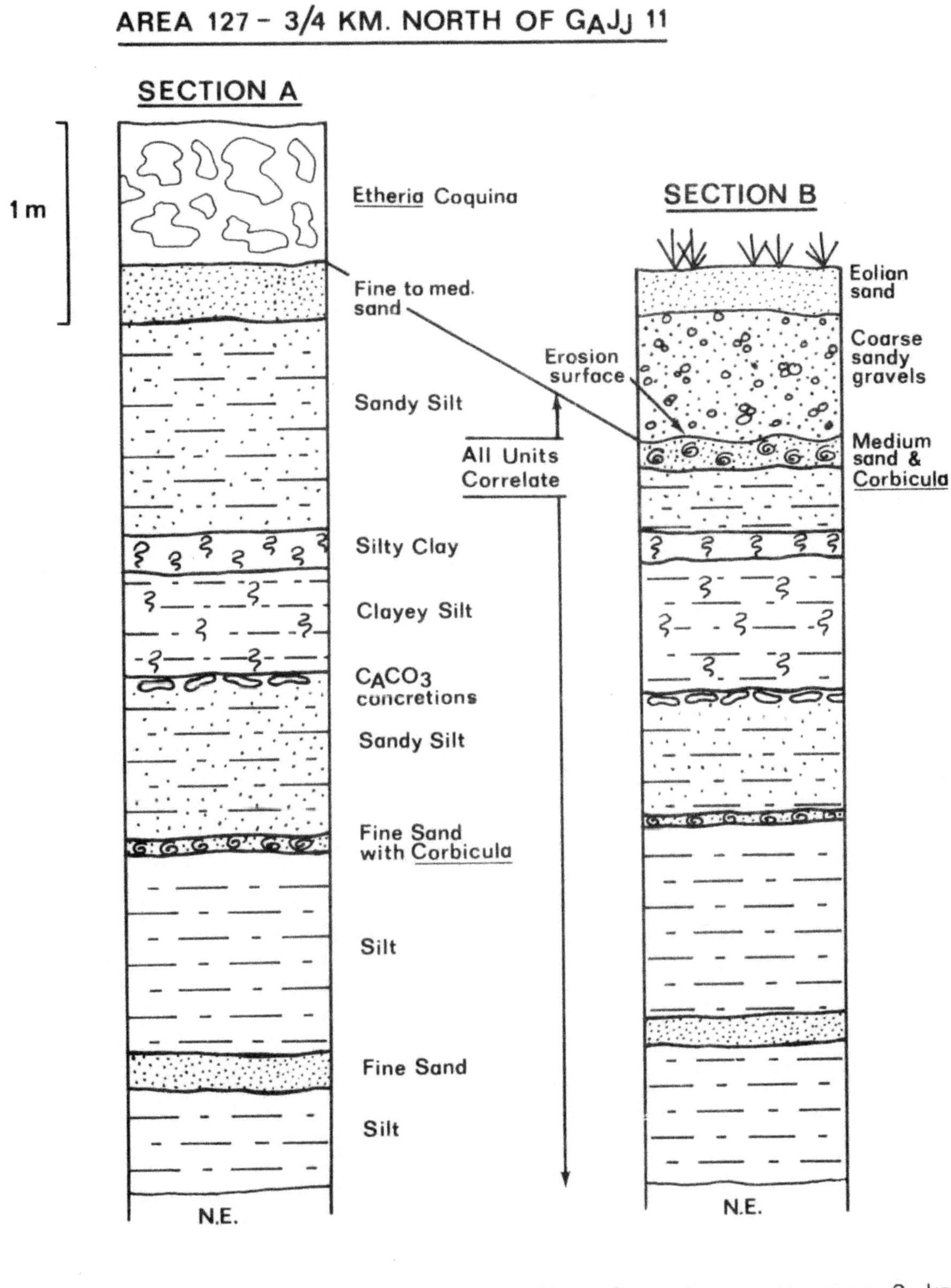

Figure 26. GaJj 11. Geological sections from two exposures 2 kms north of GaJj 11. Sections are similar except for the topmost units.

al. 1982) from its previous 75-80 m height. The Etheria reef was then exposed and the cemented mass of molluscs died.

A second interpretation proposes the low sand mound at GaJj 11 represented former beach sediments that were eroded and redeposited by stream action. This model suggests the Etheria reef also formed on an offfshore sandbar during the maximum lake transgression. However, the archaeological site would then postdate the formation of the oyster reef. During the major regression of approximately 7000-6500 BP the reef would have been exposed, the oysters died, base level was lowered and incision of the lacustrine sediments by stream channels commenced. Artifactual remains, perhaps from a nearby former beach shoreline, were then incorporated into the fluvial sediments and deposited on the surface of the Plio-Pleistocene and fine grained Holocene lake sediments. The broken Etheria shells at the base of the archaeological section would, therefore, represent reworked portions of the earlier oyster reef.

SAMPLING

A systematic checkerboard sampling strategy was utilized at GaJj 11 (Redman 1974). The area to be sampled was subdivided into 12 blocks of 30, five meter square units. Within each 30 unit block, 3 units were collected, beginning, arbitrarily, with the first unit in every other row (Fig. 27). Hence:

potential sampling units:	372
units sampled:	42
11% of the population sampled	

In addition, 5 units were collected within the excavation area. The total number of surface units collected was therefore 47 or 13% of the population.

A substantial number of bone harpoons, stone artifacts and fragmentary parts of several human skeletons lay ouside the collecting units but within the outlying margins of the site. These pieces are individually plotted on the site plan.

EXCAVATION PROCEDURES

An excavation grid, 2 m by 5 m, was established, and, except where natural stratigraphic boundaries were encountered, excavation proceeded in arbitrary 10 cm levels. After excavating 3 one meter squares to a depth of nearly 50 cms, it became apparent that the artifacts were in a disturbed context and had been size sorted into a succession, of coarse sand and gravel lenses. Further excavation was discontinued and work began on an 18 m trench to ascertain the stratigraphic and depositional nature of the ridge.

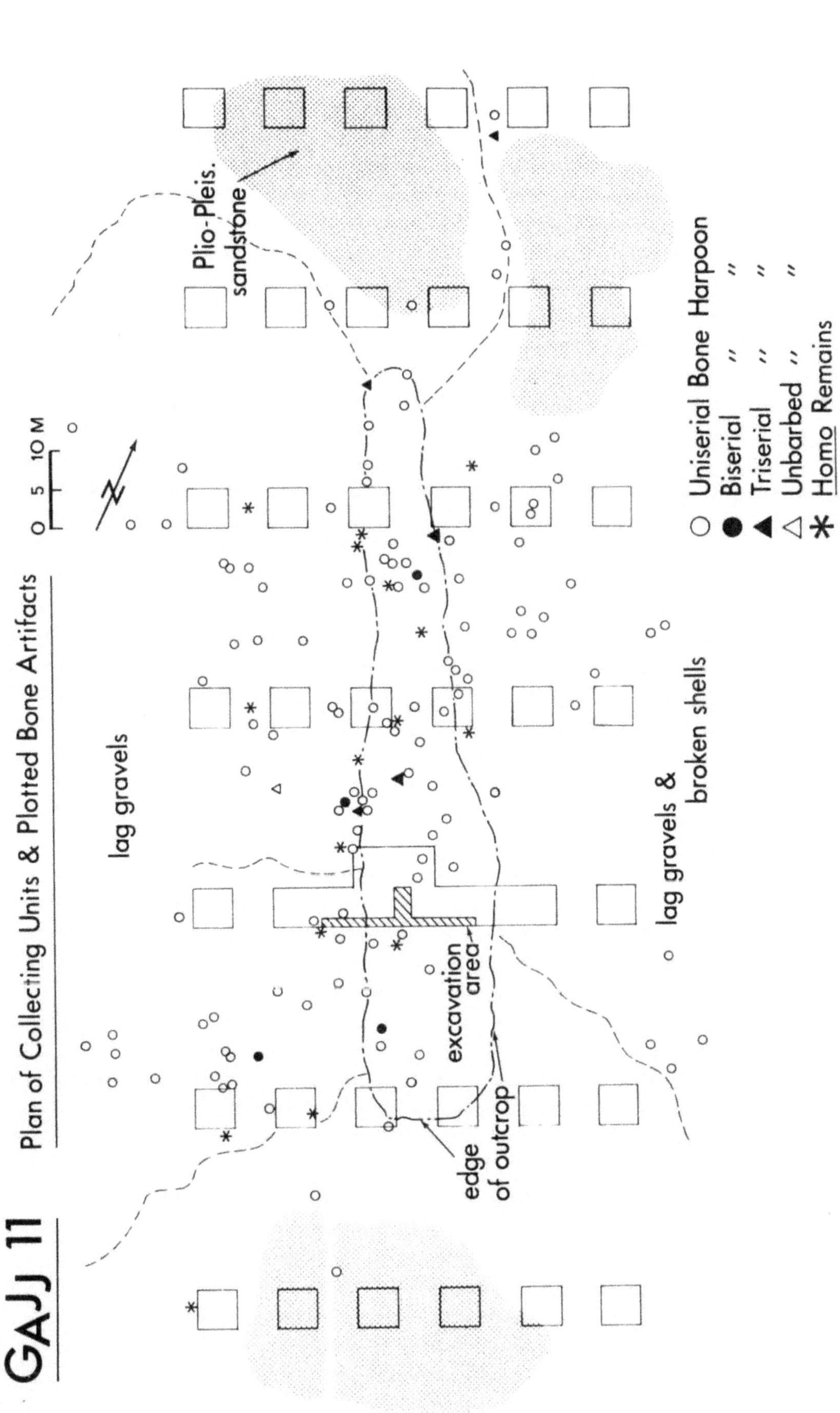

Figure 27. GaJj 11. Plan of sampling units and plotted bone artifacts.

CULTURAL MATERIAL

Table 21 presents an inventory of the collected material by sampling unit. Individually plotted pieces are also tabulated.

Stone Artifacts (Tables 22-23)

Relatively few stone artifacts were collected from the sampling units. Chert was the preferred raw material followed by lava and quartz. Several microliths, *outil écaillés*, core/choppers and a limace were identified. Four core types were represented in low numbers.

Bone Artifacts

Over 140 bone artifacts were collected from the surface including uniserially, biserially and triserially barbed harpoon heads (Figs. 28-29). Thirteen percent of the bone artifacts were collected from the sampling units. The remainder lay outside the sampling units and were recorded on the site plan. All of the 62 harpoon bases were notched. Basal grooved circular incisions wer not identified.

Five broken triserial harpoon heads were collected. In cross section, the artifacts revealed a fourth unbarbed ridge (Fig. 29, e-g). Other triserially barbed bone harpoons have been reported from Daima in the Lake Chad region of Nigeria (Connah 1976, Plate III) and were associated with radiocarbon dates of nearly 2700 BP. The artifacts from GaJj 11 are the only triserial harpoons known to the author from either Northern or Eastern Africa. Whether the three-sided harpoons were used in specialized subsistence activities, such as spearing hippopotami or large Nile Perch, or represented individual or group stylistic preference in artifact form is presently not known.

Faunal Remains

A substantial quantity of faunal material was collected from the sampling units. The great majority of the sample consisted of fish vertebrae and fish spines. Selected fish and hippo bones with articular surfaces were removed from further identification. Importantly, terrestrial mammals were not identified from the preliminary sorting and subsequent identification. The remaining faunal refuse was counted, weighed by sampling unit and returned to the site.

Although broken *Etheria* shells occurred within the site's lower horizons, direct evidence for their consumption was lacking. The mollusc is occasionally eaten, however, as reported from Zaire (Pilsbry and BeQuaert 1927:454).

Human Remains

A large collection of fragmentary but well preserved human remains was discovered by the writer and later by Karen Bell of the National Museum. She reports that at least seven individuals were represented and, judging from the mandibles and maxillae, the individuals varied in size and age. The teeth, especially the molars, showed

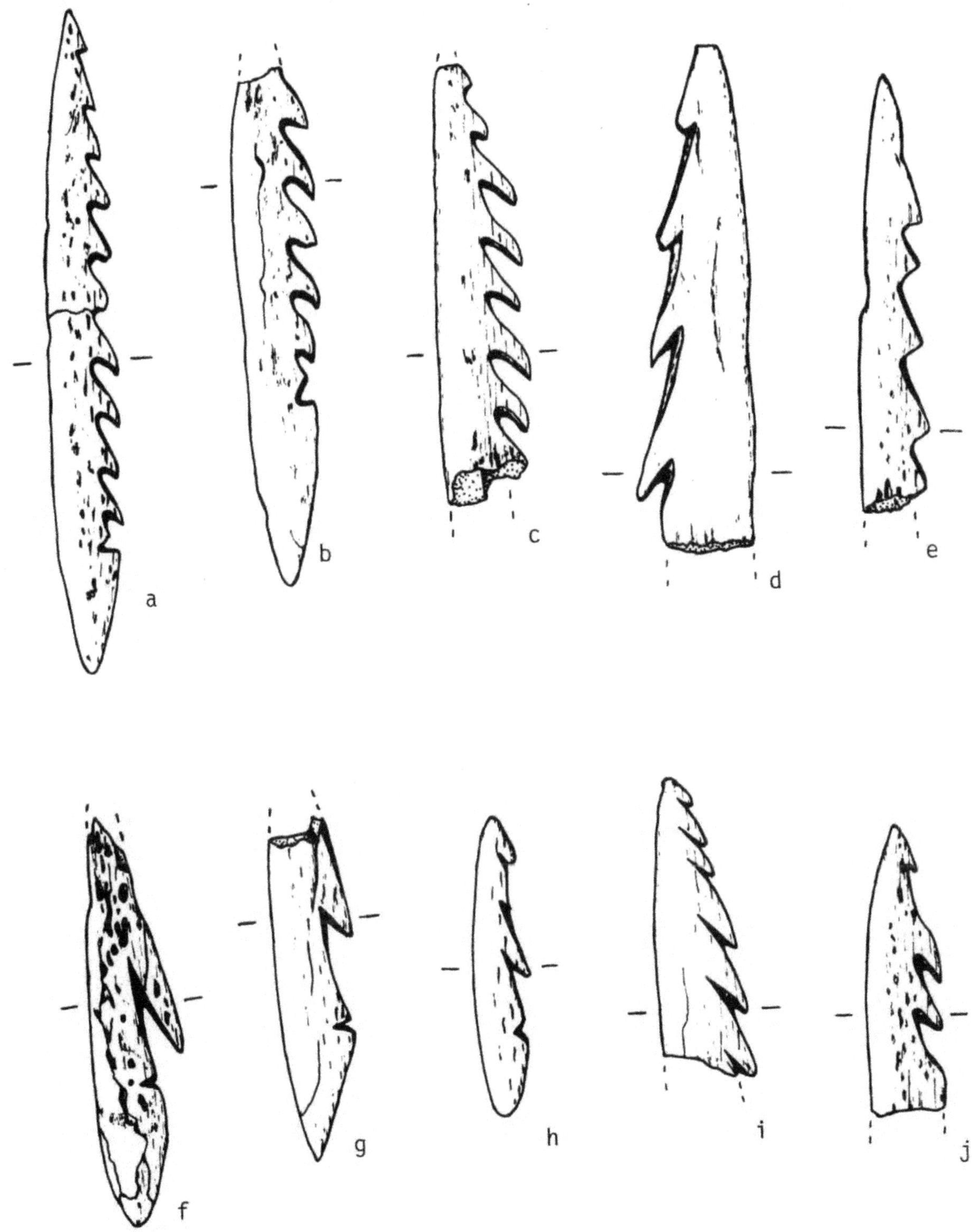

Figure 28. GaJj 11. Bone artifacts.
(a-j) uniserially barbed harpoons. Natural size.

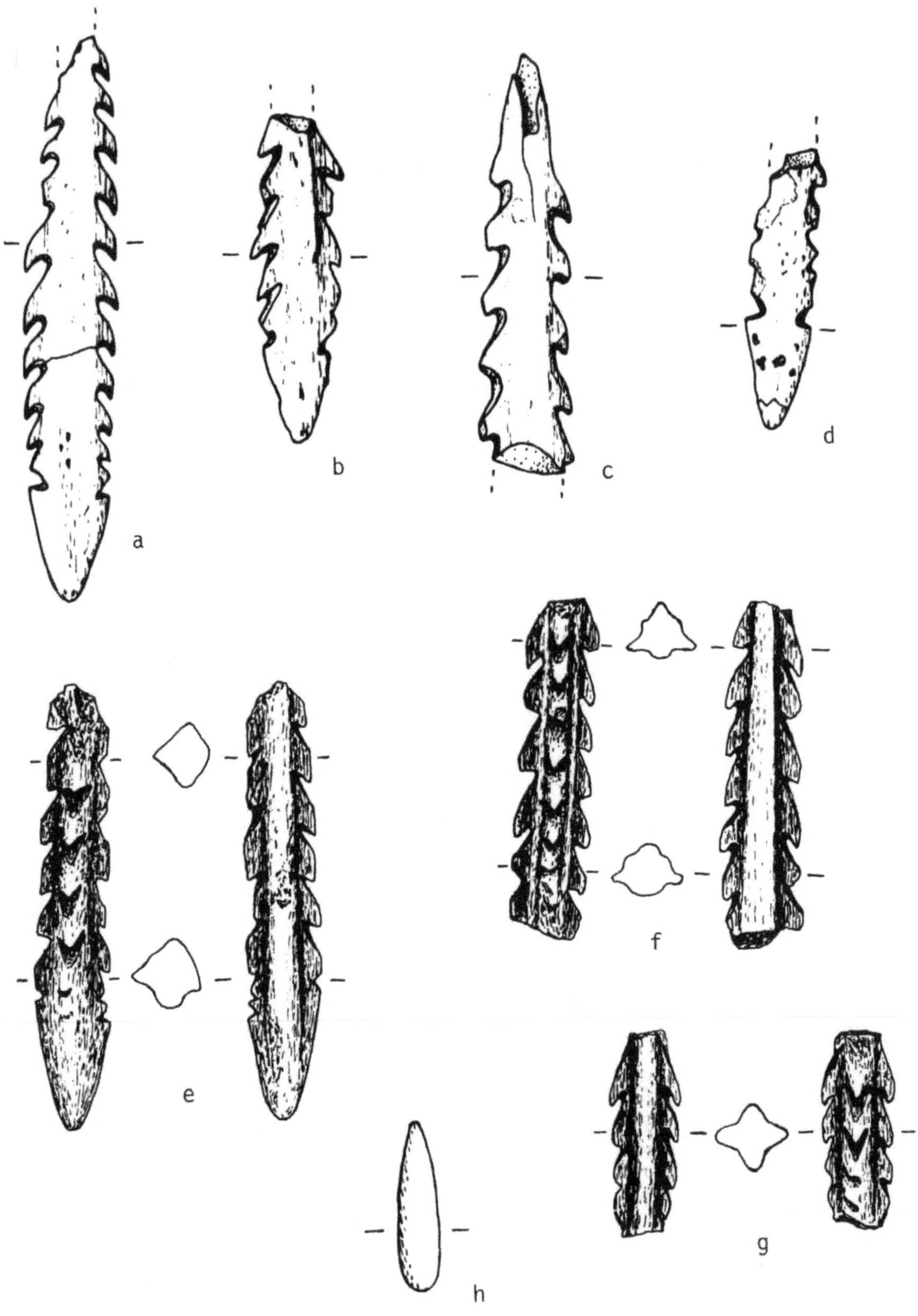

Figure 29. GaJj 11. Bone artifacts.
(a-d) biserial harpoons; (e-g) triserial harpoons; (h) barbless.
Natural size.

considerable amounts of wear. In several cases, dentine was exposed over the entire occlusal surface of the molars. Cranial fragments indicated the individuals had very thick vaulted skulls.

Dating

Three C^{14} age determinations have been received from snail (Melanoides and Mutela emina), freshwater osyter (Etheria elliptica) and fish bone samples (Table 2). Small amounts of charcoal were collected from the upper levels of the geological trench and submitted to laboratories at the Universities of Pennsylvania and Helsinki, Finland. Unfortunately, the charcoal sample proved too small to run a statistically valid age determination.

There was a discrepancy of 1200-1000 years between the shell and fish bone dates. Comparable dating results from Dongodien also indicated wide variance between shell and bone dates. While both shell dates indicated an age of approximately 9000 years for the age of the occupation, the single fish bone date suggests an age of over 1000 years later. Without further dating evidence, a cautionary posture should be taken and the site is therefore thought to date between 8000-9000 BP.

Two Etheria samples from the reef immediately to the north of the archaeological site both yielded an anomalous 'modern' age. The Helsinki laboratory reported clear "traces of atomic bomb radiocarbon in the sample" and that it is difficult to believe "the modern age could be due to contamination only." This perplexing situation is currently under investigation.

PRELIMINARY INTERPRETATION

The reconstruction of human land use patterns at GaJj 11 is largely dependent on resolving the site's paleogeographic setting. One interpretation, favored by the author, suggests the site was located on the surface of an offshore sand and gravel bar and was occupied prior to formation of the Etheria reef. Ths site may have served as a temporary fishing station for groups camped in more sedentary fishing-hunting-gathering settlements along the high beach shoreline. Alternatively, the offshore bar may have been occupied by human groups who lacked a large landbase connection. In the recent past, El Molo fishing settlements were located entirely on two probably similar offshore sandbars in Allia Bay (von Hohnel 1894:132). A second interpretation suggests the site was associated with a regression of the high lake stand, subsequent stream erosion and redeposition of the culture materials from a nearby shoreline based fishing occupation.

The site appears to have been occupied between 8000-9000 BP. The accumulation of large quantities of barbed bone harpoons, fish bones and the absence of terrestrial fauna suggests the site was utilized primarily as a fishing station. Pottery was not present at the site although decorated and undercorated sherds occurred nearby at GaJj 12. Occupation features such as hearths or windbreak foundations were not identified. The recovery of triserially barbed bone harpoon heads are the first known examples from the early Holocene fishing sites of Northern and Eastern Africa.

Table 21
GaJj 11

INVENTORY OF SURFACE FINDS BY SAMPLING UNIT

Unit	1	2	3	4	5	6	7	8	9	10	11	12	13	14	15	16	17	18	19	20	21	22	23	24	25	26
Stone:																										
Microliths	-	-	-	-	-	-	-	-	-	1	-	-	-	-	-	-	-	-	-	-	-	-	1	-	-	-
Scrapers	-	-	-	-	-	-	-	-	-	-	-	-	-	-	-	-	-	-	-	-	-	-	-	-	-	-
Outils écaillés	-	-	-	-	1	1	-	-	-	-	-	-	-	-	-	-	-	-	-	-	-	-	-	-	-	-
Bec	-	-	-	-	-	-	-	-	-	1	-	-	-	-	-	-	-	-	-	-	-	-	-	-	-	-
Core/chopper	1	-	-	-	-	-	-	-	2	-	-	-	-	-	-	1	-	-	-	-	-	-	-	-	-	-
Limace	-	-	-	-	-	-	-	-	-	-	-	-	-	-	-	-	-	-	-	-	-	-	-	-	-	-
Cas. trimm.	-	-	-	-	-	-	-	-	-	1	-	-	-	-	-	-	-	-	-	-	-	-	-	-	-	-
Whole flakes	-	-	-	3	6	9	3	4	1	12	9	1	3	3	1	4	2	1	-	2	2	6	3	3	6	1
Broken flakes	-	-	-	-	3	3	1	-	-	18	4	-	-	2	-	2	-	-	2	-	4	3	3	1	1	1
Angular frags.	1	2	1	10	6	10	3	2	2	6	6	1	-	4	1	3	1	-	-	4	1	1	-	2	1	-
Cores	-	-	-	-	1	2	-	-	-	4	1	-	-	-	-	-	-	-	-	-	-	1	-	-	-	-
Bone harpoons:																										
Uniserial	-	-	-	-	-	-	-	-	-	1	-	-	1	-	-	-	-	1	-	-	2	1	-	2	-	-
Biserial	-	-	-	-	-	-	-	-	-	-	-	-	-	-	-	-	-	-	-	-	-	-	-	-	-	-
Triserial	-	-	-	-	-	-	-	-	-	-	-	-	-	-	-	-	-	-	-	-	-	-	-	-	-	-
Barbless	-	-	-	-	-	-	-	-	-	-	-	-	-	-	-	-	-	-	-	-	-	-	-	-	-	-
NID	-	-	-	-	-	-	-	-	-	-	-	-	-	-	-	-	-	-	-	-	-	-	-	-	-	-
Pottery:																										
Undecorated	-	-	-	-	-	-	-	-	-	-	-	-	-	-	-	-	-	-	-	-	-	-	-	-	-	-
Bone frags. (NID):																										
Mammal	-	-	-	-	-	-	-	-	-	-	-	-	-	-	-	-	-	-	-	-	-	-	-	-	-	-
Fish	-	-	-	-	-	-	-	3	-	-	-	-	-	-	-	-	-	-	-	2	-	-	-	-	-	-
Turtle	-	-	-	-	-	-	-	1	-	-	-	-	-	-	-	-	-	-	-	1	-	-	-	-	-	-
Crocodile	-	-	-	-	-	-	-	-	-	-	-	-	-	-	-	-	-	-	-	-	-	-	-	-	-	-
Bones w/art. sur.	-	-	-	-	-	-	-	-	-	1	-	-	-	-	-	-	-	-	-	-	-	-	-	1	1	-
Teeth:																										
Mammal	-	-	-	-	-	-	-	-	-	-	-	2	-	-	-	-	-	-	-	1	-	-	-	-	-	-
Homo sapiens	-	-	-	-	1	-	-	1	-	-	-	-	-	-	-	-	-	-	-	1	2	-	-	-	-	-
Raw counts:																										
Stone	2	2	1	13	15	25	7	6	5	33	20	2	3	9	2	10	3	1	2	6	7	11	7	6	8	2
Bone/teeth	-	-	-	-	1	-	-	4	-	1	-	2	-	-	-	-	-	-	-	5	2	-	-	1	1	-

Table 21 continued

GaJj 11

INVENTORY OF SURFACE FINDS BY SAMPLING UNIT

Unit	27	28	29	30	31	32	33	34	35	36	37	38	39	40	41	42	43	44	45	46	47	Sampled Units (N)	Plotted (N)	Assemblage Total (N)
Stone:																								
Microliths	-	-	-	1	-	-	-	-	-	-	-	-	-	-	-	-	1	-	-	-	-	4	2	6
Scrapers	-	-	-	-	-	-	-	-	-	-	-	-	-	-	-	-	-	-	-	-	-	-	-	-
Outils écaillés	-	-	-	-	1	-	-	-	-	-	-	-	-	-	-	-	-	-	-	-	-	3	-	3
Bec	-	-	-	-	-	-	-	-	-	-	-	-	-	-	-	-	-	-	-	-	-	1	-	1
Core/chopper	-	-	-	-	-	-	-	-	-	-	-	-	-	-	-	-	-	-	-	-	-	4	-	4
Limace	-	-	-	-	-	-	-	-	-	-	-	1	-	-	-	-	-	-	-	-	-	1	-	1
Cas. trimm.	-	-	-	-	-	-	-	-	-	-	-	-	-	-	-	-	-	-	-	-	-	1	-	1
Whole flakes	6	8	8	1	1	4	5	14	4	1	6	6	2	3	2	2	8	8	-	4	-	180	-	180
Broken flakes	9	4	4	1	-	3	7	3	6	2	4	-	3	1	2	-	5	6	1	2	2	113	-	113
Angular frags.	4	2	1	-	-	2	8	8	4	2	4	7	1	-	1	2	2	12	2	1	2	131	-	131
Cores	4	-	1	-	-	1	-	-	-	1	-	-	1	-	-	-	-	-	-	-	-	17	1	18
Bone harpoons:																								
Uniserial	1	-	3	-	-	-	1	1	-	-	-	-	-	-	-	-	2	2	-	-	-	18	105	123
Biserial	-	-	-	-	-	-	-	-	-	-	-	-	-	-	-	-	-	-	-	-	-	-	4	4
Triserial	-	-	-	-	-	-	-	-	-	-	-	-	-	-	-	-	-	-	-	-	-	-	5	5
Barbless	-	-	-	-	-	-	-	-	-	-	-	-	-	-	-	-	-	-	-	-	-	-	1	1
NID	-	-	-	-	-	-	-	-	-	-	-	-	-	-	-	-	-	-	-	-	-	-	8	8
Pottery:																								
Undecorated	-	-	-	-	-	-	-	-	-	-	-	-	-	-	-	-	-	-	-	-	-	-	1	1
Bone frags. (NID):																								
Mammal	-	-	-	-	-	-	-	1	-	-	-	-	-	-	-	-	-	-	-	-	-	1	-	1
Fish	-	-	-	-	-	-	-	1	-	-	-	-	-	-	-	-	-	-	14	-	-	20	-	20
Turtle	-	-	-	-	-	-	-	-	-	-	-	-	-	-	-	-	-	-	1	1	1	5	-	5
Crocodile	-	-	-	-	-	-	-	-	-	-	-	-	-	-	-	-	-	-	2	-	-	2	-	2
Bones w/art. sur.	-	1	-	-	-	-	-	-	-	-	-	-	-	-	-	-	-	1	-	2	2	9	-	9
Teeth:																								
Mammal	-	-	-	-	-	-	-	1	-	-	-	-	-	-	-	-	-	-	1	-	-	5	1	6
Homo sapiens	4	-	1	-	-	-	-	-	-	-	-	1	-	-	-	-	-	-	1	-	4	16	37	53
Raw counts:																								
Stone	23	12	14	3	2	10	20	25	14	6	14	14	7	4	5	4	16	26	3	7	4	455	3	458
Bone/teeth	4	1	1	-	-	-	-	3	-	-	-	1	-	-	-	-	-	1	19	3	3	48	38	86

Table 22

GaJj 11

FREQUENCY OF CORE TYPES

	Chert	Lava	Quartz	Obsidian	Total (N)	Total (%)
Single platform	5	1	-	-	6	33.3
Double: alternate	2	-	-	-	2	11.1
other	-	-	-	-	-	-
Multi	7	1	-	-	8	44.5
Disc	-	-	-	-	-	-
Casual	2	-	-	-	2	11.1
Fragmentary	-	-	-	-	-	-
Raw material (N)	16	2	-	-	18	100.0
(%)	88.9	11.1	-	-		

Table 23

GaJj 11

FREQUENCY OF MAJOR ARTIFACT CLASSES

	Chert	Lava	Quartz	Obsidian	Total (N)	Total (%)
Trimmed pieces	11	5	-	-	16	3.5
Cores	16	2	-	-	18	3.9
UTFF	299	121	3	1	424	92.6
Raw material (N)	326	128	3	1	458	100.0
(%)	71.2	27.9	0.7	0.2		

INTRODUCTION

The site of GaJj 12 was located in Area 127 approximately 3/4 km east of GaJj 11 (Fig. 8). Abundant lithic artifacts, bone harpoons, pottery and faunal remains were exposed on the eroding flanks of a discontinuous outcrop of Galana Boi sediments. Two areas of the site were sampled and geological sections were drawn. Three altimeter readings indicated the site lay 74-78 m above the 1976 July level.

STRATIGRAPHY

The Holocene exposures at GaJj 12 were nearly 500 m in length and consisted of three discontinuous outcrops. The sediments were composed of a fluctuating series of diatomaceous silts, coarse sands and fine gravels. Calcuim carbonate concretions were a regular component of the fine grained silt and sandy units. Molluscs were abundant. They occurred as distinct shell bands and as dispersed fragments throughout several sandy horizons. To the west, the finer grained sediments increased in thickness and could be traced for nearly a kilometer. Artifactual material was eroding from separate horizons of coarse sand and sandy silt at both outcrops A and B. The stratigraphic context of artifacts at outcrop C remains unclear.

Outcrop A (Fig. 30)

Figure 30 reveals a series of fine to coarse sands, diatomaceous silts and shell bands. Cultural remains were present within the uppermost coarse sands as well as the underlying horizon of $CaCO_3$ cemented diatomaceous silt. Broken fragments of *Melanoides* and *Corbicula* shells were plentiful within this latter unit. The section had a cumulative thickness of 2.25 m and was sampled.

Outcrop B (Fig. 31)

Two horizons each of sand and diatomaceous silt were present at this outcrop. Artifactual material was eroding from the uppermost coarse sands as well as the underlying consolidated sandy silt. Molluscs, in the form of scattered broken shell fragments, were distributed throughout the upper three levels. The section had a cumulative thickness of over three meters and was sampled.

Outcrop C (Fig. 31)

A small isolated patch of coarse sand with broken mollusc shells occurred 40 m north of outcrop B. This sand unit was overlain by a second lens of very coarse sandy gravels. At the base of outcrop C a complete uniserial harpoon, several decorated pottery sherds and a small collection of obsidian chippage were exposed. It was not possible to assign a definite stratigraphic provenience to these artifacts as the cultural material could have been derived from either sand horizon.

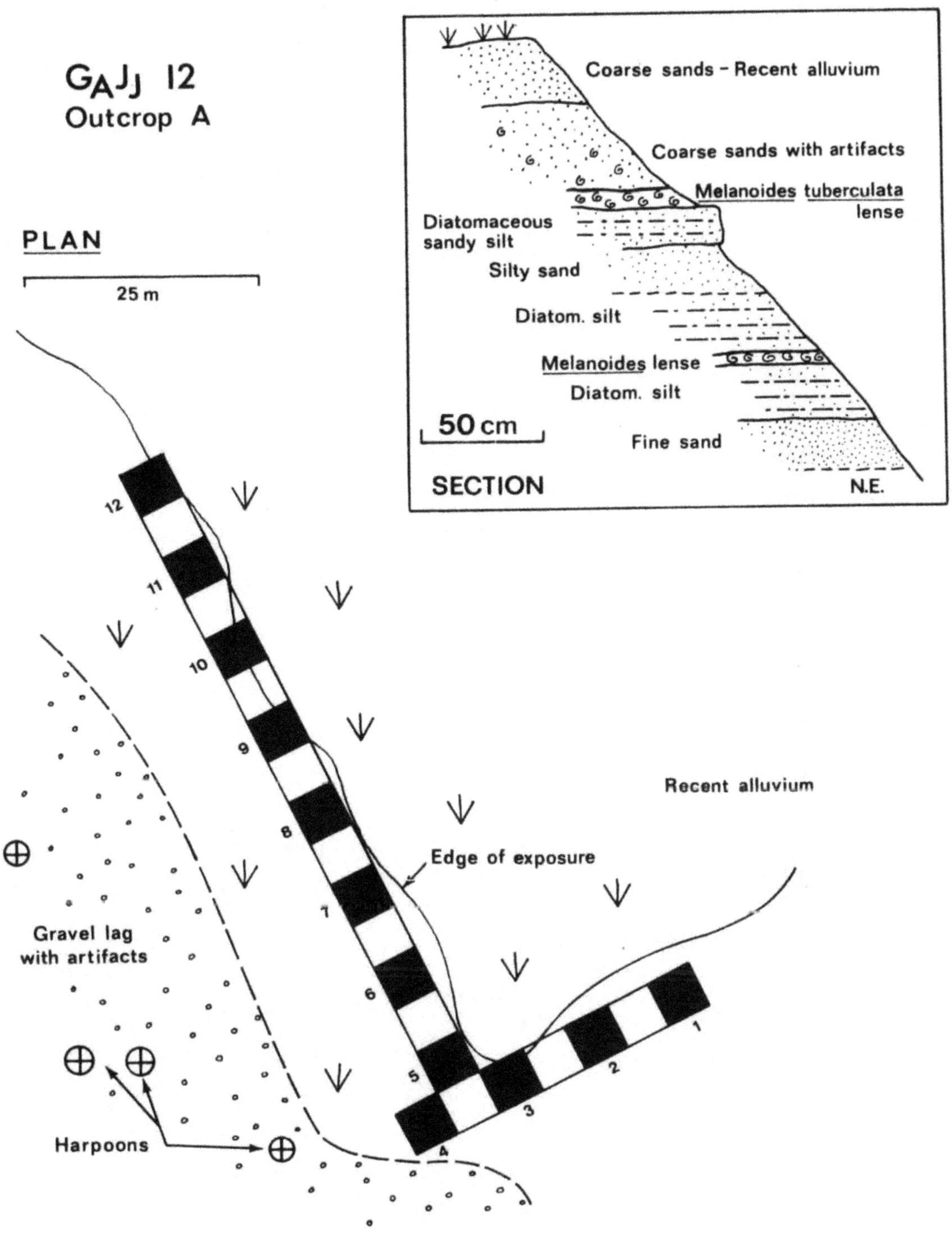

Figure 30. GaJj 12. Sampling units and stratigraphic section at Outcrop A.

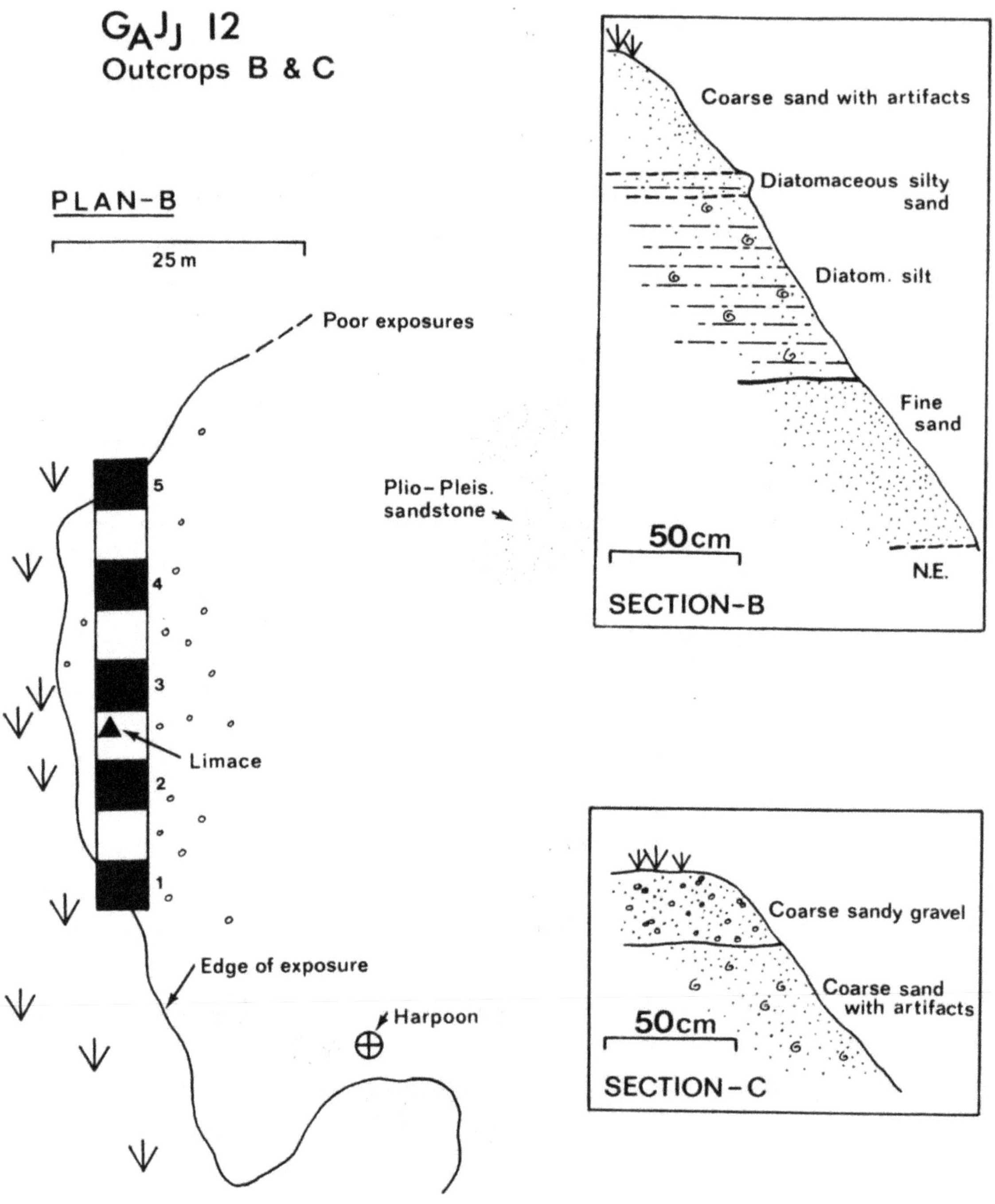

Figure 31. GaJj 12. Sampling units and stratigraphic sections at Outcrops B and C.

Summary

Artifactual material was eroding out of two distinct archaeological horizons at GaJj 12. The uppermost coarse sands at outcrops A and B are thought to represent beach deposits while the underlying consolidated diatomaceous sandy silt is believed to represent deposition along a beach shoreline that was subsequently submerged and covered by fine grained sediments. Figure 32 presents a provisional correlation of the main artifact bearing horizons.

SAMPLING

Artifacts exposed by erosion were sampled from outcrops A and B. For each outcrop, a linear transect sampling grid was laid out parallel to the orientation of the exposure. Alternate sampling units beginning with unit one were collected. Seventeen total units were sampled, twelve from outcrop A and five from outcrop B.

CULTURAL MATERIAL

Table 24 provides an inventory of surface finds by sampling unit. Artifacts from outcrop C and plotted pieces from outcrops A and B are also tabulated.

Stone Artifacts (Tables 25-27; Figs. 33-34)

The lithic artifacts revealed a number of interesting features. Lava accounted for 48% of the total sample. Except for the site of FxJj 12 North, this figure represented the highest percentage of non crypto-crystalline silica raw materials utilized among the shoreline fishing settlements. Microlithic implements comprised the majority of the shaped stone tools followed by outils écaillés, core/choppers and scrapers. Crescents were the most common microlithic type. A variety of core forms were present, including bipolar fragments manufactured from small quartz pebbles. These quartz bipolar fragments comprised nearly 20% of the core sample. A single unifacially trimmed limace was recovered from outcrop B (Fig. 34, a). Similar artifacts have been recovered near the Holocene deposits at GaJj 11 and to the west of GaJj 12. The pieces were manufactured from a fine grained greenish-gray lava. Their morphology bears a striking resemblance to examples of unifacial core/choppers from the Karari Industry (Harris 1978).

Bone Artifacts

A total of 8 uniserial bone harpoons was collected. Three fragments showed basal notching (Fig. 35, b-d) while the only complete specimen had a combination of basal grooved incisions and notching (Fig. 35, a).

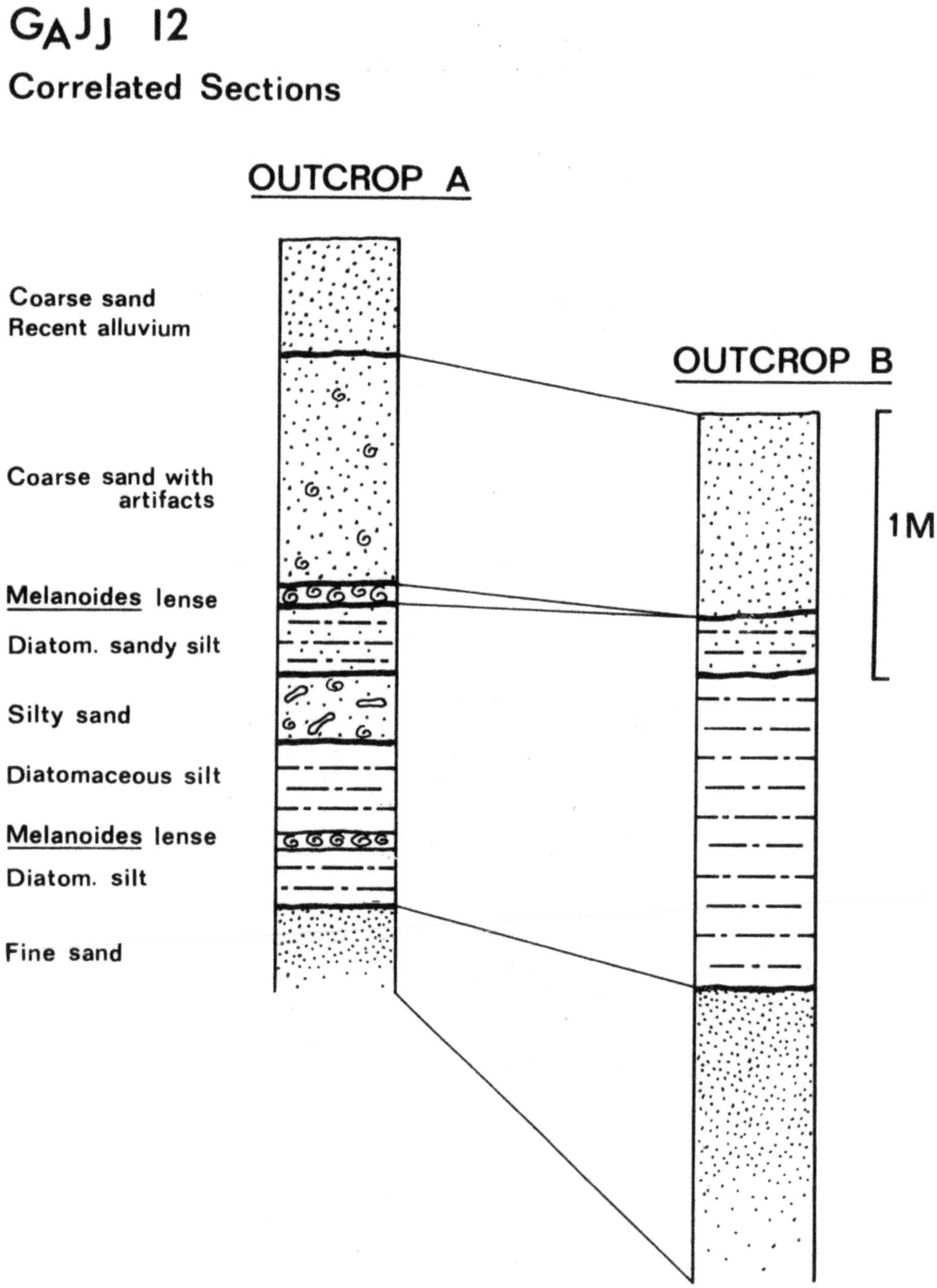

Figure 32. GaJj 12. Correlated stratigraphic sections from Outcrops A and B.

Pottery

Twenty-four undecorated and seventeen decorated sherds were collected. Figure 35, e-g illustrates three decorated sherds from outcrop A. The sherds were tempered with coarse sand and tended to be friable. A series of small impressions, usually set in rows, were a diagnostic feature of the specimens. Figure 35, h-l illustrates a second set of sherds from outcrop C. These sherds showed two series of small curving impressions which were separated by shallow grooved lines. Several of the rim sherds had tapered undercorated lips which suggested open mouthed vessels. Both sets of decorated sherds showed affinity to material from GaJj 2 and FxJj 12 North. However, the stratigraphic provenience of the material from outcrop C remains unclear. The sherds may represent pottery fragments derived from a lens of coarse gravel that mantles much of the Galana Boi deposits in Area 127 and thus would postdate the 75-80 m high lake transgression. The sherds from outcrop C have been included in the total ceramic inventory but their relationship to the artifacts from outcrops A and B is unclear.

Faunal

The faunal collection was fragmentary, water abraided, and well mineralized. Fish, reptile and mammal bones were recovered. Mammal remains were especially abundant and outnumbered fish bones by a five to one margin. Preliminary field indentifications include hippopotamus, warthog, zebra and small to medium sized bovids.

PRELIMINARY INTERPRETATION

The site of GaJj 12 was situated on the high Holocene beach. Artifactual remains, including barbed bone harpoons and decorated pottery, were sampled from several well eroded horizons of coarse sands and semi-consolidated silty sands. Archaeological features were not recognized. The artifacts are believed to represent a mixture of at least two distinct occupations whose inhabitants shared similar technological and economic traditions. Unfortunately, the culture material from the occupation episodes could not be separated, nor could the time interval between them be determined.

To the west of the site, lay a sequence of fine grained sediments which were deposited in a large quiet water embayment. A series of low sandy hummocks lay further westward and may have originally formed offshore sand and gravel bars. One or more of the archaeological horizons may represent more permanent settlements whose occupants maintained temporary fishing camps along the nearby offshore sandbar at GaJj 11.

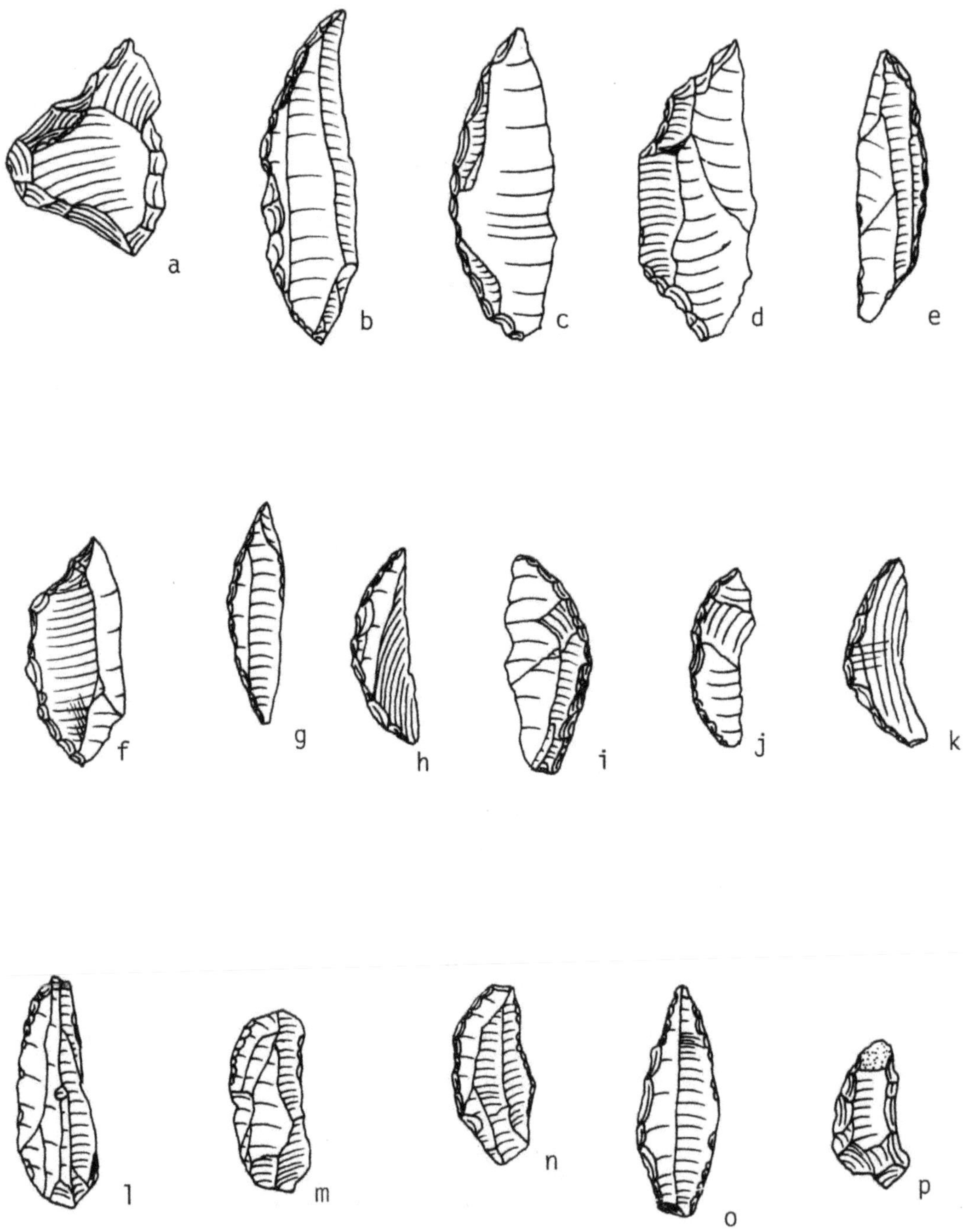

Figure 33. GaJj 12. Backed Tools.
(a) trapezoid; (b-k) crescents; (l-m) curved backed flakes; (o-p) double backed flakes. Natural size.

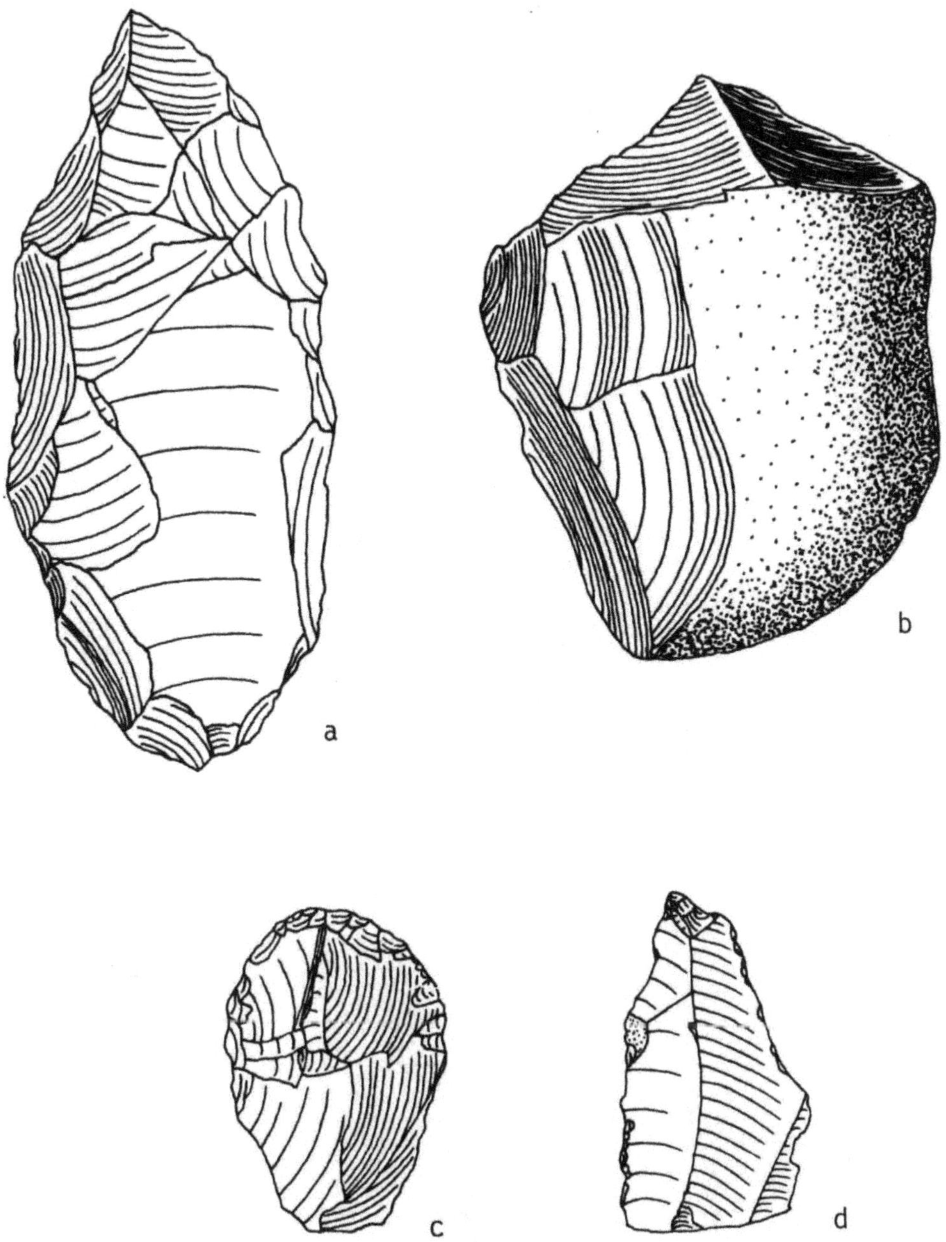

Figure 34. GaJj 12. Stone artifacts.
(a) core/chopper; (b) limace; (c) end scraper; (d) bec. Natural size.

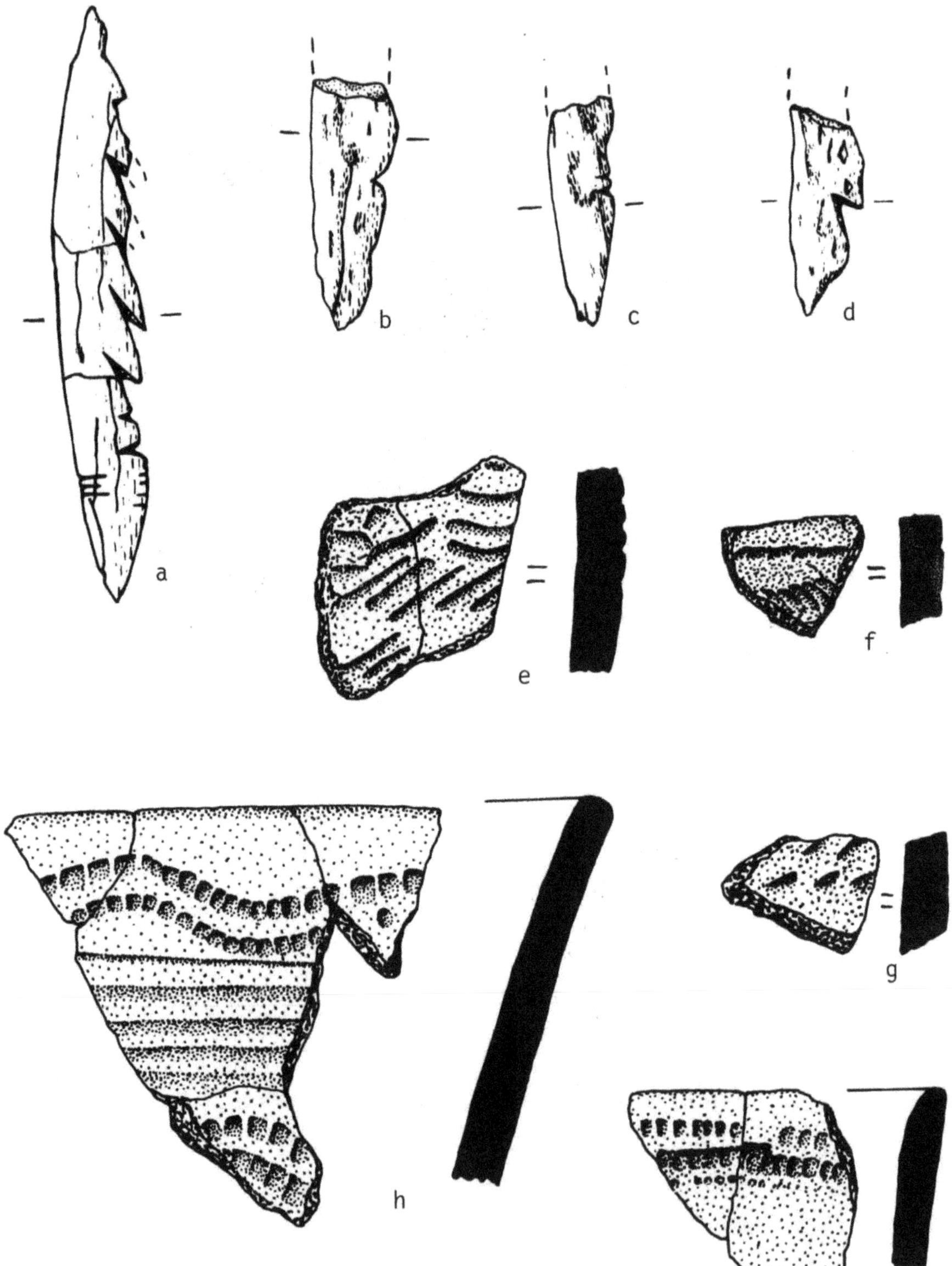

Figure 35. GaJj 12. Bone artifacts and pottery sherds.
Outcrop A (b-d) uniserial harpoons; Outcrop C (a) uniserial harpoon; Outcrop A (e-g) decorated body sherds; Outcrop C (h-i) decorated rim sherds. Natural size.

Table 24
GaJj 12

INVENTORY OF SURFACE FINDS BY SAMPLING UNIT

Unit	Outcrop A 1	2	3	4	5	6	7	8	9	10	11	12	(N)	Outcrop B 13	14	15	16	17	(N)	Outcrop C	Plotted Outcrop A	Plotted Outcrop B	Assemblage Total (N)
Stone:																							
Microliths	19	-	4	7	9	-	2	2	-	-	1	-	44	3	2	3	4	-	12	2	-	-	58
Scrapers	2	-	2	3	-	1	-	-	-	-	2	-	10	-	-	1	1	1	3	-	-	-	13
Outils écaillés	6	-	-	1	-	-	2	1	-	-	3	-	13	-	-	12	1	-	13	-	-	-	26
Bec	-	-	-	-	1	-	-	-	-	-	-	-	1	-	-	-	-	-	-	-	-	-	1
Burin	-	-	-	-	-	-	-	-	-	-	-	-	-	-	-	-	1	-	1	-	-	-	1
Awl	-	-	-	-	-	-	-	-	-	-	-	-	-	-	-	1	-	-	1	-	-	-	1
Limace	-	-	-	-	-	-	-	-	-	-	-	-	-	1	-	-	-	-	1	-	-	-	1
Core/chopper	3	1	1	-	-	-	-	-	1	3	1	-	10	-	-	3	1	-	4	-	-	-	14
Cas. trimm.	2	-	-	1	-	-	-	-	1	-	1	-	5	-	1	2	1	3	7	-	-	-	12
Cas. retouch	-	-	2	5	1	-	-	-	-	-	2	-	10	-	-	2	-	-	2	-	-	-	12
Whole flakes	37	8	20	49	9	-	5	6	22	30	6	6	198	8	5	47	18	12	90	-	-	-	288
Broken flakes	77	9	28	109	21	1	17	10	34	27	9	5	347	21	7	72	34	15	149	2	-	-	498
Angular frags.	36	10	13	49	10	1	10	13	14	14	2	7	179	8	11	35	15	10	79	3	-	-	261
Cores	9	2	3	40	3	-	4	1	4	6	4	1	77	2	2	8	2	4	20	1	-	8	106
Bone harpoons:																							
Uniserial	-	-	1	-	-	1	-	-	-	-	-	-	2	-	-	-	-	-	-	1	4	1	8
Pottery:																							
Decorated	-	4	-	-	-	-	-	-	-	-	-	-	4	-	-	-	-	-	-	8	-	-	17
Undecorated	14	1	1	-	-	-	-	-	-	-	-	-	16	-	-	-	-	-	-	13	-	-	24
Bone frags. (NID):																							
Mammal	38	8	15	86	12	1	9	10	14	5	10	5	213	5	13	15	2	2	35	-	-	-	248
Fish	4	-	-	18	7	1	-	-	1	-	-	-	31	1	-	14	-	-	15	-	-	-	46
Turtle	16	-	-	2	1	1	1	-	-	1	-	-	22	1	-	4	-	-	5	-	-	-	27
Crocodile	1	-	-	8	-	-	-	-	-	-	-	-	9	-	-	-	-	-	-	-	-	-	9
Bones w/art. sur.	3	1	1	1	1	-	2	1	1	-	-	-	11	1	-	1	-	-	2	-	-	-	46
Teeth:																							
Mammal	2	-	5	7	3	1	-	-	1	-	1	-	20	-	-	2	1	-	3	-	1	-	24
Ostrich eggshell	2	-	-	4	-	-	-	-	-	-	-	-	6	-	-	-	-	-	-	7	-	-	13
Raw counts:																							
Stone	191	30	73	264	54	3	40	33	76	80	31	19	892	43	28	186	78	45	382	8	-	8	1292
Bone/teeth	66	9	21	122	24	4	12	11	17	6	11	5	306	8	13	26	3	2	42	-	1	-	348

Table 25

GaJj 12

FREQUENCY OF TRIMMED PIECES

	Outcrop A Chert	Outcrop A Lava	Outcrop A Quartz	(N)	Outcrop B Chert	Outcrop B Lava	Outcrop B Quartz	(N)	Outcrop C Chert	Outcrop C Obsidian	(N)	Assemblage Total (N)	Assemblage Total (%)
Microliths:												(58)	(41.2)
Triangle	-	-	-	-	-	-	-	-	-	-	-		
Trapezoid	-	-	-	-	-	-	-	-	-	-	-	-	-
Crescent	14	-	-	14	5	-	-	5	-	-	-	19	13.7
Curved backed flake	10	-	-	10	4	-	-	4	-	-	-	14	10.1
Double backed flake	1	-	-	1	1	-	-	1	-	-	-	2	1.4
Truncations	1	-	-	1	-	-	-	-	-	-	-	1	0.7
Miscellaneous backed	4	-	-	4	-	-	-	-	-	-	-	4	2.9
Fragmentary	14	-	-	14	2	-	-	2	1	1	2	18	13.0
Scrapers:												(12)	(8.6)
Straight side	1	1	-	2	1	-	-	1	-	-	-	3	2.2
Convex side	1	-	-	1	-	-	-	-	-	-	-	1	0.7
Double side	1	1	-	2	-	-	-	-	-	-	-	2	1.4
Circular	1	-	-	1	-	-	-	-	-	-	-	1	0.7
End	2	-	-	2	1	-	-	1	-	-	-	3	2.2
Notch	2	-	-	2	1	-	-	1	-	-	-	3	2.2
Outils écaillés	9	-	4	13	8	-	5	13	-	-	-	26	18.7
Bec	1	-	-	1	-	-	-	-	-	-	-	1	0.7
Burin	-	1	-	1	-	-	-	-	-	-	-	1	0.7
Awl	-	-	-	-	1	-	-	1	-	-	-	1	0.7
Limace	-	-	-	-	-	1	-	1	-	-	-	1	0.7
Core/chopper	-	10	-	10	-	4	-	4	-	-	-	14	10.1
Cas. trimm.	5	-	-	5	3	4	-	7	-	-	-	12	8.6
Cas. retouch	2	8	-	10	2	-	-	2	-	-	-	12	8.6
Raw material (N)	69	21	4	94	29	9	5	43	1	1	2	139	100.0

Table 26

GaJj 12 Surface

FREQUENCY OF CORE TYPES

	Chert	Lava	Quartz	Obsidian	Total (N)	Total (%)
Single platform	6	1	-	-	7	6.6
Double: alternate	14	-	20	1	35	33.0
other	7	-	-	-	7	6.6
Multi	24	5	-	-	29	27.4
Disc	5	1	-	-	6	5.7
Casual	12	2	-	-	14	13.2
Fragmentary	5	3	-	-	8	7.5
Raw material (N)	73	12	20	1	106	100.0
(%)	68.9	11.3	18.9	0.9		

Table 27

GaJj 12 Surface

FREQUENCY OF MAJOR ARTIFACT CLASSES

	Chert	Lava	Quartz	Obsidian	Total (N)	Total (%)
Trimmed pieces	99	30	9	1	139	10.8
Cores	73	12	20	1	106	8.2
UTFF	450	571	19	7	1047	81.0
Raw material (N)	622	613	48	9	1292	100.0
(%)	48.1	47.4	3.8	0.7		

INTRODUCTION

The site of FxJj 12 was located in Area 116 on a ridgetop overlooking Laga Aberegaya (Fig. 8). The Galana Boi deposits consisted primarily of silty sands and coarse sands with gravels of sub-littoral and littoral origin. Cultural remains, including microliths, core/choppers, bone harpoons and faunal refuse, were exposed by erosion along the full length of the deflated ridge summit. In 1971 a team of archaeologists from the University of California conducted a series of geological and excavation trenches at the site. Cultural material from one of their excavation trenches was subsequently analyzed by myself. The site was systematically sampled during the 1976 field season.

STRATIGRAPHY

The Galana Boi deposits consisted of a relatively narrow horizontal strip of sediments perched on top of Lower Member beds of the Koobi Fora Formation. The section contained 2.25 m of Holocene sediments which overlay Plio-Pleistocene indurated sandstones and silty clays. The Holocene beds were composed of muddy coarse sands, diatomaceous silty sands, shell coquina and several horizons of coarse gravelly sands.

A major objective of the 1976 field season was to expand earlier excavations and more precisely define the depositional nature of the various sandy horizons. Consequently, one of the previous excavations (trench B) was partially re-opened. Figure 36 has been redrawn from field sections originally drafted by Drs. G. Isaac and J. W. K. Harris.

The association of diatomaceous silty sands, clean coarse gravels and a coquina of Melandoides and Corbicula in horizons D and E indicate off-shore to shoreline depositional conditions at a time when Lake Turkana stood at a very high level. Layers A-C were deposited under similar conditions. The suite of Holocene units in trench B can best be interpreted as representing beach deposits that accumulated and underwent small shoreline flucuations during a major transgressive stand.

SAMPLING

In response to the relatively long and narrow configuration of the outcrop, a linear transect sampling strategy was chosen (Fig. 37). The outcrop was subdivided into 10 sampling blocks. Each block measured 25 m long and 5 m wide. The first unit within each sampling block was selected for collection. A total of 10 units was sampled.

EXCAVATION

During 1971 three graduate students from the University of California, John W. K. Harris, John Onyango-Abuje, and Jonathan

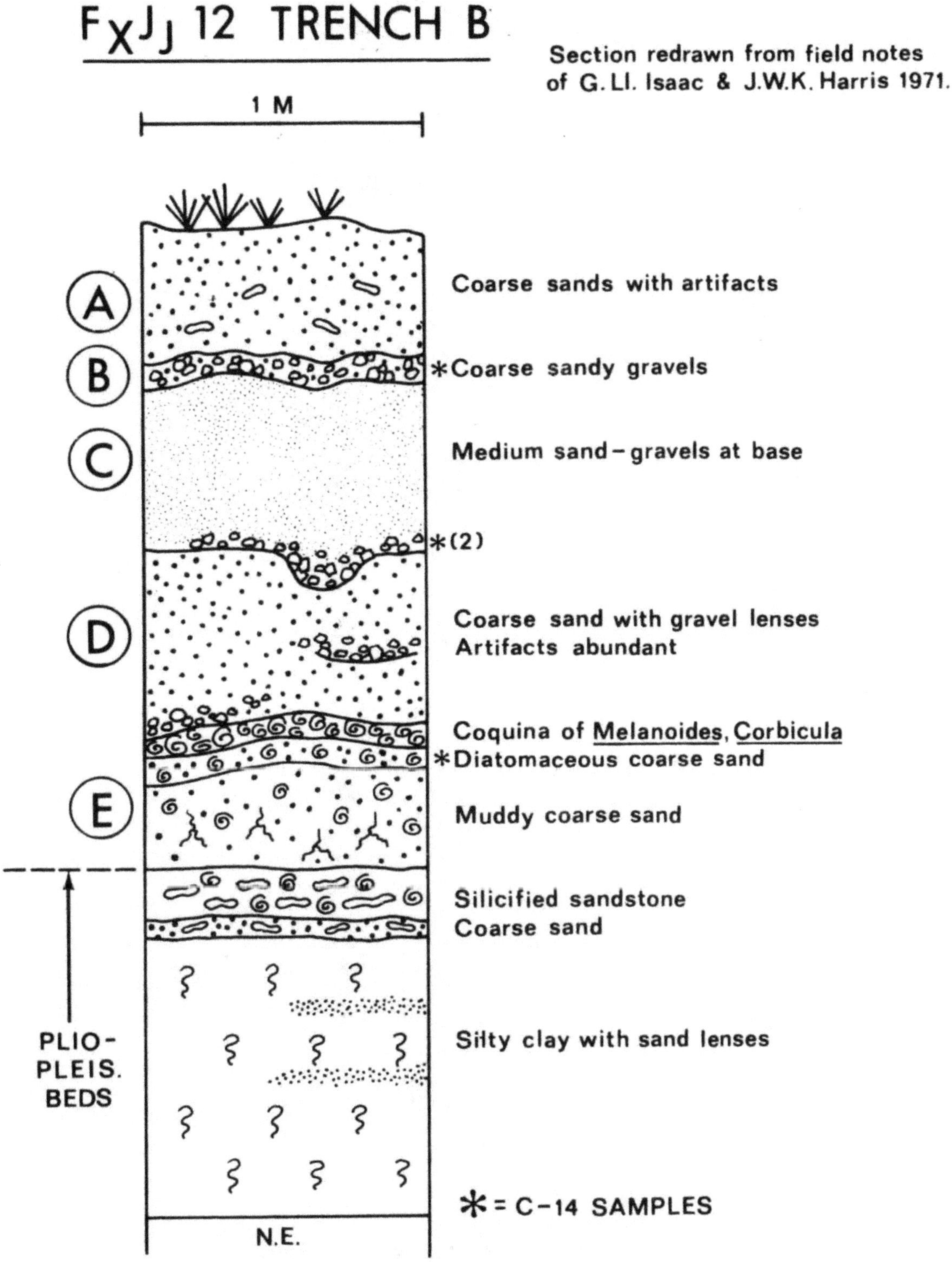

Figure 36. FxJj 12. Excavation section from Trench B. Redrawn from field notes of G. Isaac and J.W.K. Harris, 1971.

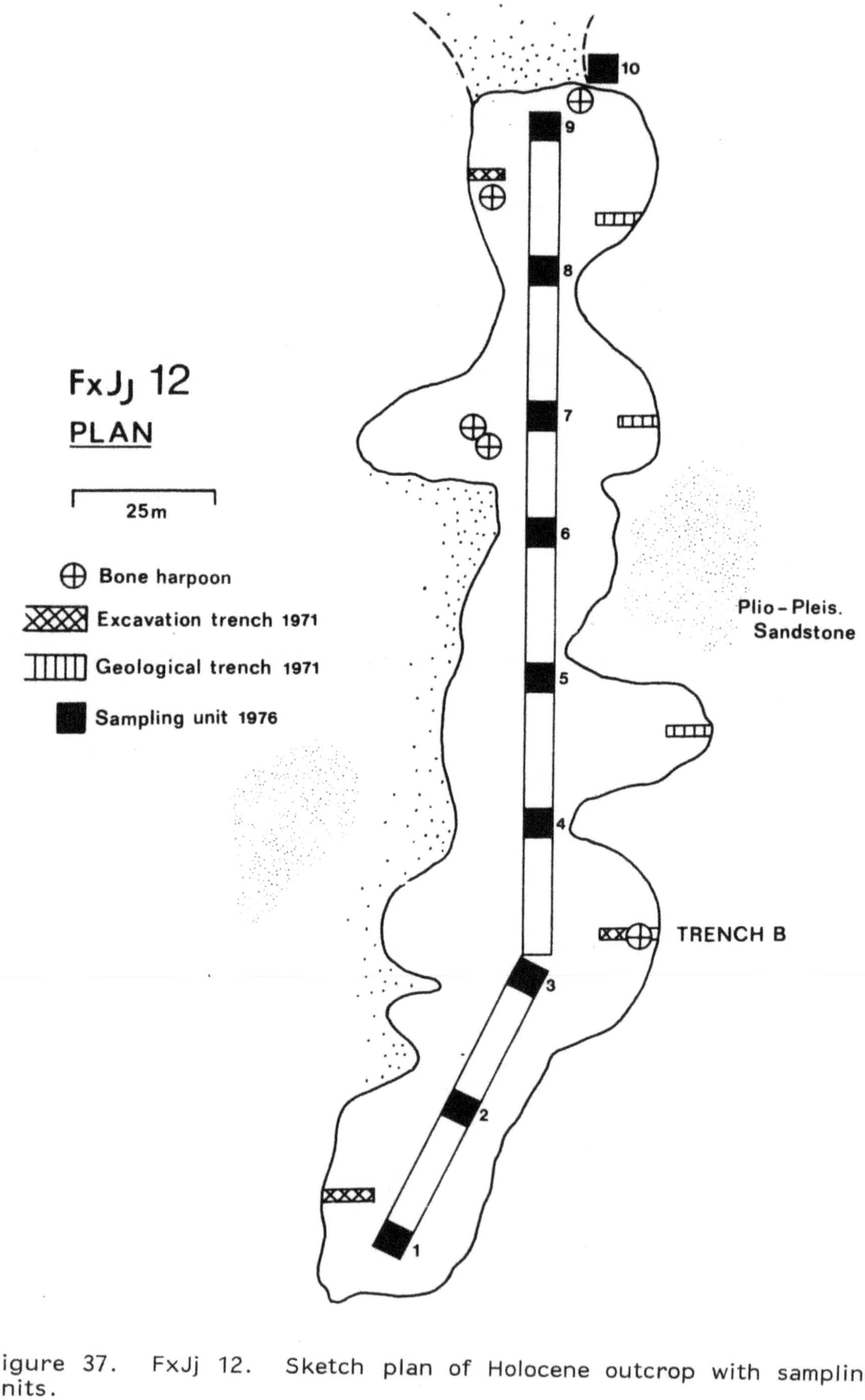

Figure 37. FxJj 12. Sketch plan of Holocene outcrop with sampling units.

Karoma, supervised separate test excavations at FxJj 12. A thorough search of archaeological collections at the National Museum in Nairobi, failed to turn up artifacts from two of the excavations. Consequently, only the excavated material from Harris' test trench (B) is reported in this monograph.

Test trench B measured two meters in width and was excavated in a series of seven levels. Because of the unconsolidated nature of the sandy horizons and the difficulty of following distinct bedding surfaces, the trench was originally excavated in a series of arbitrary 10 cm spits. Cultural remains were, as a result of this decision, excavated from a combination of arbitrary and natural stratigraphic divisions. The majority of the excavated finds were recovered from units A-C.

CULTURAL REMAINS: SURFACE

Table 28 presents an inventory of the artifactual remains by sampling unit. Artifacts and faunal remains collected during the 1971 field season are also tabulated.

Stone Artifacts (Tables 29-31; Fig. 38)

Lava accounted for over 40% of the raw material. Microliths were the most common shaped implement followed by scrapers, *outils écaillés* and core/choppers. A bifacially flaked chert pick was also recovered (Fig. 38, e). Curved backed flakes were well represented within the microlithic tools. Double platform cores accounted for nearly half of the surface collection of cores. Most of the debitage was derived from small chert pebbles and large lava cobbles.

Bone Artifacts

Six uniserially barbed bone harpoons were collected from the surface. One specimen was unbroken (Fig. 38, f) while the remainder were fragmentary. All of the harpoon specimens were considerably water worm. Two basal fragments showed evidence of notching while a third example had circular incisions.

Faunal Remains

A large and well preserved series of fish, reptile and mammal remains were identified. The full mammal bone assemblage is currently being analyzed. Preliminary identifications of wildebeest (*Connochaetes taurinus*), Hippopotamus (*Hippopotamus* sp.) and cheetah (*Acinonyx* sp.) have been made.

CULTURAL MATERIAL: EXCAVATION

Table 32 presents an inventory of the recovered material by excavation spit.

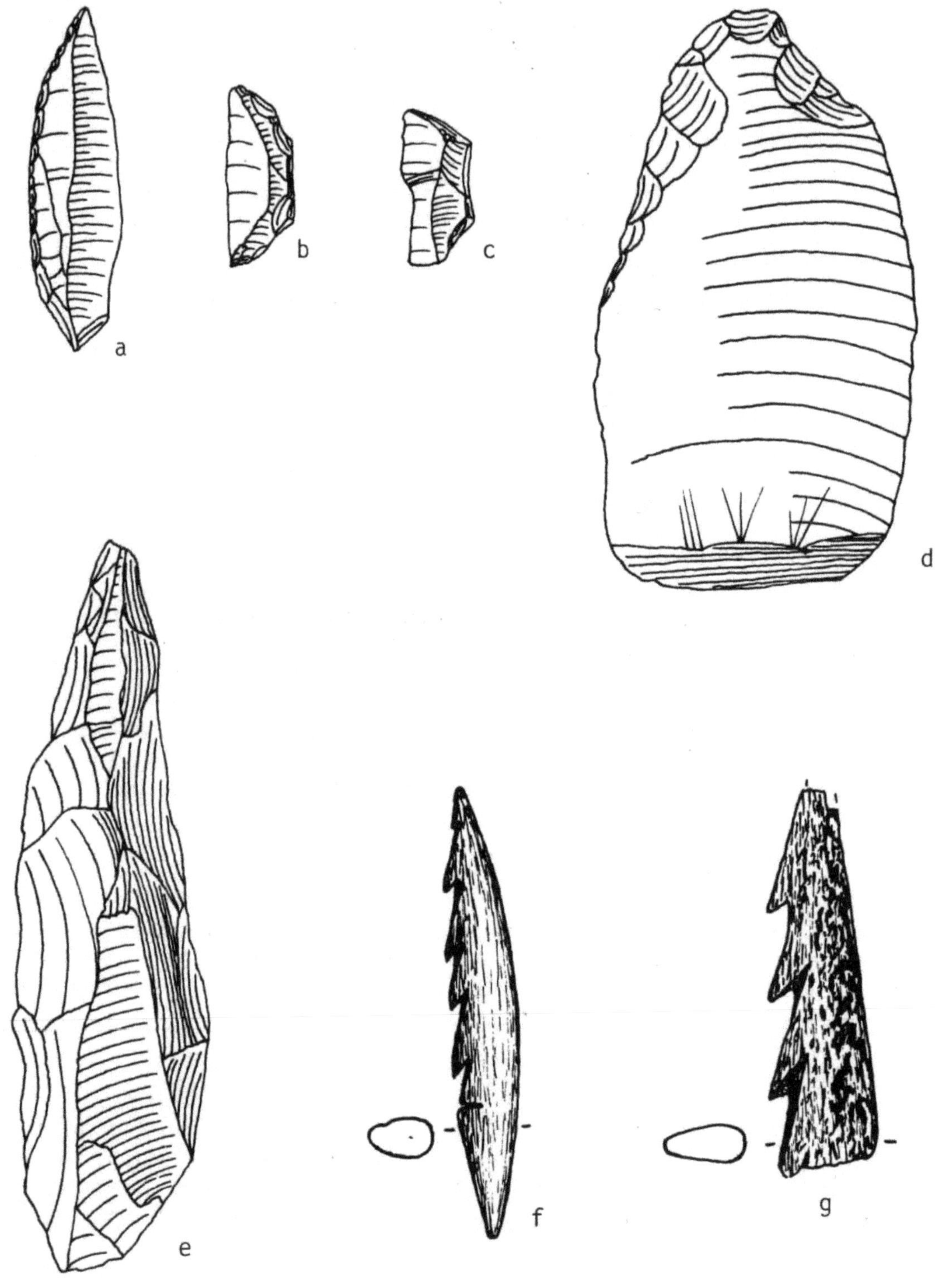

Figure 38. FxJj 12. Stone and bone artifacts: Surface.
(a-c) crescents; (d) convergent scraper; (e) pick; (f-g) uniserial harpoons. Natural size.

Stone Artifacts (Tables 33-35; Fig. 39)

A relatively small sample of lithic artifacts was recovered from trench B. Microliths and outils écaillés were the most common shaped implement. Two core/choppers were collected from coarse gravels within unit C. Their morphology was similar to surface examples collected from FxJj 12 and excavated pieces from GaJj 1.

Bone Artifacts

A single fragmentary tip from a uniserially barbed harpoon was identified from unit B.

Faunal Remains

Considerable quantities of faunal remains were excavated. Fish bones were numerous. Mammal post-cranial bones and teeth were well mineralized and usually showed manganese staining. The excavated faunal collection is currently undergoing analysis.

DATING

A series of mollusc shell and mammal bone samples were submitted for radiocarbon analysis. The shell samples were collected in 1971 from the lower most Holocene level within trench B. The sample was later divided into two equal lots and shipped to different radiocarbon laboratories. The samples yielded ages of:

9660 ± 235 BP	unit E	Geochron Labs.
9540 ± 260 BP	unit E	Radiocarbon Ltd.

A second series of samples consisting of mammal bone was submitted from units A-C and unit D within trench B. Both samples were dated on their bone apatite fraction. The mammal bone samples yielded ages of:

9355 ± 235 BP	units A-C	Geochron Labs.
9395 ± 270 BP	unit D	Geochron Labs.

PRELIMINARY INTERPRETATION

The association of microlithic stone tools, core/choppers, abundant fish and mammal remains and barbed bone harpoon heads indicates FxJj 12 represented a large fishing-hunting settlement located on the high Holocene transgressive shoreline. The Galana Boi units were deposited under beach conditions and displayed a fluctuating sequence of sand and gravel horizons, mollusc shells and associated artifacts. To the east of the site was a flat lying surface of dark brown silty clay with vertical cracking and associated pisoliths. This unit, probably a developed soil, is thought to represent a back beach facies and to have been laterally equivalent to the coarse beach sands at the site.

The absence of pottery suggests the site may have functioned, perhaps like GaJj 11, as a short term encampment. The site's inhabitants may have been familiar with pottery making techniques but chose not to manufacture or utilize ceramic vessels at the temporary settlememt. Alternatively, the absence of pottery may indicate the existence of an aceramic phase within the group of early Holocene fishing settlements. This important consideration is discussed further in Chapter VIII.

The series of radiocarbon determinations indicates that by 9500-10,000 BP the Holocene lake had reached its highest transgressive stand and that scatters of characteristic Later Stone Age microlithic tools, excavated from the sites lowermost horizons, date to this period. While bone harpoons were not recovered from the lower levels, the large quantity of fish bone refuse argues strongly for their likely occurrence. The main archaeological horizons (A-D) date to 8400 BP. Bone harpoons were a consistent feature of these layers.

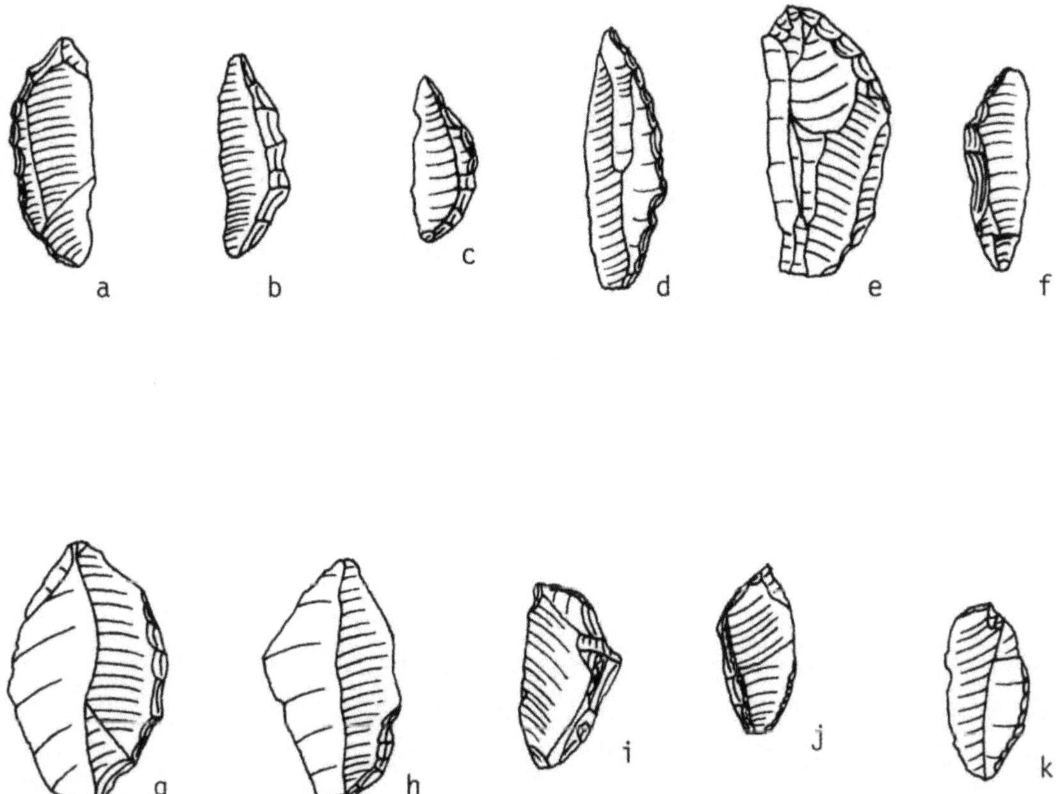

Figure 39. FxJj 12. Stone artifacts: Excavation.
(a-c) crescents; (d-k) curved backed flakes. Natural size.

Table 28
FaJj 12

INVENTORY OF SURFACE FINDS BY SAMPLING UNIT

Unit	1	2	3	4	5	6	7	8	9	10	11	12	Assemblage Total (N)
Stone:													
Microliths	2	3	3	2	2	-	-	2	2	-	1	3	20
Scrapers	1	-	-	-	-	2	2	1	1	-	-	1	8
Outils écaillés	1	-	-	-	2	2	-	1	-	-	-	1	7
Bec	-	-	-	-	1	-	-	-	-	-	1	-	2
Burin	-	2	-	-	-	-	-	-	-	-	-	-	2
Core/chopper	1	1	-	-	-	-	3	1	-	-	-	-	6
Pick	-	-	-	-	-	-	-	-	-	-	-	1	1
Cas. trimm.	-	2	1	-	2	-	-	1	-	-	-	1	7
Cas. retouch	-	-	-	-	-	-	1	-	-	-	2	-	3
Whole flakes	11	18	12	6	13	5	-	6	6	1	9	13	100
Broken flakes	14	25	24	16	20	7	5	6	7	1	41	41	207
Angular frags.	2	17	25	7	9	4	6	4	4	-	18	23	119
Cores	5	2	10	3	6	3	3	4	2	1	8	10	57
Bone harpoons:													
Uniserial	-	-	-	-	-	-	1	-	-	-	-	-	6
Bone frags. (NID):													
Mammal	48	16	8	12	20	5	19	45	27	7	42	147	396
Fish	62	31	-	13	38	4	40	99	4	10	96	632	1029
Turtle	9	-	2	2	7	-	3	15	-	-	18	50	106
Crocodile	-	-	-	-	-	-	-	-	-	-	-	-	-
Bones w/art. sur.	13	4	3	-	9	-	15	22	3	2	9	32	119
Horn core	-	-	-	-	-	-	-	-	-	1	-	-	4
Teeth:													
Mammal	3	1	-	-	-	-	-	-	4	4	3	9	26
Fish	-	-	-	-	-	-	-	-	-	-	-	-	-
Raw counts:													
Stone	37	70	75	34	55	23	20	26	22	3	80	91	539
Bone/teeth:													
Mammal	64	21	11	12	29	5	34	67	34	14	54	188	545
Fish	71	31	2	15	45	4	43	114	4	10	114	682	-

Table 29

FxJj 12: Surface

FREQUENCY OF TRIMMED PIECES

	Chert	Lava	Total (N)	Total (%)
Microliths:			(20)	(35.7)
Trapezoid	1	-	1	1.8
Crescent	2	-	2	3.6
Curved backed flake	7	-	7	12.5
Truncations	3	-	3	5.4
Fragmentary	7	-	7	12.5
Scrapers:			(8)	(14.3)
Side	1	2	3	5.4
End	1	1	2	3.6
Notch	1	1	2	3.6
Convergent	-	1	1	1.8
Multi	1	-	1	1.8
Other	1	-	1	1.8
Outils écaillés	7	-	7	12.5
Bec	2	-	2	3.6
Burin	2	-	2	3.6
Core/chopper	-	7	7	12.5
Pick	1	-	1	1.8
Cas. trim.	4	2	6	10.7
Cas. retouch	3	-	3	5.4
Raw material (N)	43	14	57	100.0

Table 30

FxJj 12: Surface

FREQUENCY OF CORE TYPES

	Chert	Lava	Quartz	Obsidian	Total (N)	Total (%)
Single platform	4	2	-	-	6	10.5
Double: alternate	17	-	4	-	21	36.9
other	1	-	-	-	1	1.7
Multi	11	-	-	-	11	19.4
Disc	-	2	-	-	2	3.5
Casual	7	-	-	1	8	14.0
Fragmentary	6	1	1	-	8	14.0
Raw material (N)	46	5	5	1	57	100.0
(%)	80.6	8.8	8.8	1.8		

Table 31

FxJj 12: Surface

FREQUENCY OF MAJOR ARTIFACT CLASSES

	Chert	Lava	Quartz	Obsidian	Total (N)	Total (%)
Trimmed pieces	43	14	-	-	57	10.9
Cores	46	5	5	1	57	10.9
UTFF	190	204	14	-	408	78.2
Raw material (N)	279	223	19	1	522	100.0
(%)	53.4	42.8	3.5	0.3		

Table 32

FxJj 12

INVENTORY OF FINDS BY EXCAVATION SPITS 1971 EXCAVATION

Spits	1	2	3	4	5	6	7	8	9	10	11	12	13	14	15	16	17	18	19	20	21	22	23	Total (N)
Stone:																								
Microliths	2	1	3	-	2	-	1	4	2	-	1	-	1	-	1	1	-	-	-	-	-	-	-	19
Scrapers	-	-	-	-	-	-	-	-	-	-	-	-	-	-	-	-	-	-	-	-	-	-	-	-
Outils écaillés	-	-	-	2	1	-	1	-	-	-	1	-	1	-	-	-	-	-	-	-	-	-	-	6
Core/choppers	-	-	-	-	-	1	-	1	-	-	-	-	-	-	-	-	-	-	-	-	-	-	-	2
Cas. trimm.	-	-	-	-	-	-	-	-	1	-	-	1	-	-	-	-	-	-	-	-	-	-	-	2
Whole flakes	1	3	3	4	5	-	6	4	3	8	3	2	1	-	-	2	-	-	-	-	-	-	1	46
Broken flakes	3	4	6	13	10	10	14	11	8	9	5	4	6	-	-	3	-	-	-	-	-	-	-	106
Angular frags.	1	8	5	6	4	5	8	9	9	19	5	8	-	1	1	2	-	1	-	-	-	1	2	95
Cores	-	-	-	1	1	3	2	3	4	3	1	-	-	-	-	1	-	-	-	-	-	-	-	19
Bone harpoon:																								
Uniserial	-	-	-	-	-	-	-	1	-	-	-	-	-	-	-	-	-	-	-	-	-	-	-	1
Bone frags. (NID):																								
Mammal	4	15	17	9	32	14	12	24	12	35	39	56	5	7	7	32	1	1	-	-	4	2	5	333
Fish	76	304	174	178	264	47	55	41	82	106	21	26	41	10	32	198	33	33	-	3	4	11	30	1769
Turtle	-	7	3	4	8	-	2	-	3	3	3	6	4	1	2	14	3	1	-	-	1	3	1	69
Crocodile	-	-	-	-	-	-	-	-	-	-	-	-	1	-	-	-	-	-	-	-	-	-	-	1
Snake/lizard	-	-	-	-	-	-	-	-	-	-	-	1	-	-	-	-	-	-	-	-	-	-	-	1
Bones w/art. sur.	1	-	-	4	-	3	3	2	2	9	5	12	-	2	7	-	-	-	-	-	-	-	1	51
Teeth:																								
Mammal	-	4	-	2	1	-	2	-	5	8	2	2	1	1	-	1	1	-	-	-	-	-	-	30
Fish	-	1	-	-	-	-	-	-	-	-	-	-	-	-	-	-	-	-	-	-	-	-	-	1
Ostrich eggshell	-	-	-	-	-	-	-	-	-	-	-	-	-	-	-	-	-	-	-	-	-	-	2	2
Raw counts:																								
Stone	7	16	17	26	23	19	32	32	27	39	16	15	9	1	2	9	-	1	-	-	-	1	3	295
Bone/teeth:																								
Mammal	5	19	17	15	41	17	17	26	17	52	46	68	6	10	7	40	2	1	-	-	4	2	6	418
Fish	76	305	174	178	264	47	55	41	82	106	21	26	41	10	32	198	33	33	-	3	4	11	30	1770

Table 33

FxJj 12: Excavation

FREQUENCY OF TRIMMED PIECES

	Chert	Lava	Quartz	Total (N)	(%)
Microliths:				(19)	(65.5)
Crescent	4	-	-	4	13.8
Curved backed flakes	6	-	-	6	20.7
Truncations	1	-	-	1	3.4
Fragmentary	8	-	-	8	27.6
Outils écaillés	3	-	3	6	20.7
Core/chopper	-	2	-	2	6.9
Cas. trimm.	2	-	-	2	6.9
Raw material (N)	24	2	3	29	100.0

Table 34

FxJj 12: Excavation

FREQUENCY OF CORE TYPES

	Chert	Lava	Quartz	Obsidian	Total (N)	Total (%)
Single platform	-	-	-	-	-	-
Double: alternate	10	-	2	-	12	63.1
other	1	-	-	-	1	5.3
Multi	3	-	-	-	3	15.8
Disc	-	-	-	-	-	-
Casual	1	-	-	-	1	5.3
Fragmentary	1	-	1	-	2	10.5
Raw material (N)	16	-	3	-	19	100.0
(%)	84.2	-	15.8	-		

Table 35

FxJj 12: Excavation

FREQUENCY OF MAJOR ARTIFACT CLASSES

	Chert	Lava	Quartz	Obsidian	Total (N)	Total (%)
Trimmed pieces	24	2	3	-	29	9.8
Cores	16	-	3	-	19	6.4
UTFF	148	87	12	-	247	83.4
Raw material (N)	188	89	18	-	295	100.0
(%)	63.7	30.2	6.1	-		

INTRODUCTION

The site of FxJj 12 North was located in Area 117 approximately 400 kms north of FxJj 12 (Fig. 8). Large numbers of stone artifacts, bone harpoons and faunal remains had been exposed by erosion. Wavy-line pottery and fragments of a human skeleton were collected. The site was sampled and a geological trench was excavated. A single altimeter reading of 105 m indicates the probability of localized tectonic activity.

STRATIGRAPHY

The Galana Boi sediments at FxJj 12 North consisted of a series of silty clays, fine silty sands and coarse sands with gravels (Fig. 40). These units lay unconformably on Plio-Pleistocene dark brown silty clays. The lowermost Holocene levels included a fluctuating sequence of coarse beach sands and silty clays. The section coarsened upward and was composed of several poorly sorted sand horizons with a fine silty matrix. Molluscs, both whole and fragmentary, were abundant in the lower but absent in the upper sand levels.

Cultural materials were exposed along the eroded margins of two archaeological horizons. The topmost coarse sands (bed A) contained dispersed numbers of microlithic stone tools and pottery sherds. This horizon formed a loose mantle over the underlying consolidated sediments and is thought to have a mixed fluvial/eolian origin. The main archaeological level (bed B) occurred immediately below the topmost sandy unit. Wavy-line pottery, stone artifacts and fragments of a human skeleton from this level were discovered *in situ* along the outcrop's edge. Surface mixing of the two horizons has most probably occurred. However, the great majority of the surface finds were derived from bed B.

The principal archaeological horizon, as at FxJj 12, can be traced to the east where it interfingered with a surface of dark brown silty clay. This latter widespread surface probably represents an area of backbeach mudflats that was laterally equivalent to the section's upper coarse beach sands.

The altimeter readings from FxJj 12 and FxJj 12 North indicated that in Areas 116 and 117 the maximum Holocene lake transgression lay 95-105 meters above the present day level of Lake Turkana. Raynolds (1973) also reported an altimeter reading of 100 m from FxJj 12. These measurements are in conflict with altimeter figures from four other high shoreline occupation sites (Table 1). Three alternative interpretations for this apparent anomaly are suggested. First, all three measurements may be in error. While possible, this interpretation is considered unlikely. The independent readings of Raynolds and myself, separated by four field seasons, argues against this position. Second, the high Holocene transgression may have reached an elevation greater than 80 m. However, a considerable number of altimeter readings from many parts of the Basin (Butzer 1971, 1980; Phillipson 1977a; Robbins 1972, 1974; Whitworth 1965a and my own

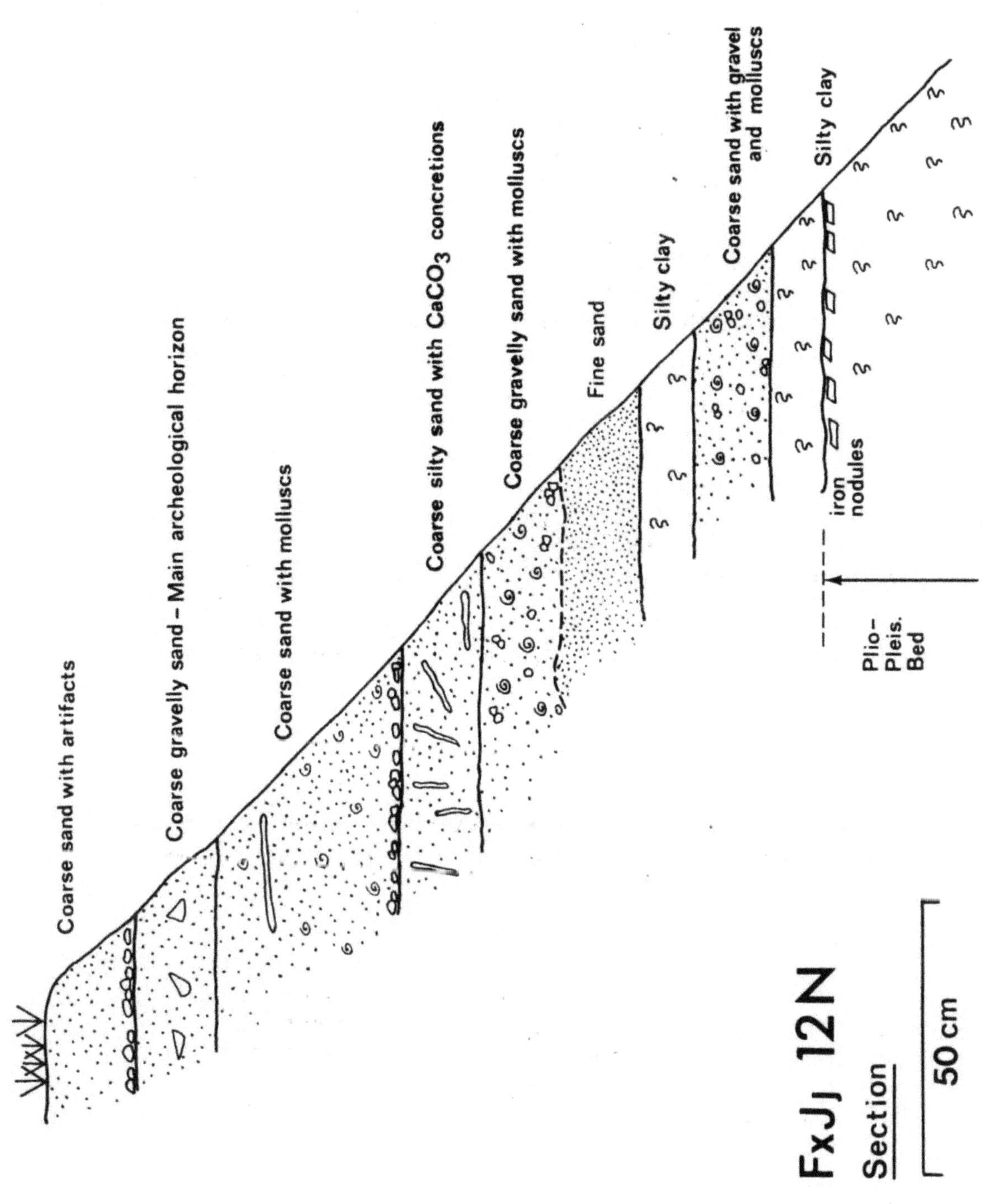

Figure 40. FxJj 12 North. Principal geological section.

data) all show culmination at 75-80 m. Third, Areas 116 and 117 may have undergone relatively small but significant earth movements within the last 10,000 years. This interpretation gains support from observations by Vondra and Bowen (1978:403). They reported faulting of Galana Boi beds in the Kokoi horst area of up to 120 m. Areas 116 and 117 are close to or border on the horst. It appears, therefore, during parts of the Holocene the Kokoi horst area was tectonically unstable and probably resulted in the uplift and faulting of nearby Galana Boi outcrops. The Kokoi may be undergoing further uplift which would indicate tectonic activity continues to play an important role in the Basin (Vondra and Bowen 1978). Additional Holocene post-depositional movements were observed in Area 127. A small fault, with maximum displacement of 1½ m, occurred to the west of GaJj 12. Further Holocene faulting activity has been reported on the south-western side of the lake from the Kabua Beds (Walsh and Dodson 1969).

SAMPLING

A well exposed surface with high densities of stone artifacts and bone refuse was selected for sampling (Fig. 41). Because of the relatively long and narrow configuration of the outcrop, a linear transect approach, similar to the method used at FxJj 12, was chosen. A 200 m transect was laid out and divided into ten 20 m blocks. Within each block, beginning, arbitrarily, with the first unit, a single 5 m square sampling unit was collected. A total of 10 units was sampled.

CULTURAL MATERIAL

An inventory of the surface finds by sampling unit is presented in Table 36. Cultural remains collected during the 1971 and 1974-75 field seasons are also recorded.

Stone Artifacts (Tables 37-39; Figs. 42-43)

Lava accounted for over 50% of the raw material. This figure represented the highest percentage of non cyrpto-crystalline silica reported from the 75-80 m shoreline fishing and hunting settlements. Microliths, scrapers and *outils écaillés* were present in nearly equal numbers. Core/choppers accounted for nearly 23% of the shaped stone artifacts.

Bone Artifacts

A total of nine uniserially barbed bone harpoon heads was recovered from the surface. One specimen was complete (Fig. 44, a) while the remainder were broken and heavily water abraided. All three basal portions had been notched.

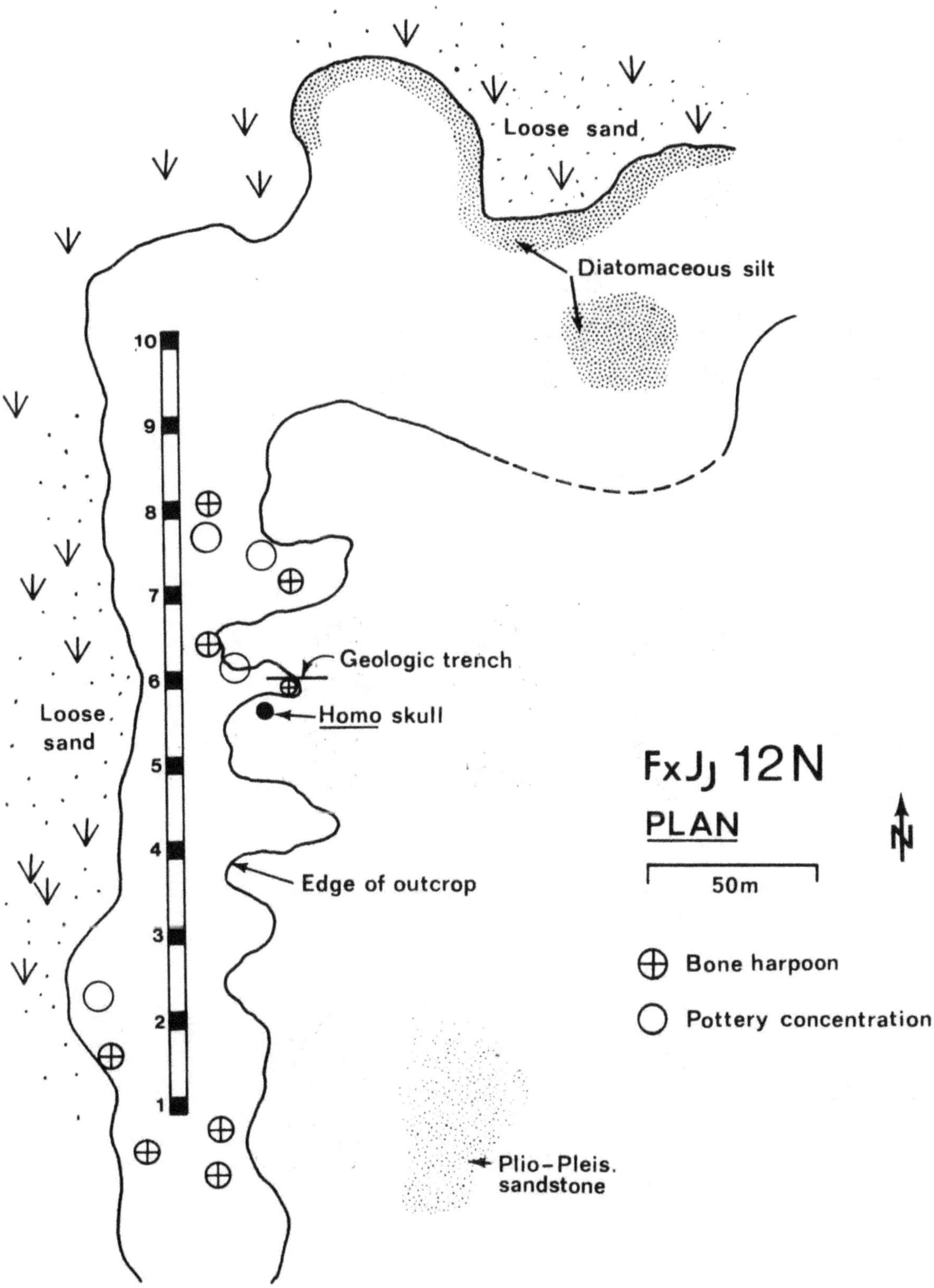

Figure 41. FxJj 12 North. Sketch plan of Holocene outcrop with sampling units.

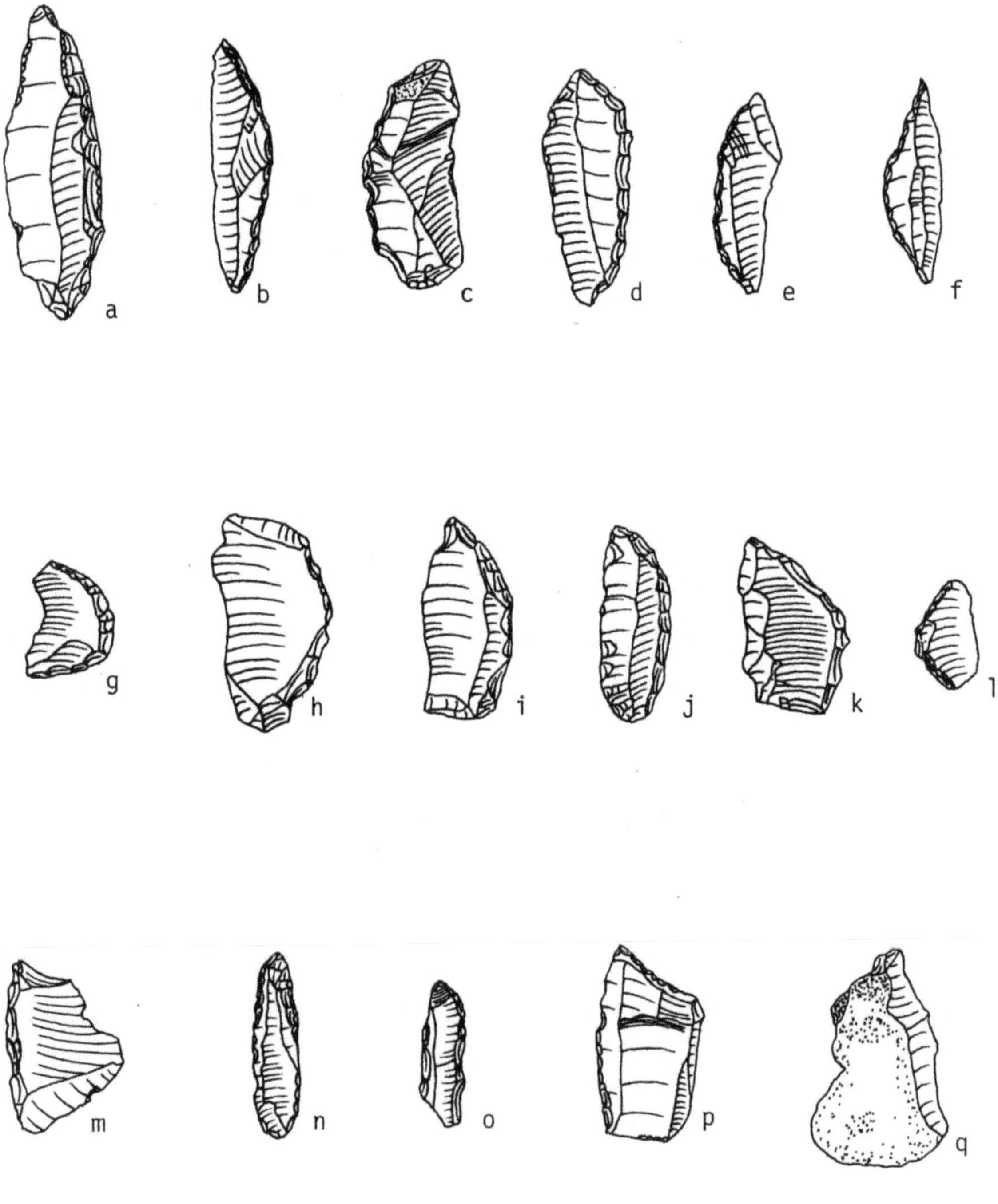

Figure 42. FxJj 12 North. Stone artifacts.
(a-g) crescents; (h-l) curved backed flakes; (m) straight backed flake; (n-o) micro-percoirs; (p-q) becs. Natural size.

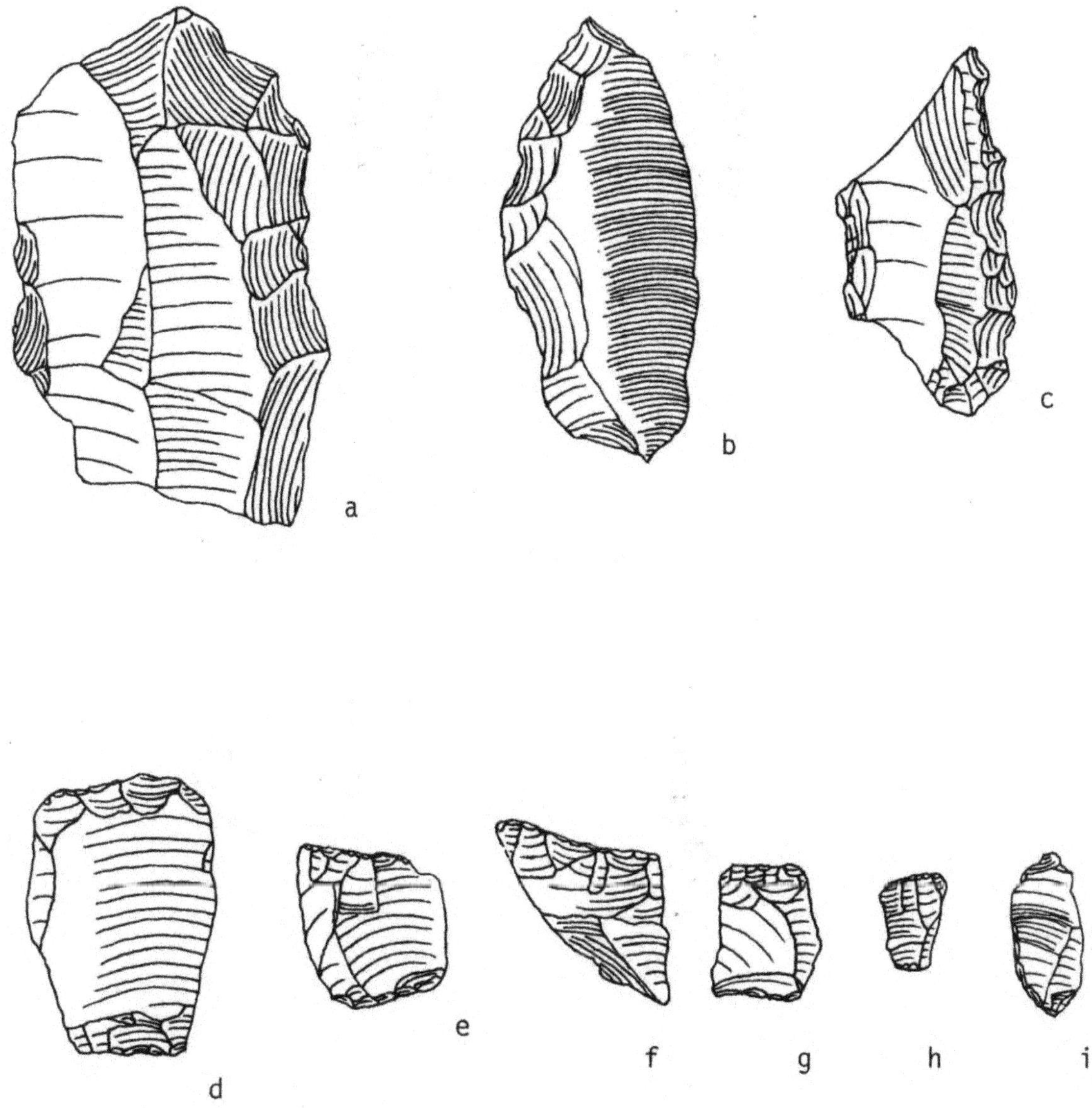

Figure 43. FxJj 12 North. Stone Artifacts.
(a-c) side scrapers; (d-i) outils écaillés. Natural size.

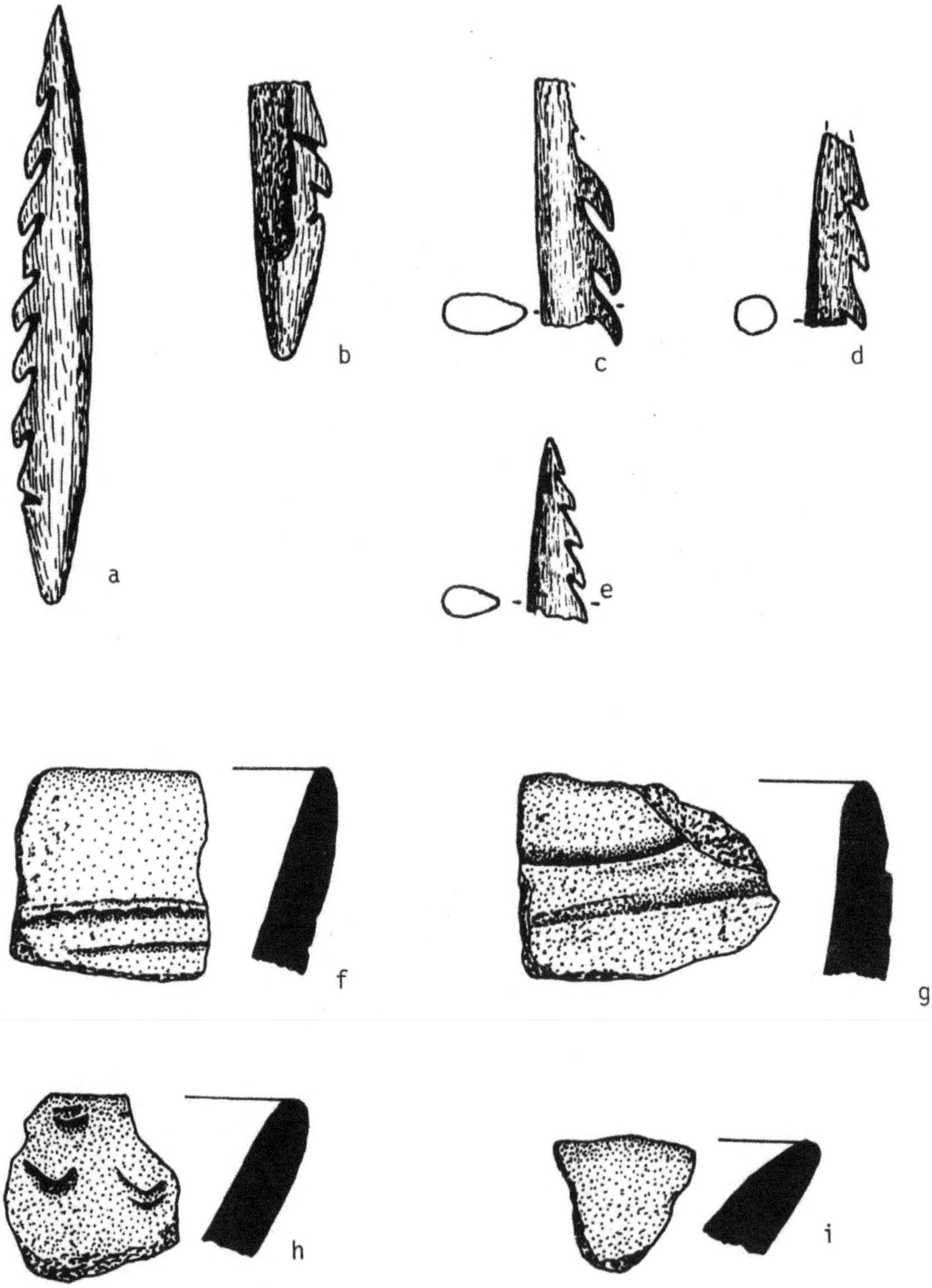

Figure 44. FxJj 12 North. Bone artifacts and pottery sherds. (a-e) uniserial harpoons; (f-h) decorated rim sherds; (i) undecorated rim. Natural size.

Pottery (Table 40; Figs. 44-46)

A total of 200 sherds were collected from the surface. Significantly, nine wavy-line sherds were identified, four of which were photographed *in situ* from bed B. The specimens were strikingly similar to sherds first collected at Khartoum (Arkel 1949:81-84) and since reported from numerous sites in North Africa (Hays 1975). Surface finds, usually single sherds, of wavy-line pottery have also been made within the Turkana Basin (Whitworth 1965b; Robbins 1972, 1974; F. H. Brown 1975; Phillipson 1977a). The site of FxJj 12 North is believed to represent the first *in situ* collection of wavy-line pottery directly associated with a fishing settlement from the Turkana Basin.

The sherds were dark in color (5Y3/2-5Y2/1), extremely friable, comparatively thick walled (7-10 mm) and tempered with ample amounts of coarse gravelly sand. The distinctive external design consisted of a varied number of closely space, shallow, undulating furrows or wavy-lines. The undulating lines may have been produced with a catfish spine (Arkel 1949; Phillipson 1977b:46). Using commercial clay tempered with coarse sand and a pectoral spine of a catfish (*Clarius* sp.), the author successfully reproduced the wavy-line decorative pattern. Other instruments however, such as barbed harpoon heads, may also have been employed.

A second series of decorated sherds was associated with the wavy-line specimens. This group was characterized by multiple rows of circular or lenticular impressions (Fig. 46, a-e). Sherds with a similar decoration occurred at the fishing settlements of GaJj 2 and GaJj 12. The examples from FxJj 12 North, like the wavy-line sherds, were dark in color, very crumbly and tempered with coarse gravelly sand. Significantly, *both* decorative features were combined on several sherds (Fig. 45, d-e).

Six rim sherds were collected, including four decorated specimens. Three of the rims had inwardly sloping profiles and tapered lips suggesting the presence of open mouthed vessels (Fig. 44, f-h, i). The absence of wavy-lines and circular rows of impressions among the rim sherds suggests these decorative features were limited to the main body of the vessel.

The largest proportion of surface sherds were concentrated in discrete patches along the outcrop's edge. These patches may have originally represented the remains of whole pots or perhaps refuse dumping areas that included numbers of broken sherds. Three sherds were rejoined form one area (Fig. 46, a).

To the northwest of sampling unit 2 a small collection of sherds was recovered at the interface of beds A and B. In character of paste, temper and decoration, these sherds were different from the wavy-line pottery. Their external surfaces were decorated with closely spaced rows of triangular impressions (Fig. 46, f-g) and their internal surfaces were decorated with shallow grooving or internal scoring not unlike the excavated sherds from Dongodien. Twenty out of 22 collected sherds had internal scoring, including specimens that

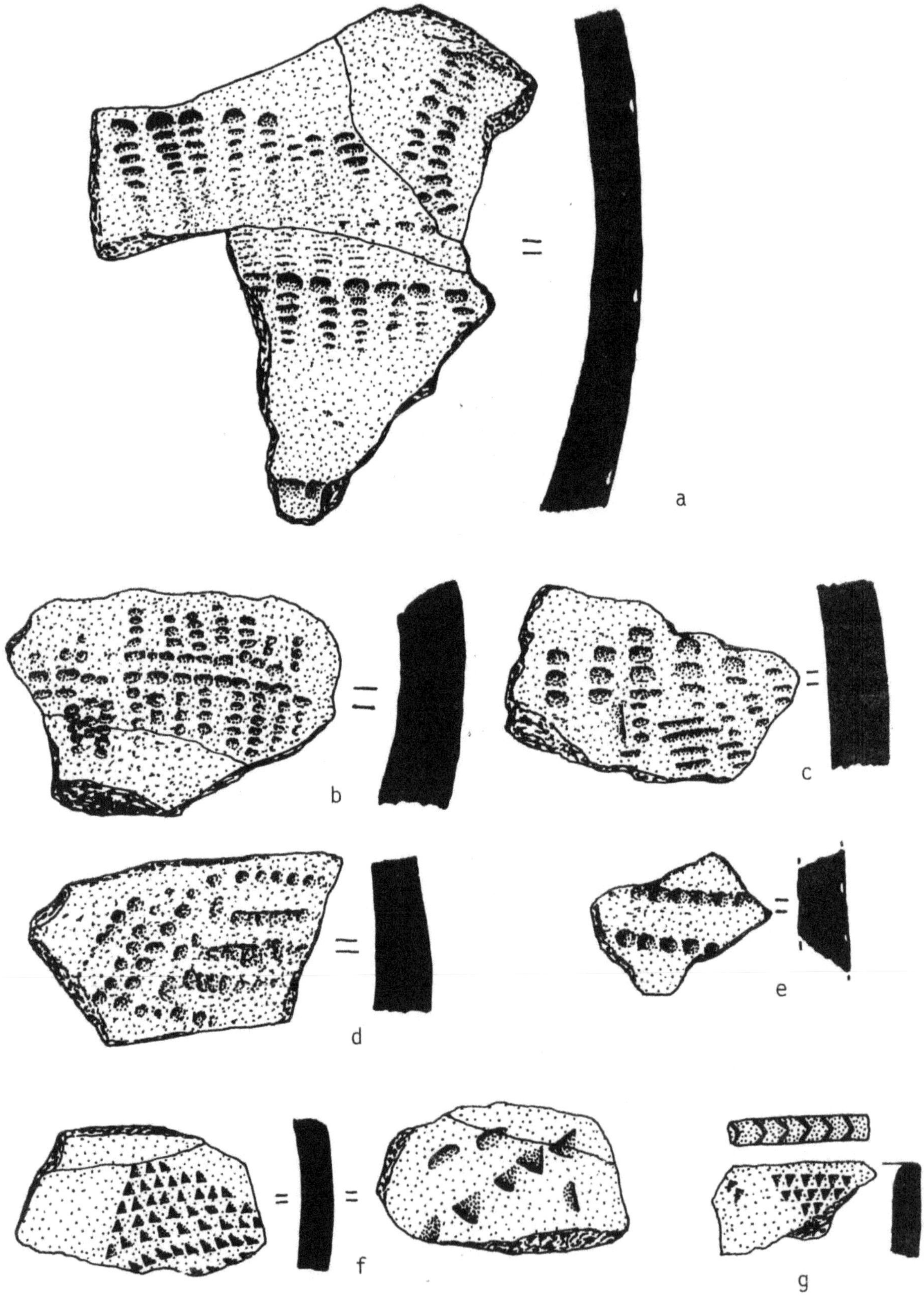

Figure 45. FxJj 12 North. Pottery.
(a-f) wavy-line body sherds. Natural size.

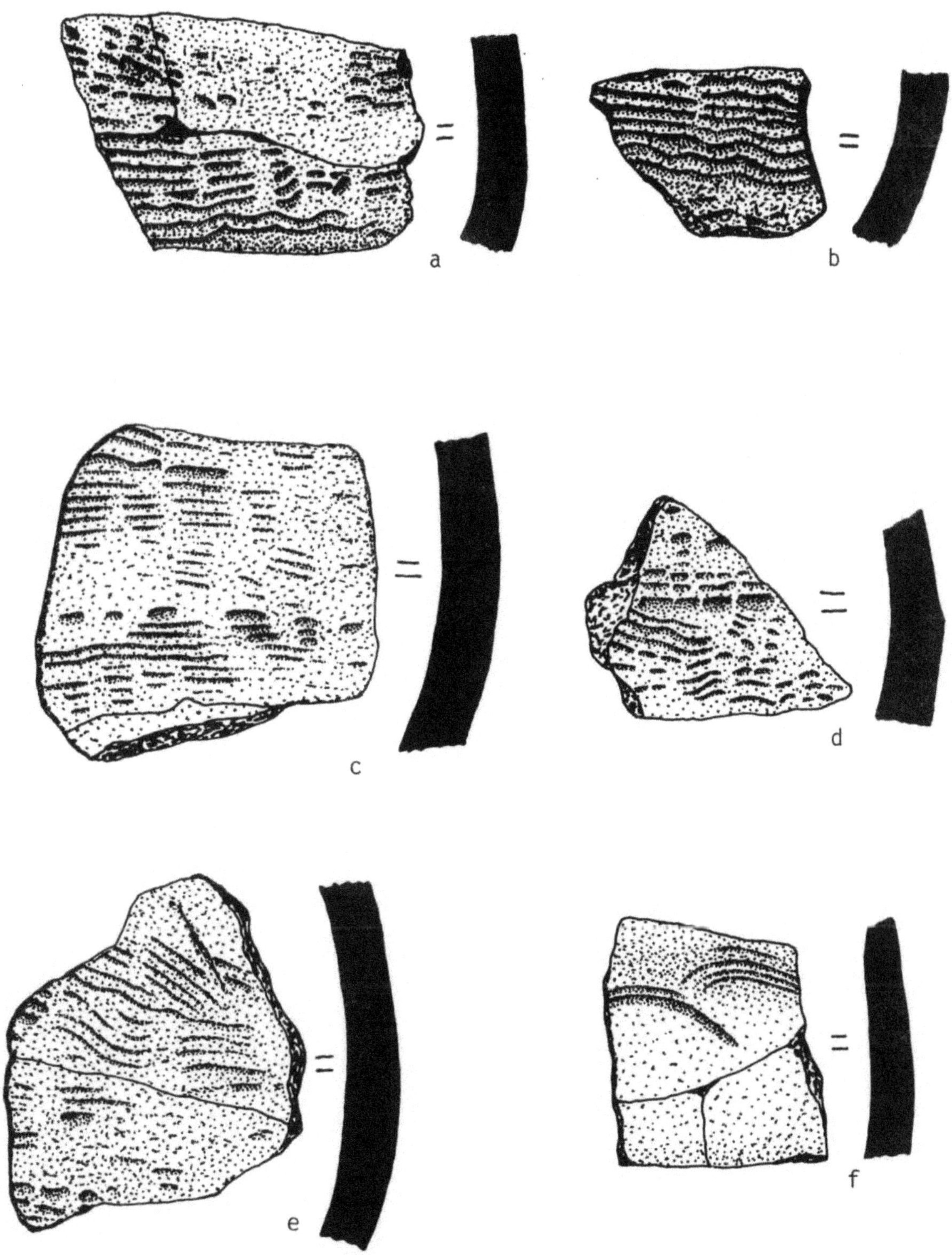

Figure 46. FxJj 12 North. Pottery.
(a-e) decorated body sherds; (f) internally scored sherds; (g) rim.
Natural size.

lacked external decoration. The pottery's location and combination of technical and stylistic features strongly suggests the assemblage was derived from bed A and was not associated with the wavy-line pottery from bed B.

Faunal Remains

A small collection of fragmentary and water abraided fish, reptile and mammal remains was recovered from the sampled units. While the complete assemblage of mammal remains has not been analyzed, preliminary field identification by the writer indicated several species of Bovidae and Warthog (*Phacochoerus* sp.) were present.

Human Remains

Well preserved fragments of a human skeleton, including a right humerus shaft, were discovered *in situ* within bed B. A partial calvaria, right maxilla and most of the body of a mandible were collected within several meters of the humerus. The combination of very small mastoid processes and a relatively small, slender-shafted humerus suggests the individual was a female. The sutures in the skull were fused and the tooth crowns were quite worn. No cause of death was evident.

DATING

Fragments of the human skeleton were submitted for radiocarbon analysis. The sampled yielded a date of:

3245 ± 155 BP Geochron Labs.

This date is not considered, for several reasons, to accurately reflect the age of the main archaeological horizon. First, the site's location along the high 75-80 m beach terrace indicates an early to middle Holocene age. Second, the skeleton may well represent a later intrusive burial. The radiocarbon date would therefore represent the valid age of burial but not the correct age of the archaeological site. Third, the technical difficulties of dating bone apatite are considerable (Taylor 1982; Owen et al. 1982). The possibility exists that the skeleton was associated with wavy-line pottery and bone harpoons but the radiocarbon date is much too recent. Unfortunately, neither fragments of charcoal nor mollusc shell were identified within bed B. At this moment, the age of the main archaeological horizon and the possible association of the human skeleton remain a frustrating uncertainty. The collection of additional C^{14} samples from FxJj 12 North has the highest priority in future field seasons.

PRELIMINARY INTERPRETATION

Stone artifacts, bone harpoons, wavy-line pottery and faunal remains were collected from a level of coarse beach sand along the high Holocene transgression in Area 117. Wavy-line pottery was relatively abundant and apparently represents the first occurrence, within the Turkana Basin, of this pottery style from a well defined stratigraphic context. The presence of two contrasting pottery styles with different distributions over the site raises the possibility that more than one phase of occupation was represented.

Analysis of altimeter readings from Areas 116 and 117 suggests the Kokoi horst area was tectonically active during parts of the Holocene. The single radiocarbon age determination may or may not be a valid date for the human skeleton, but it is not believed to be a reliable date for the lake shoreline deposits and the main archaeological horizon. Unfortunately, this horizon remains inadequately dated. Collection of suitable dating materials will have top priority in future field work.

Table 36

FxJj 12 North

INVENTORY OF SURFACE FINDS BY SAMPLING UNIT

Unit	1	2	3	4	5	6	7	8	9	10	11	Site Survey 1971	Site Survey 1974-1975	Plotted 1976	Assemblage Total (N)
Stone:															
Microliths	2	2	-	1	9	10	-	1	1	1	1	-	8	-	36
Scrapers	2	3	2	1	2	7	2	5	-	5	1	1	1	-	32
Outils écaillés	-	1	-	-	2	9	-	2	-	1	1	8	-	-	24
Bec	-	-	-	-	-	1	-	1	-	-	-	-	-	-	2
Core/chopper	1	1	4	3	2	-	13	1	-	4	4	-	-	-	33
Cas. trimm.	3	2	2	2	-	-	3	-	1	2	-	4	-	-	19
Whole flakes	20	20	15	15	50	91	56	31	6	74	51	14	-	-	443
Broken flakes	20	35	43	34	101	143	141	82	15	172	87	17	-	-	890
Angular frags.	12	12	17	20	70	109	101	47	7	96	56	15	-	-	562
Cores	5	15	3	2	20	24	29	16	2	14	28	50	-	-	209
Hammerstones	-	-	-	2	3	-	-	2	-	-	2	-	1	-	10
Bone harpoons:															
Uniserial	-	1	-	-	-	-	-	-	-	-	-	-	-	8	9
Pottery:															
Decorated	1	-	-	1	-	-	1	-	-	8	-	-	-	35	46
Undecorated	1	1	-	2	-	-	3	1	-	6	-	-	-	140	154
Bone frags. (NID):															
Mammal	1	3	4	-	8	21	12	15	3	26	8	-	-	-	85
Fish	17	17	8	9	-	5	7	34	1	10	20	-	-	-	128
Turtle	1	1	-	-	-	1	-	1	-	2	1	-	-	-	7
Crocodile	-	-	-	-	-	1	-	-	-	-	-	-	-	-	1
Bones w/art. sur.	-	2	-	3	1	-	2	1	-	-	-	-	-	-	9
Teeth:															
Mammal	1	-	-	-	1	4	5	1	-	4	2	-	-	-	18
Fish	1	-	-	-	-	1	-	1	-	-	-	-	-	-	3
Raw counts:															
Stone	65	91	86	80	259	394	345	188	32	369	231	109	10	-	2259
Bone/teeth	21	23	12	12	10	33	26	53	4	42	31	-	-	-	267

Table 37

FxJj 12 North: Surface

FREQUENCY OF TRIMMED PIECES

	Chert	Lava	Obsidian	Total (N)	Total (%)
Microliths:				(36)	(24.7)
Crescents	9	-	2	11	7.5
Curved backed flakes	11	-	-	11	7.5
Straight backed flakes	1	-	-	1	0.7
Double backed micro-percoir	2	-	-	2	1.4
Fragmentary	11	-	-	11	7.5
Scrapers:				(32)	(21.9)
Side	11	7	-	18	12.4
Double side	2	4	-	6	4.1
Convex	-	2	-	2	1.4
Notch	-	4	-	4	2.7
Circular	2	-	-	2	1.4
Outils écaillés	22	1	-	23	16.4
On a crescent	1	-	-	1	
Bec	2	-	-	2	1.4
Core/chopper	-	33	-	33	22.6
Cas. trimm.	8	11	-	19	13.0
Raw material (N)	82	62	2	146	100.0

Table 38
FxJj 12 North: Surface

FREQUENCY OF CORE TYPES

	Chert	Lava	Quartz	Obsidian	Total (N)	Total (%)
Single platform	30	7	1	-	38	18.2
Double: alternate	41	-	2	-	43	20.7
other	22	1	-	-	23	11.0
Multi	40	2	2	-	44	21.1
Disc	2	1	-	-	3	1.4
Casual	38	1	2	-	41	20.5
Fragmentary	9	8	-	-	17	8.1
Raw material (N)	182	20	7	-	209	100.0
(%)	87.1	9.6	3.3	-		

Table 39
FxJj 12 North: Surface

FREQUENCY OF MAJOR ARTIFACT CLASSES

	Chert	Lava	Quartz	Obsidian	Total (N)	Total (%)
Trimmed pieces	82	62	-	2	146	6.5
Cores	182	20	7	-	209	9.3
UTFF	730	1142	32	-	1904	84.2
Raw material (N)	994	1224	39	2	2259	100.0
(%)	44.0	54.2	1.7	0.1		

Table 40

FxJj 12 North: Surface

INVENTORY OF POTTERY SHERDS

Bed A: Internally Scored			Bed B: Main Arch. Horizon		
Rim:	decorated	3	Rim:	decorated	4
Body:	decorated	17		undecorated	2
	undecorated	2	Body:	'wavy-line'	9
				impressed	28
				other	4
				undecorated	130
Total (N)		22	Total (N)		178

INTRODUCTION

The site of GaJi 3 was located in Area 102 (Fig. 8). Small numbers of stone artifacts, bone harpoons, pottery sherds and a high density of fish bone fragments were exposed by erosion along the margins of an outcrop of Galana Boi sediments. The site was sampled and an excavation was undertaken. Two altimeter readings indicated the main archaeological horizon lay 55-56 m above the June 1976 level of Lake Turkana.

STRATIGRAPHY

Galana Boi deposits were well exposed for nearly 60 m along the margins of the east-west trending Koobi Fora Ridge (Fig. 47). The deposits had a cumulative thickness of just over two meters and rested unconformably on Upper Member units of the Koobi Fora Formation. The main stratigraphic divisions of a composite excavation section are illustrated in Figure 48. Silty clays, shell coquina, silty sands and coarse sands with gravels were present. Molluscs occurred throughout the section. They formed conspicuous shell aggregations at the interface of units B and C and between units D and C. Melanoides, Corbicula, Cleopatra and Caelatura were well represented in the upper shell band. Articulated Mutela shells were present near the top of unit B. Cultural material was recovered primarily from the silty sands in unit B. Additional artifacts were collected 30 m to the west along the interface between units D and C.

SAMPLING

A large area, eight meters by ten meters square, along the exposed margins of the Holocene outcrop, was selected for sampling. The area to be sampled was subdivided into one meter square units and fourty squares were collected. Each unit was scraped, sieved and all cultural remains were saved. Over 325 large fish elements were plotted on the field contour plan.

EXCAVATION

A partially eroded sequence of lacustrine sediments, immediately above the sampling area, was selected for excavaiton. Rich quantities of fish bone and a uniserially barbed bone harpoon head were eroding out of a horizon of silty unconsolidated sands. Fifteen excavation units, each one meter square, were either completely or partially excavated. Unit A (Fig. 48) consisted of a poorly sorted coarse sand with broken mollusc shells and few artifacts. This lithology was removed as overburden until a depth of 20 cms above unit B was reached.

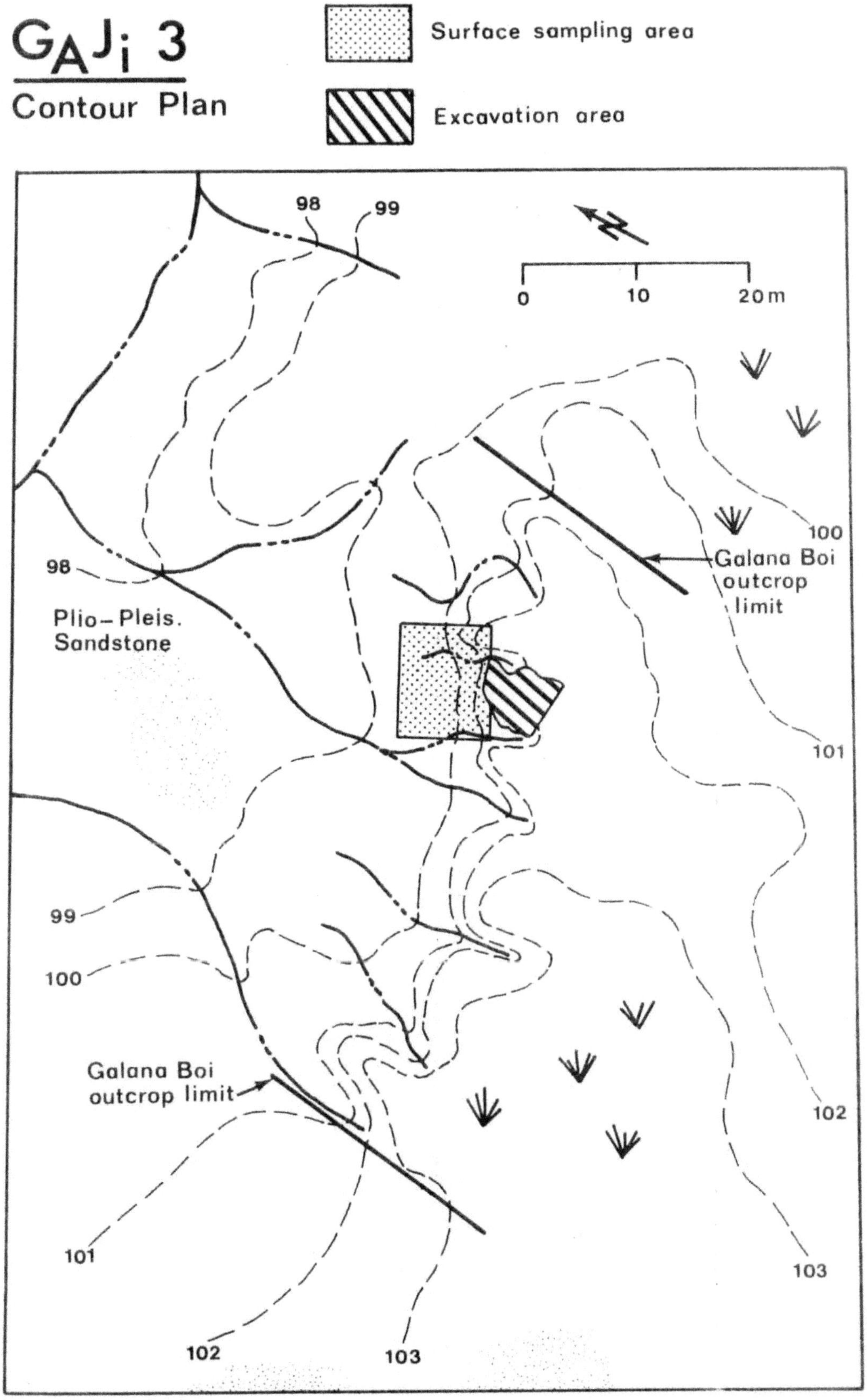

Figure 47. GaJi 3. Contoured site plan. Location of excavation and sampling units are indicated.

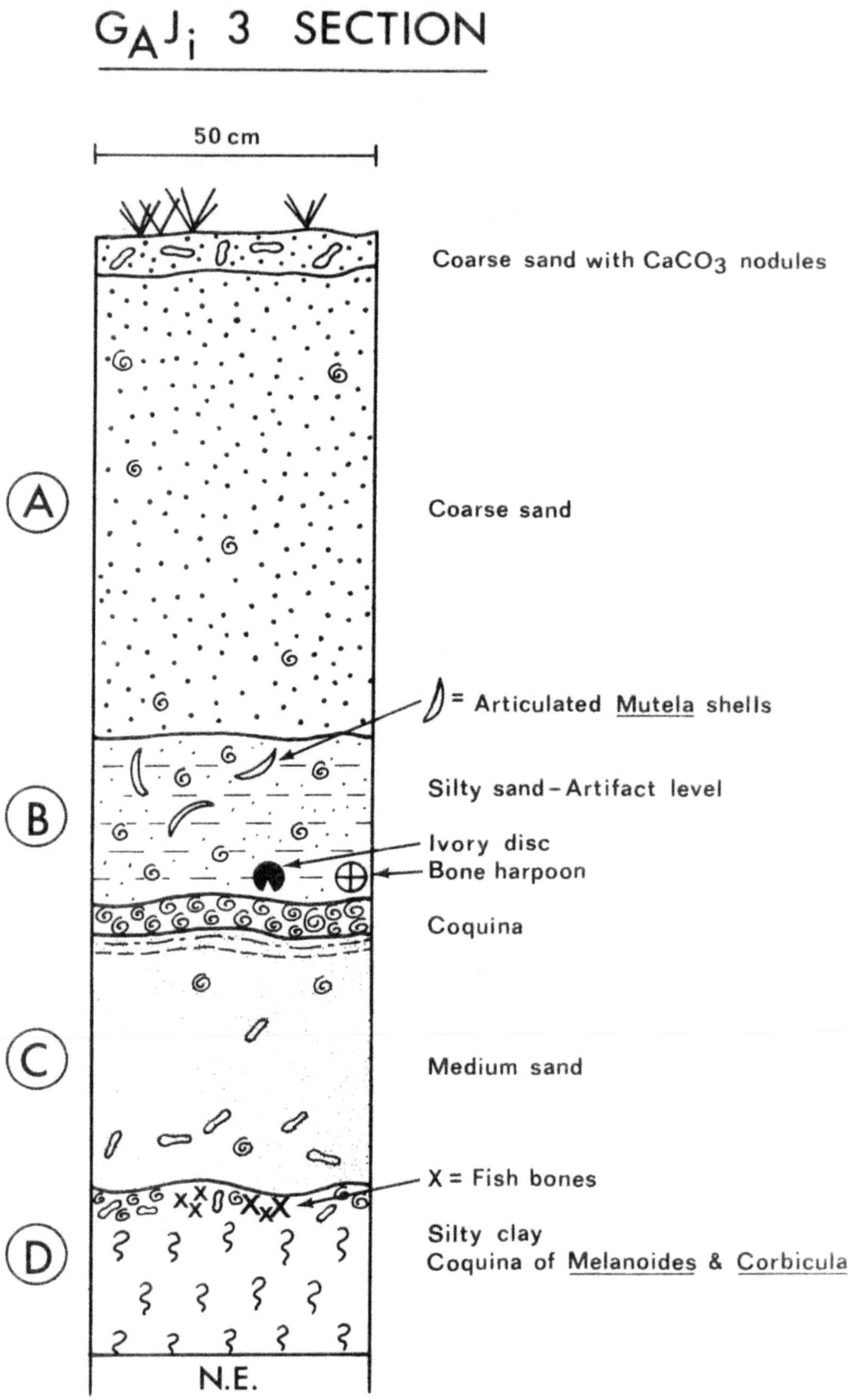

Figure 48. GaJi 3. Composite stratigraphic section. Unit B refers to the main archaeological level.

CULTURAL MATERIAL: SURFACE

Very few cultural remains were recovered from the 40 sampling units (Table 41). A total of 15 stone artifacts, including a core/chopper, outil écaillé, several sherds of pottery, uniserial bone harpoon, and an ostrich eggshell bead were collected. The bone harpoon was complete with sharply pointed barbs and showed little evidence of water abrasion (Fig. 49). A single decorated rim sherd showed cross hatching and thickening along the lip.

Fish bone accounted for the great bulk of the recovered surface material. Nile perch (Lates niloticus) and several species of catfish were provisionally identified. An atlas vertebra from a hippopotamus was also collected.

CULTURAL MATERIAL: EXCAVATION

Only a small number of stone and bone artifacts were recovered during the excavation (Table 41). Fragmentary fish bone was abundant throughout all levels of the excavation. Large fish bones were individually plotted in units B, C and D. Due to the low number of associated bone and stone artifacts, there was some uncertainty whether the recovered fish remains represented naturally or culturally related accumulations of refuse. The recovery of two barbed harpoons, a circular shell pendent, small fragments of charcoal and associated fish bones suggest the materials derived from unit B represented an occupation of low artifact density. The upper level of unit D suggests a different means of accumulation. Here, fish bone and broken mollusc shells occurred in a single compacted horizon. The partial remains of a large Nile Perch skeleton, including skull fragments, an articulated basio-cranium, several vertebrae and numerous pectoral and dorsal spines were excavated. No artifacts were directly associated with the fish skeleton. It would appear this second assemblage of fish bone accumulated by natural agencies. Thirty meters to the west, however, a broken biserial harpoon was discovered in situ at the same stratigraphic level as the excavated Nile Perch skeleton. The harpoon was clearly associated with other fish bone refuse. The material from unit D may well represent a second low artifact density-high fish bone concentration. Gifford's (1978) recent ethnoarchaeological research with the Dassanetch has demonstrated that similar localized patches of abundant fish bone refuse may be associated with few artifacts and little concentrated charcoal. However, a clear association of artifactual and faunal material from unit D remains uncorroborated.

DATING

A single age determination on fish bone from unit B yielded a date of 4560 ± 185 BP.

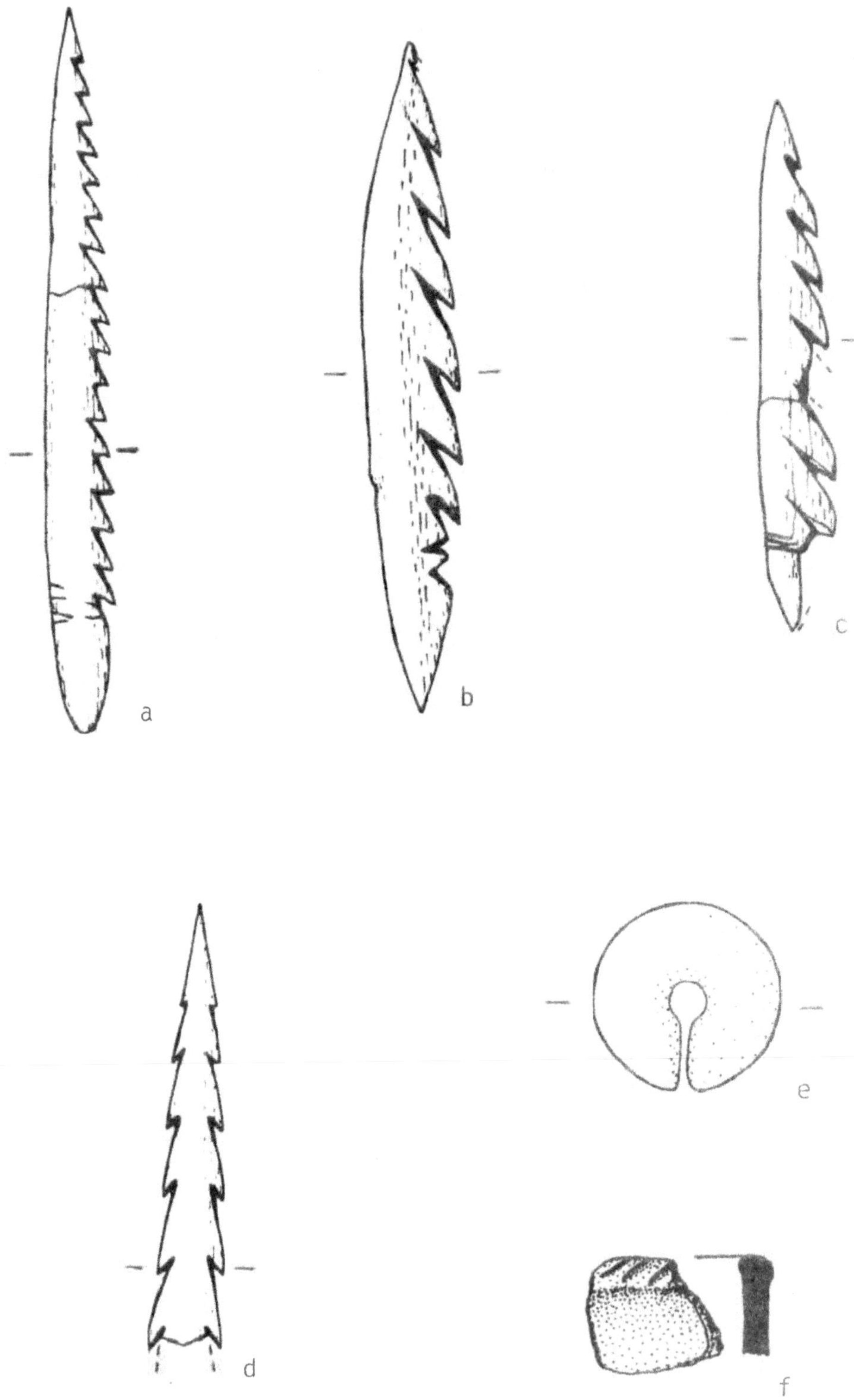

Figure 49. GaJi 3. Surface and excavated artifacts: Surface. (b-c) uniserial harpoons; (f) decorated rim sherd; Excavated. (a) uniserial harpoon; (d) biserial harpoon; (e) shell disc. Natural size.

PRELIMINARY INTERPRETATION

The site of GaJi 3 was originally selected for excavation because of a low density of artifacts and contrasting abundance of associated fish refuse. It was believed the site would furnish different settlement and subsistence patterns when compared to the series of larger fishing and hunting settlements located along the 75-80 m Holocene transgression. Unfortuntely, such comparisons were rendered difficult by the uncertain association of artifacts and faunal remains in unit D. On the other hand, excavations within unit B revealed a definite association of fish bone, bone harpoons and charcoal fragments. These artifacts occurred within a unit of silty sand that is best interpreted as beach sediments. Both horizons B and D may have represented short duration occupations primarily oriented towards fishing. The evidence from unit D was equivocal however. Alternatively, the single harpoon from this unit may have been lost during a fishing expedition which subsequently became incorporated within offshore silty clays. Robbins reported on a possible analogous situation from Lothagam (1974:204). The single radiocarbon date, in conjunction with two altimeter readings, suggests the occupation was probably associated with the retreating phase of second Holocene transgressive-regressive cycle.

Table 41

GaJi 3

INVENTORY OF SURFACE AND EXCAVATION FINDS

	Surface	Excavation	Total (N)
Stone:			
Microliths	-	-	-
Scrapers	-	-	-
Outils écaillés	1	-	1
Whole flakes	1	-	1
Broken flakes	1	-	1
Angular frags.	9	2	11
Cores	3	-	3
Bone harpoons:			
Uniserial	2	2	4
Biserial	1	1	2
Pottery:			
Decorated	1	-	1
Undecorated	8	-	8
Bone frags: (NID)			
Mammal	2	-	2
Fish	*	*	*
Turtle	6	-	6
Bones w/artic. sur.	1	-	1
Teeth:			
Mammal	1	-	1
Fish	17	36	53
Beads:			
Ostrich eggshell	1	-	1
Shell pierced disc	-	1	1

* records missing

CHAPTER IV

PASTORAL NEOLITHIC SITES

The classificatory term Pastoral Neolithic has been proposed to describe:

> . . . cultures of eastern Africa which (1) relied substantially on domesticated stock for their livelihood, (2) used pottery, and (3) employed typical Later Stone Age technologies for the manufacture of edged tools. (Bower and Nelson 1978: 562)

The term is a useful contribution to the cultural-historical nomenclature of East Africa especially as it helps identify a widespread and distinctive set of similar cultural traditions.

Chapter IV presents three site reports that describe middle Holocene Pastoral Neolithic settlements at East Turkana. Two of the sites were situated along former beach shorelines while the third occupation was located along the margins of a major inland river drainage. A comprehensive faunal analysis has been completed for each occupation. Ten radiocarbon determinations have been received from the three sites.

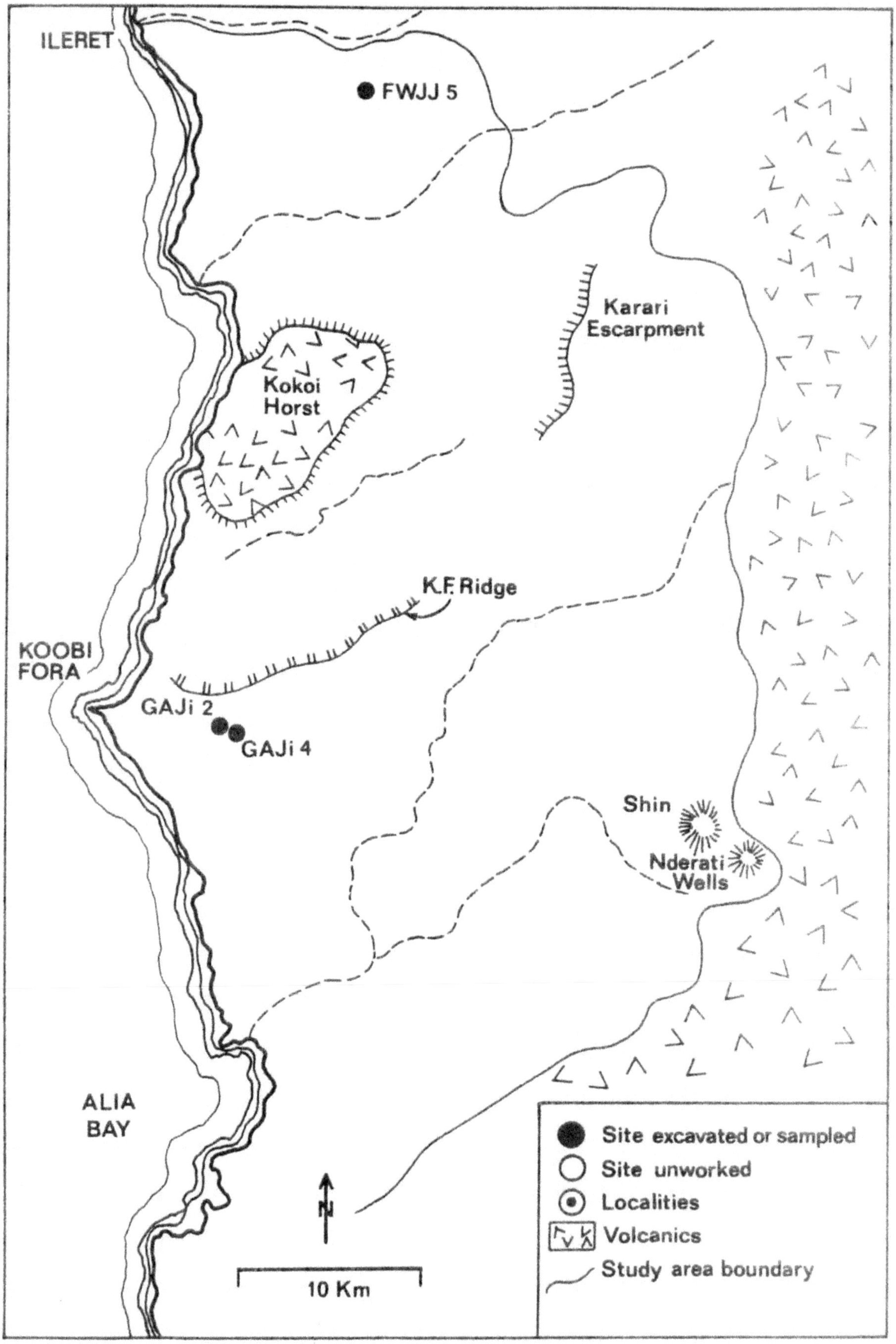

Figure 50. Location of excavated Pastoral Neolithic sites.

INTRODUCTION

The site of GaJi 2 was located in Area 102 near the top of the Koobi Fora Ridge (Fig. 50) (Plate 2). Stone artifacts, pottery, fire-cracked rock and faunal remains were exposed by erosion along the margins of a charcoal stained horizon of fine grained beach sand. An area of charcoal concentration, possibly a cooking or pottery firing feature, was carefully excavated and cross sectioned. Two excavations were undertaken. Domestic caprines and domestic cattle were identified. A series of radiocarbon dates indicated the site was occupied by 4000 years BP. Three altimeter readings suggested the site lay 43-46 m above the June 1976 lake level.

STRATIGRAPHY

The extensive Galana Boi beds in Area 102 extended inland over 10 kilometers and had a maximum thickness of between 32-35 meters (Raynolds 1973; Vondra and Bowen 1978). The deposits lay uncomformably on Upper Member units of the Koobi Fora Formation.

The principal archaeological horizon occurred within a thick sequence of littoral and fluvial sands stratigraphically above and to the west of Dongodien (GaJi 4). Figure 51 illustrates the main geological strata of the excavation section. Units A and B consisted of a pale brown to light orange unconsolidated, poorly sorted coarse sand. The units showed a distinct fining upward gradation. Broken mollusc shells occurred throughout each level. At the base of unit B a concentrated lens of coarse gravels and broken mollusc shells was present. In places, this lens truncated the underlying unit C (Fig. 51, West Face). This cut and fill erosional contact was well defined and could be traced along the exposures for over three kilometers.

The relatively few cultural remains from the upper part of unit B and unit A (Upper Horizon) were usually water worn, broken and exhibited a thick patina of calcium carbonate. These artifacts are thought to be redeposited from an unknown number of former occupation surfaces. The numerous fragments of charcoal, ash and broken bone which occurred at the very base of unit B represented reworking of the underlying occupation surface.

Unit C (Lower Horizon) was composed of a distinctive pale to dark gray, clean, medium coarse beach sand. Abundant fragments of charcoal, ash and mollusc shell (*Melanoides*, *Caelatura*, *Bellamyra* and *Corbicula*) were present. Cultural remains, including microliths, pottery, beads and bone refuse were recovered. Unit C represented the main archaeological level.

Unit D consisted of a well sorted light gray fine littoral sand with abundant heavy minerals. No artifacts were collected from this level.

Summary

The excavation section may be interpreted as follows: the site

Plate 2. Area 102: Galana Boi exposures.

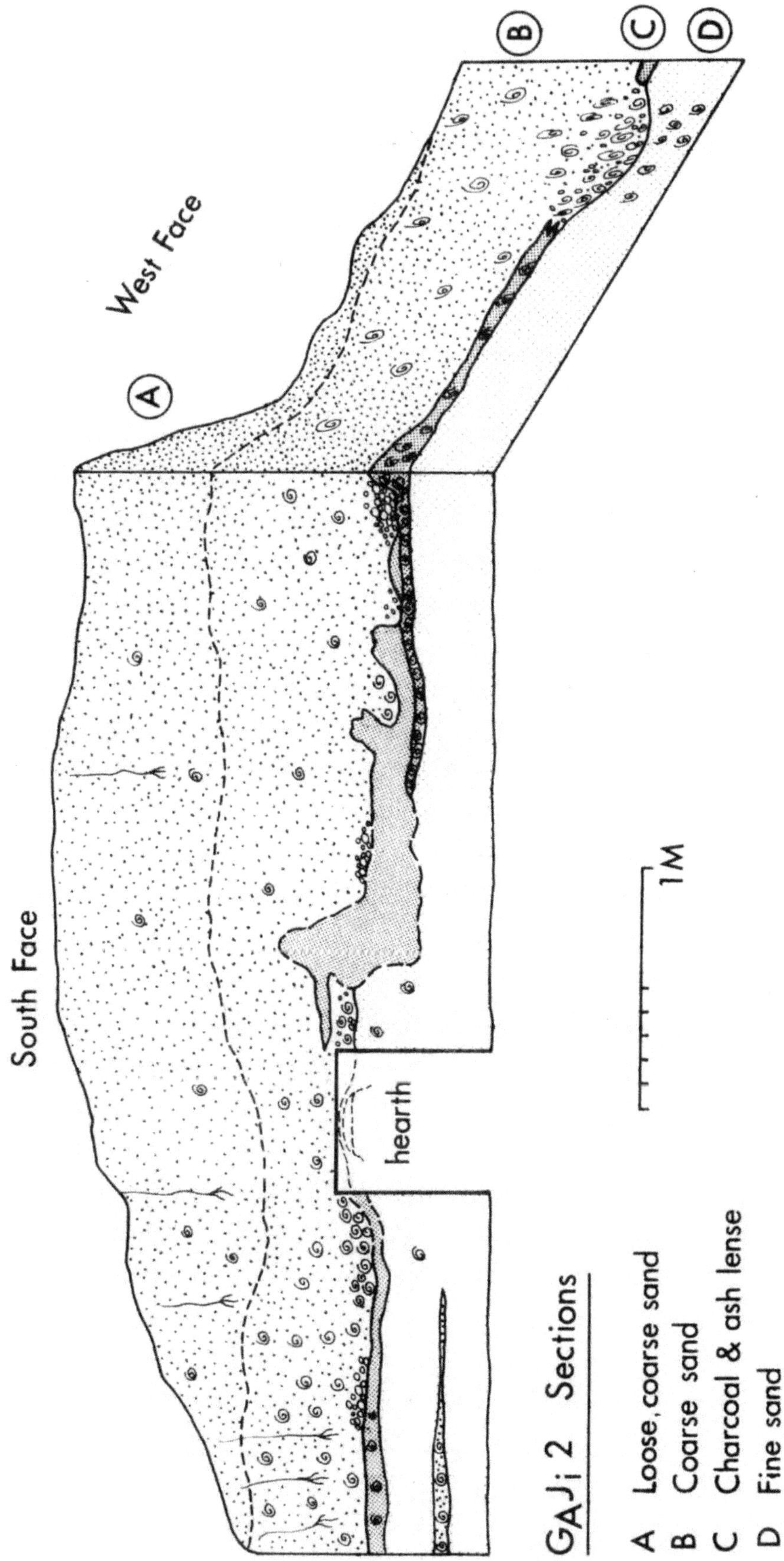

Figure 51. GaJi 2. Profile of excavation sections showing concentrations of ash, charcoal and hearth area.

was established on the surface of a clean medium coarse beach sand, probably close to the paleolake shoreline. At some point later in time, the Holocene lake underwent major recession. Burial and subsequent erosion of the soft unconsolidated littoral sediments by river or distributary channels began. Portions of the upper surface of unit C were disturbed and at least one area was truncated by the resurgence of channel cut and fill activity. Fluvial sediments, with a mixture of cultural remains from several former occupation surfaces, now accumulated to a thickness of 1-2 m. Recent eolian sands mantled the entire exposure.

EXCAVATION

Two areas of the site were chosen for excavation. Both the Upper and Lower Horizons were removed from Excavation 1. Due to time restrictions, Excavation 2 was limited to only sections of the Upper Horizon.

For Excavation 1, a five meter by two meter area was gridded into 10, one meter excavation squares. Units A and B were removed with picks and shovels until a level of 20 cms above the unit B/C interface was reached. The overburden was screened and all cultural remains were saved. Within unit C, excavation proceeded in 5 and 10 cm spits. Isolated patches of charcoal and ash, microliths, pottery and faunal remains were carefully plotted.

Four large plastic bags of charcoal, ash and sand were collected from the Lower Horizon and analyzed for possible carbonized botanical remains. Under the direction of Vickie Morse and Corrine Glesne, a make-shift flotation apparatus was set up along the modern beach shoreline. The sediment was immersed in water, examined for floating debris, and then seived through a fine wire mesh. Unfortunately, repeated efforts did not yield botanical remains.

CULTURAL REMAINS

An inventory of surface and excavated cultural remains is presented in Table 42. The following summary, due to the reworked aspect of the Upper Horizon, concentrates on the artifacts from the Lower Horizon.

Stone (Tables 43-45)

Relatively few stone artifacts were recovered from the Lower Horizon. Chert accounted for nearly 75% of the raw material followed by obsidian and lava. Shaped stone artifacts were scarce; only two microliths and three *outils écaillés* were collected. Debitage consisted primarily of small flakes and flake blades. Nine large fragments of fire cracked rock were associated with patches of charcoal and ash. The rocks were probably transported to the site from a nearby outcrop of bioclastic cemented sandstone.

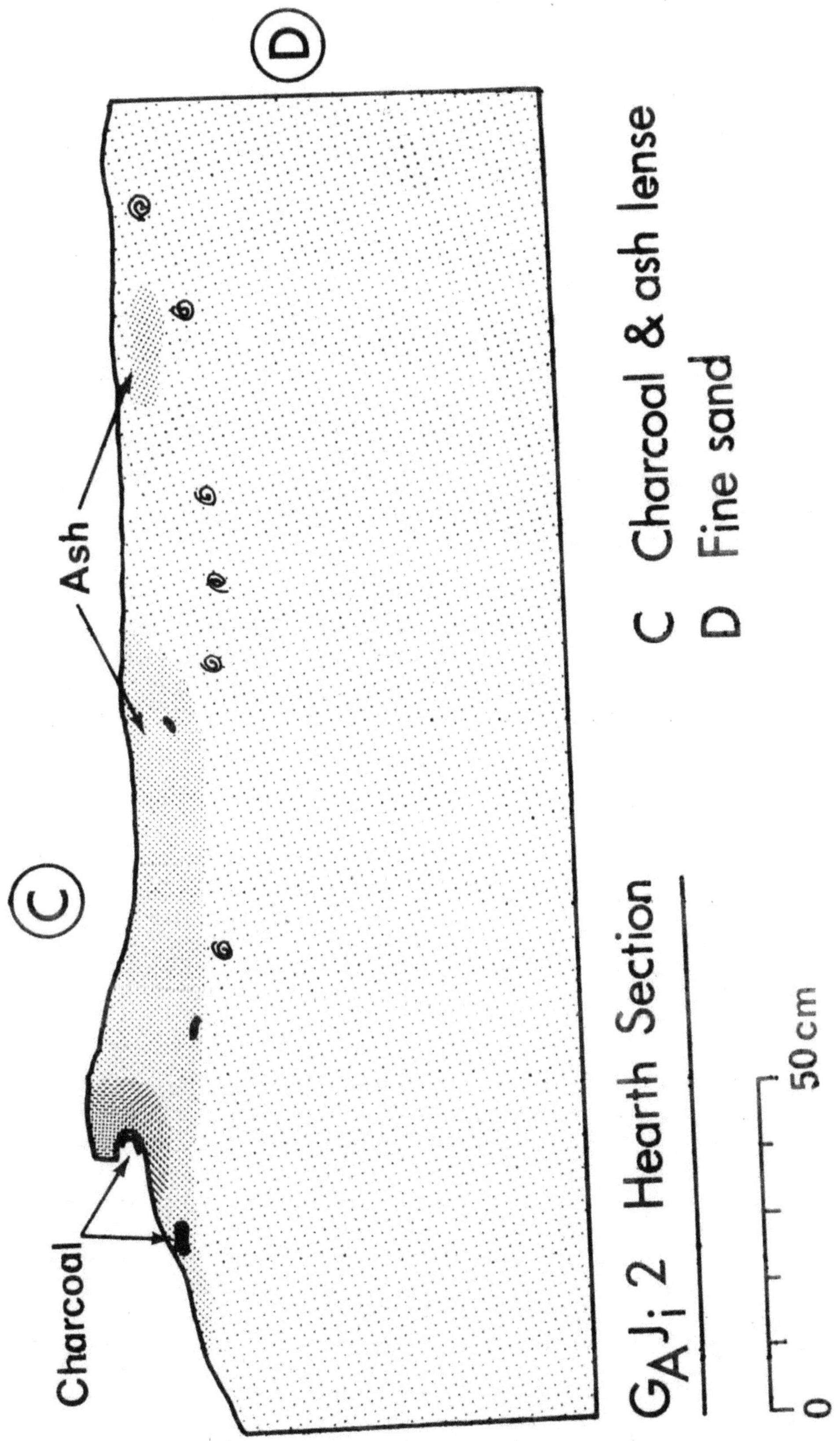

Figure 55. GaJi 2. Profile of east face showing hearth area.

Pottery: Upper Horizon

Over 50 pottery sherds were collected from the reworked Upper Horizon. The excavated sample was characterized by comb stamping and incised lines. A representative sample of decorated sherds, including a rim sherd of Nderit Ware, is illustrated in Figure 53, a-e.

Pottery: Lower Horizon

The excavated sherds from the main archaeological horizon exhibited several noteworthy features. Decorated sherds were characterized by rows of geometric comb stamped figures which were sometimes accompanied by grooving and paneling. A few sherds were decorated with punctations or incised lines. Several rim sherds had flattened cross hatched lips. The profile of another broken rim suggests the original vessel had a very constricted circular mouth (Fig. 52, f), not unlike the reconstructed burial pot from GaJj 9. In marked contrast to the nearby site of Dongodien, no sherds exhibited either internal scoring, burnishing or slipping.

A small group of sherds was collected along the outcrop's edge at the contact between both archaeological horizons. One of the rim sherds was rejoined to its excavated counterpart from the Lower Horizon (Fig. 52, e). Unfortunately, the rest of the sample may be associated with either of the levels. A wide range of comb stamped, incised and impressed decorative features was identified.

Beads

Three beads were recovered including a stone bead manufactured from microcline feldspar. The specimen was well crafted, symmetrical, and drilled from both sides. The two additional beads were made from ostrich eggshell.

FEATURES

The site was originally discovered at the end of a long morning's hitherto unproductive site survey. The excavation foreman, Muteti Nume, spotted a thick horizontal band of charcoal and ash exposed along the outcrop's sandy edge. Subsequent excavation revealed a subcircular mound with a diameter of 80-135 cm and an interior depression 25-45 cm in depth. The depression was filled with charcoal stained sand and rock, ash, and artifactual refuse. Fragments of burned bone and an immature mandible of a domestic cow were also associated. A thin discontinuous horizon of ash and charcoal patches was plotted near the feature (Fig. 54). The boundaries of these patches were indistinct however, and the horizon may have originally been distributed over the entire excavated surface. A cross section of the mound is illustrated in Figure 55. The association of charcoal, burnt bone and slabs of fire cracked rock strongly suggest the feature served as a cooking area.

Faunal Remains

A total of 609 bone fragments were recovered from the Lower Horizon. Domestic caprines and domestic cattle were identified. The caprines were identified from a collection of whole teeth while the cattle remains were recognized from a combination of dental and post cranial elements. Significantly, the remains of wild bovids and other game animals were not identified. Fish bone, including burned elements, accounted for 41% of the faunal sample. Several large vertebrae were plotted and have been identified as Lates niloticus. This material has yet to be analyzed in detail. John K. Mengitch of the National Museum in Nairobi identified the mammal remains. His identifications are listed in Tables 46-47.

DATING

Three C^{14} determinations were run on charcoal and mollusc shell samples from the Lower Horizon. The charcoal sample was divided in half and sent to two radiocarbon laboratories to insure independent assessment. The samples yielded dates of:

8915 ± 140 BP	mollusc shell	Sydney University
4160 ± 110 BP	charcoal	Sydney University
3970 ± 60 BP	charcoal	University of Pennsylvania

The radiocarbon determination for the mollusc shell is thought to date the age of the shell but not the age of the occupation surface. Early Holocene shell debris had apparently been reworked and incorporated into the archaeological horizon. Mollusc samples yielded a similarly anomalous date from Dongodien.

The two independently run charcoal samples have standard deviations that overlap at one sigma and are therefore considered statistically equivalent. They confirm a date of 4000-4100 BP for the age of the occupation.

PRELIMINARY INTERPRETATION

Analysis of the faunal material from GaJi 2, confirms the presence of domestic stock in the Lake Turkana Basin by 4000 BP. The site was located close to the paleo-shoreline when the Holocene lake stood over 45 m higher than today. Domestic caprines and cattle were associated with a Later Stone Age microlithic industry and pottery sherds that share several design motifs with the decorated pottery from the Ileret Stone Bowl Site. Fishing apparently continued to supply an important portion of the diet. As bone harpoons were not recovered, the inhabitants were evidently using alternative, as yet unknown, fishing techniques. The hunting of wild game animals had substantially declined and perhaps ceased altogether.

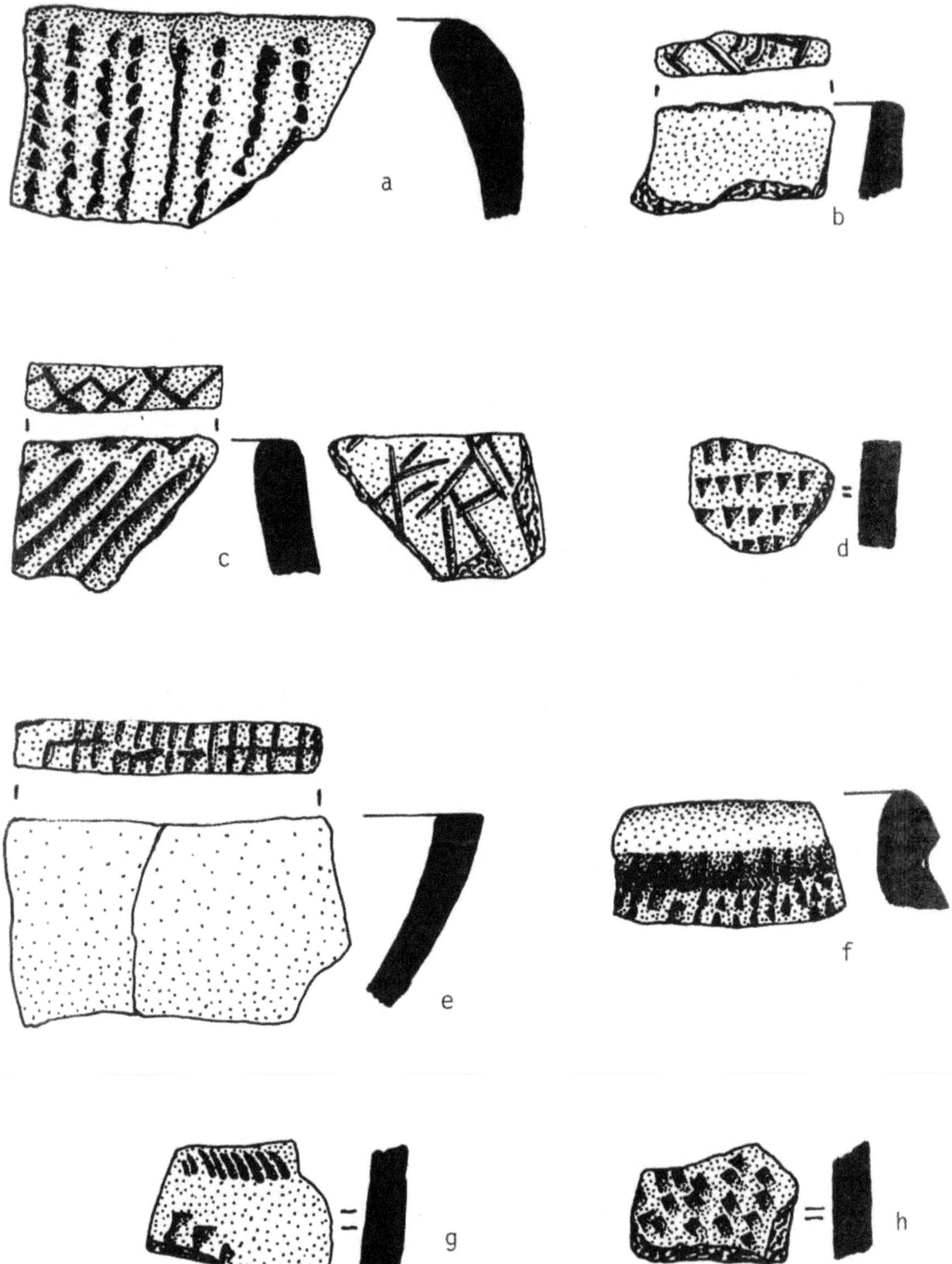

Figure 52. GaJi 2. Pottery: Excavation. Upper Horizon: (a-b) decorated rim sherds; (c) Nderit Ware rim; (d) decorated body; Lower Horizon: (e) undecorated rim; (f) decorated rim; (g-h) decorated body sherds. Natural size.

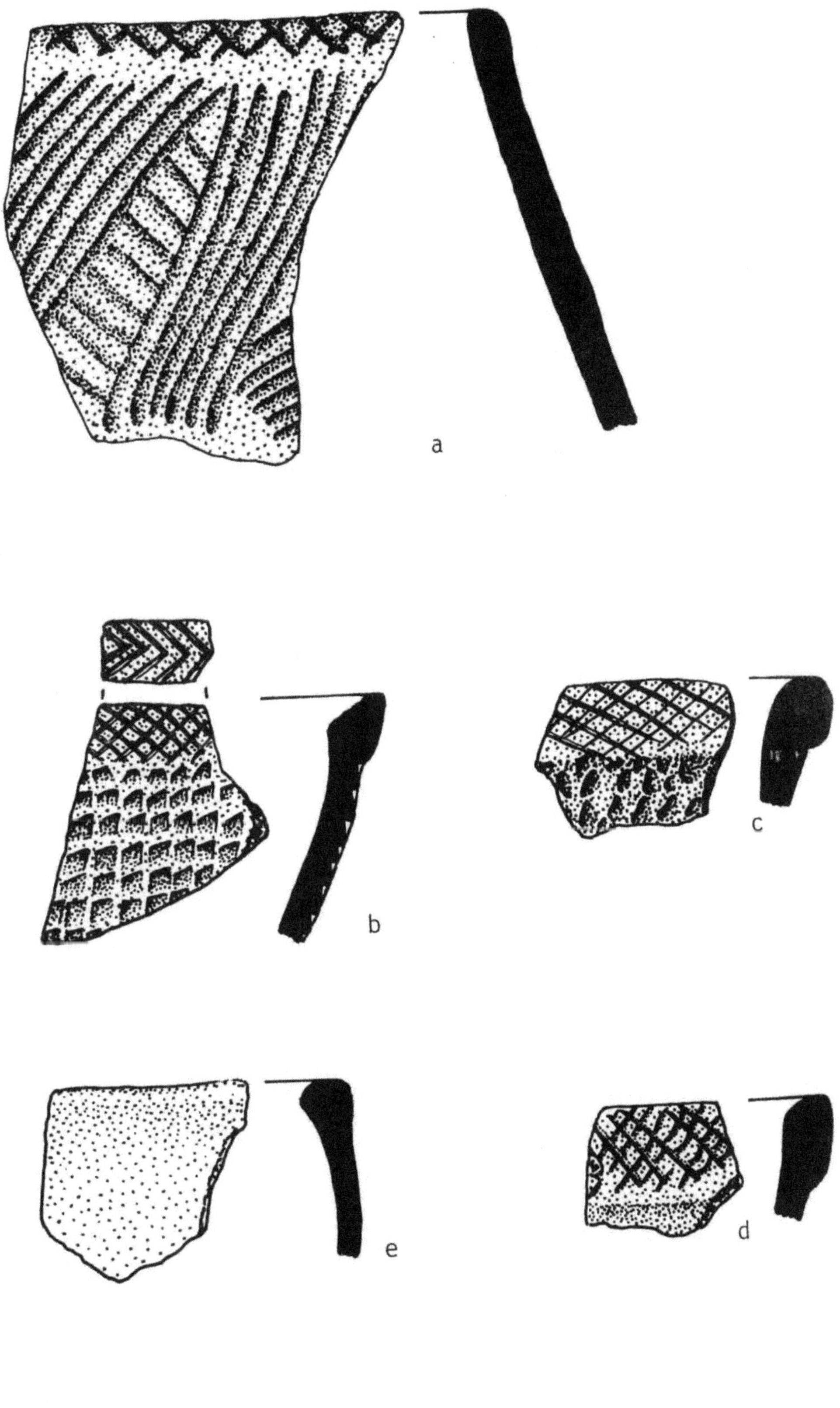

Figure 53. GaJi 2. Pottery: Surface (eroding out of outcrop's edge). (a-d) decorated rim sherds; (e) undecorated rim. Natural size.

GaJi 2: Lower Horizon

PLOT OF EXCAVATED CULTURAL REMAINS

KEY

Stone	
Stone artifact	
Fire cracked rock	
Bone/teeth	
Identifiable element: mammal	•
Fish	o
Non-identifiable: mammal	/
Pottery	
Decorated	▲
Undecorated	Δ
Concentrations of charcoal and ash	
Bovids	
Bovid	B
Medium size bovid	M
Ovicaprid	Ovicap
Other	
Ostrich eggshell bead	B ost
Stone bead	B st
Body parts	
Maxillae	mx
Tooth	th
Humerus	hu
Rib	rb
Mandible	md
Ulna	ul
Phalange	ph

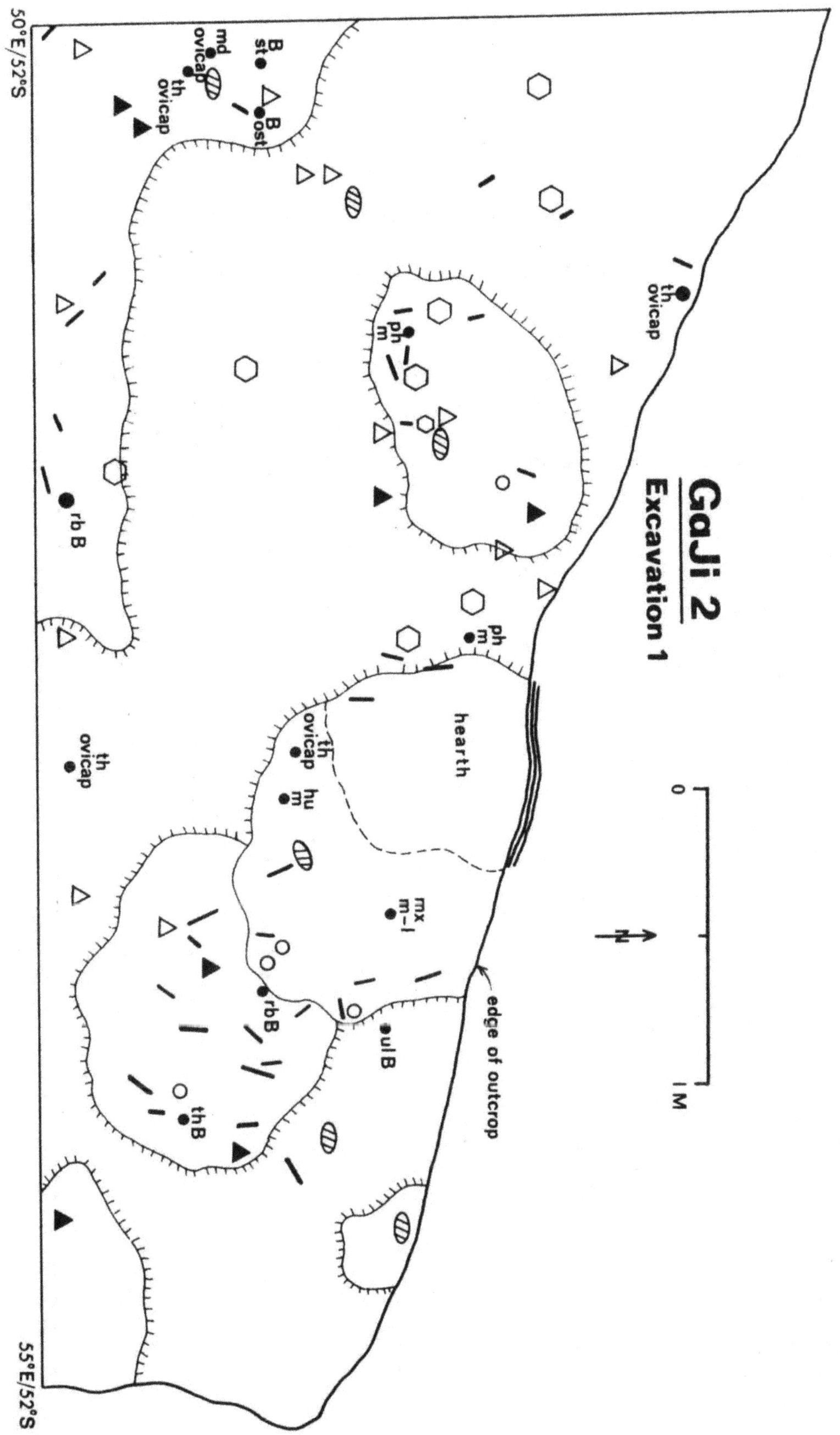

Figure 54. GaJi 2. Plot of excavated artifacts: Lower Horizon.

Table 42

GaJi 2: Surface and Excavation

INVENTORY OF CULTURAL REMAINS

	Sampling Units	Upper Horizon			Lower Horizon	Assemblage Total (N)
		Ex. 1	Ex. 2	(N)	Ex. 1	
Stone:						
Microliths	1	6	5	11	2	14
Scrapers	-	-	-	-	-	-
Outils écaillés	1	1	1	2	3	6
Whole flakes	5	6	11	17	23	45
Broken flakes	4	7	19	26	46	76
Angular frags.	5	29	24	53	49	107
Cores	1	1	-	1	-	2
Pottery:						
Rim: decorated	6	2	2	4	3	13
undecorated	1	2	-	2	2	5
Body: decorated	2	2	8	10	13	25
undecorated	17	18	17	35	44	96
Bone frags. (NID):						
Mammal	18	41	68	109	312	439
Fish	29	139	84	223	249	601
Bones w/art. sur.	3	-	6	6	17	26
Teeth:						
Mammal	-	6	3	9	29	38
Fish	-	-	-	-	1	1
Crocodile	-	-	-	-	1	1
Other:						
Bone artifact	-	-	1	1	-	1
Beads: stone	-	-	-	-	1	1
ostrich e.s.	-	1	-	-	1	1
Ostrich e.s. frags.	1	4	11	14	8	23
Fire cracked rock	-	-	-	-	9	9
Haematite	-	-	-	-	1	1
Pumice	-	-	-	-	4	4

Table 43

GaJi 2

FREQUENCY OF TRIMMED PIECES

	Upper Horizon				Lower Horizon			
	Chert	Obsidian	(N)	(%)	Chert	Obsidian	(N)	(%)
Microliths:			(11)	(84.6)			(2)	(40.0)
Crescent	2	2	4	30.8	1	-	1	20.0
Curved backed flake	1	1	2	15.4	-	-	-	-
Truncation	1	1	2	15.4	-	-	-	-
Fragmentary	2	1	3	23.0	1	-	1	20.0
Outils écaillés	1	1	2	15.4	1	2	3	60.0
Raw material (N)	7	6	13		3	2	5	

Table 44

GaJi 2 Upper Horizon

FREQUENCY OF MAJOR ARTIFACT CLASSES

	Chert	Lava	Quartz	Obsidian	Total (N)	Total (%)
Trimmed pieces	8	-	-	5	13	11.8
Cores	-	-	-	1	1	1.0
UTFF	49	1	-	46	96	87.3
Raw material (N)	57	1	-	52	110	100.0
(%)	51.8	0.9	-	47.3		

Table 45

GaJi 2 Lower Horizon

FREQUENCY OF MAJOR ARTIFACT CLASSES

	Chert	Lava	Quartz	Obsidian	Total (N)	Total (%)
Trimmed pieces	4	-	-	1	5	4.1
Cores	-	-	-	-	-	-
UTFF	86	5	-	27	118	95.9
Raw material (N)	90		-	28	123	100.0
(%)	73.2	4.0	-	22.8		

Table 46

GaJi 2: Lower Horizon

INVENTORY OF FAUNAL REMAINS BY TAXA

	N	%
MAMMAL		
Bovids	(31)	(4.9)
Caprine	8	1.3
Small bovid (caprine size)	9	1.5
Medium bovid (cattle size)	2	0.3
Size not determined	12	1.8
Carnivores		
Small carnivore	1	0.2
Non-identifiable fragments	326	53.6
REPTILE		
Crocodile	1	0.2
FISH	250	41.1
Assemblage Totals (N)	609	100.0

Table 47

GaJi 2: Lower Horizon

INVENTORY OF IDENTIFIABLE BODY PARTS (BOVIDS)

	Caprine	Small Bovid (caprine size)	Cattle	Medium Bovid (cattle size)	Bovid (Size not determined)
Body part:					
Mandible	2	-	1	-	-
Maxillae	4	-	-	-	-
Isolated teeth	2	6	-	-	5
Humerus	-	1	-	-	-
Ulna	-	-	-	-	1
Phalanges	-	-	-	1	3
Metapodials	-	1	-	-	-
Cuneiform	-	-	-	-	1
Astragalus	-	1	-	-	-
Total (N)	8	9	1	1	10

INTRODUCTION

The site of Dongodien (GaJi 4) was situated in Area 102 along the eroded flanks of the Koobi Fora Ridge (Fig. 50). Microliths, Nderit Ware pottery, faunal remains and concentrations of charcol and ash were excavated from several horizons of coarse beach sands (Plate 2). The faunal remains included domestic caprines and domestic cattle. Two excavations were conducted and the site was sampled (Fig. 56). Three altimeter readings indicated the archaeological horizon lay 44-47 m above the June 1976 level of Lake Turkana. A series of radiocarbon determinations provided firm support that the occupation horizon dates to at least 4000 BP.

STRATIGRAPHY

The Galana Boi deposits in Area 102 attained a maximum thickness of 32-35 m (Raynolds 1973; Vondra and Bowen 1978) and lay unconformably on Upper Member beds of the Koobi Fora Formation. The Holocene sediments mantled the Koobi Fora Ridge and formed a continuous exposure for nearly 6 kms. A localized Pleistocene anticlinal structure (Behrensmeyer 1970) produced a SW depositional dip of nearly 4° that affected the otherwise undisturbed Holocene deposits.

The succession of Galana Boi beds at the archaeological site consisted of dark brown silty clays, fine silts, silty sands and coarse beach sands and gravels (Fig. 57). Reworked cultural remains occurred among the three uppermost coarse sandy gravel horizons (Units 1-3). The principal archaeological level was located within Unit 5. A more detailed lithological description of this unit is outlined below.

Sub-Lithology	Description
5A	light orange clean coarse gravelly sand; forset beds contained whole and broken mollusc shells, chiefly Melanoides tuberculata; subrounded gravels and abundant bone refuse; stone artifacts and faunal remains showed distinct glossy patina; charcoal fragments concentrated at base; 20-45 cms.
5B	light orange clean medium coarse sand; small lenses of gravel; forset beds contained mollusc shells, artifacts and bone refuse; scattered charcoal fragments; numerous artifacts; 5-15 cms.
5C	pale to dark gray charcoal stained medium-fine sand; no bedding structures; few mollusc shells; distinct concentrations of ash and charcoal; artifacts very numerous especially minute obsidian debitage; 15-30 cms.

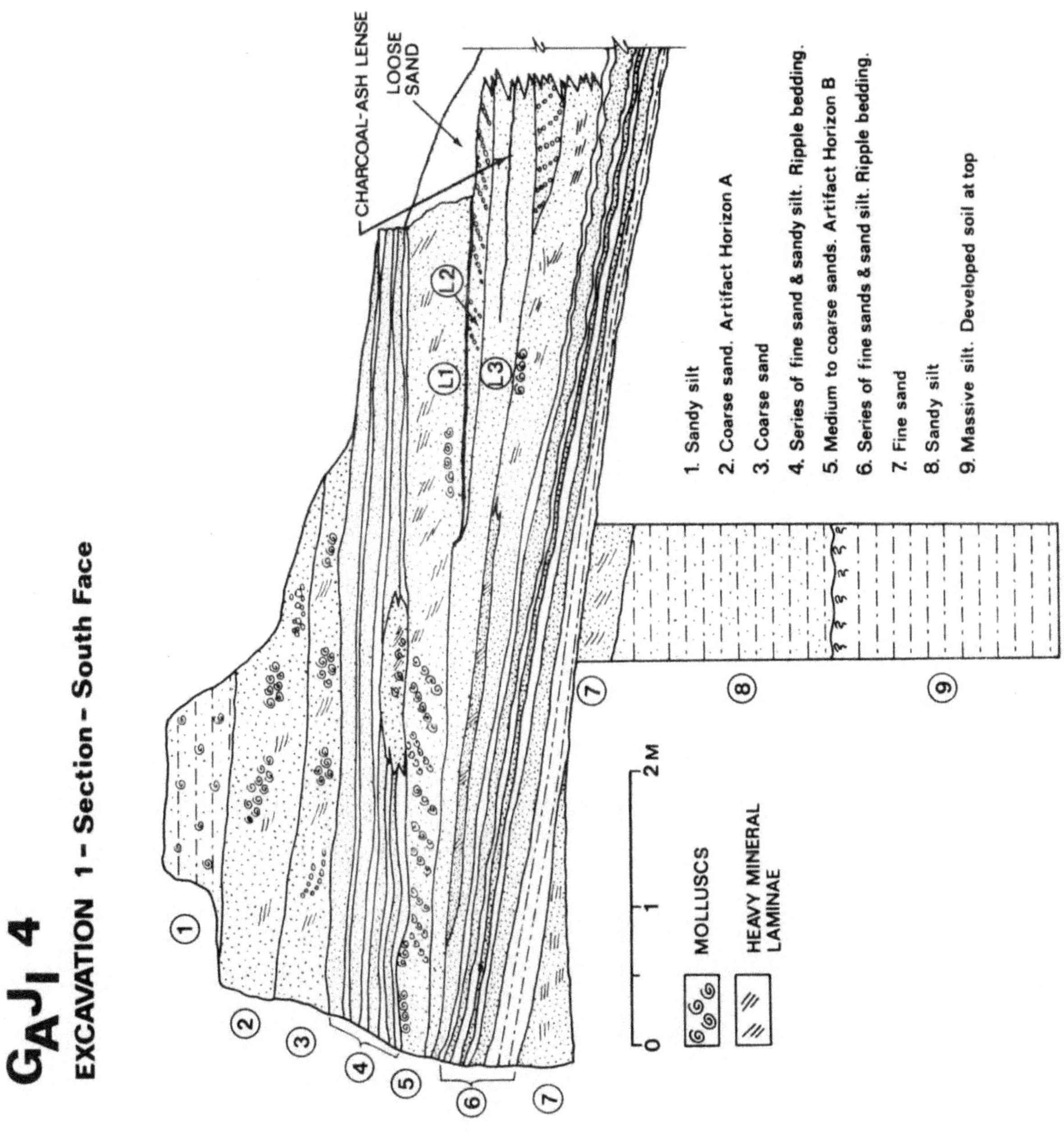

Figure 57. Dongodien. Profile of main excavation section - south face.

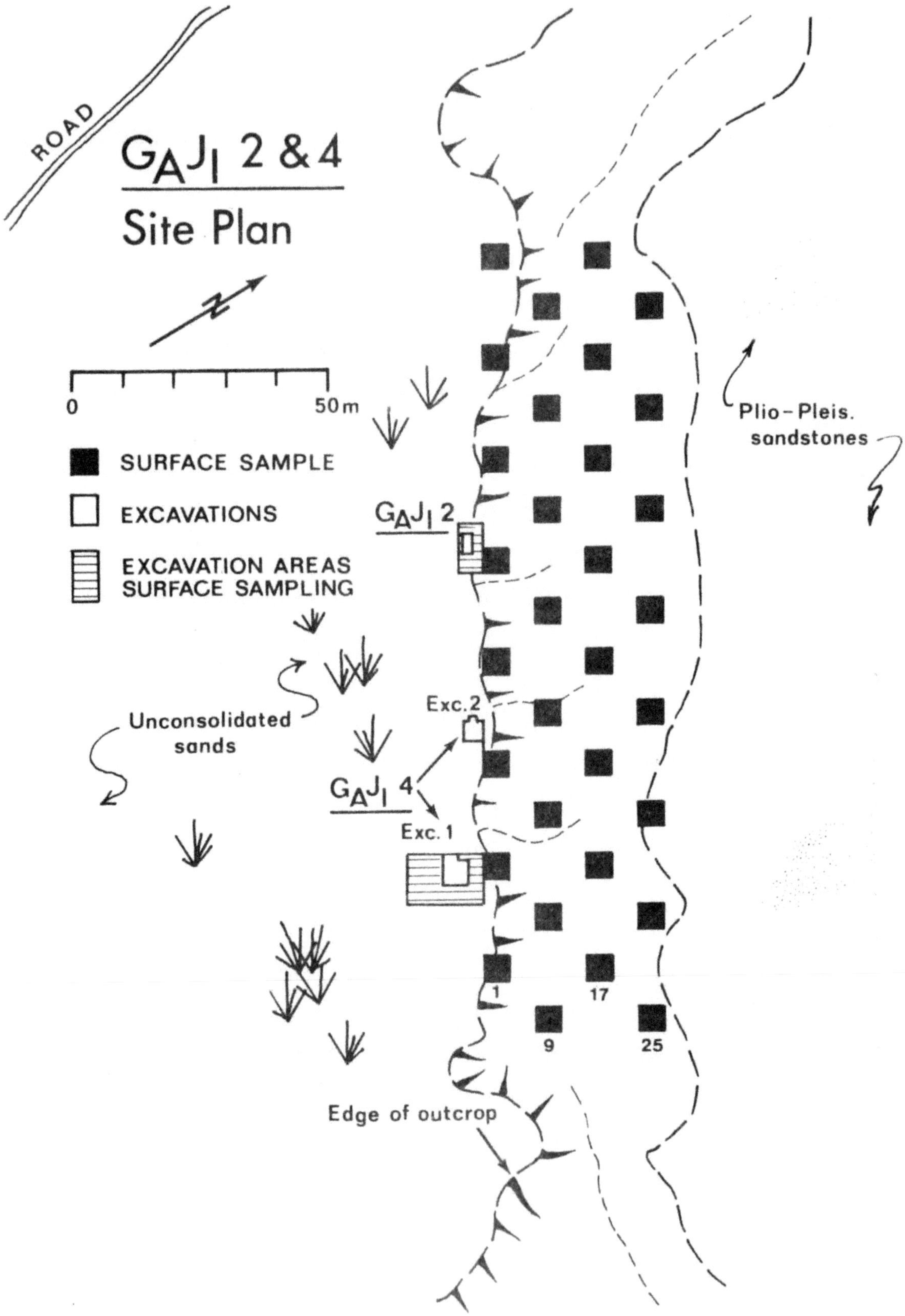

Figure 56. GaJi 2 and Dongodien (GaJi 4). Sketch plan showing location of sampling units and excavations.

The sedimentary succession is interpreted as follows: the basal exposures consisted of silty clays and fine grained (diatomaceous?) silts (Unit 9), which are indicative of offshore, sub-littoral environments. The Holocene lake had evidently begun to shallow and a series of oscillating lenses of sandy silts and silty sands (Units 6-7) were laid down. The lake continued to recede and a layer of fine to medium grained beach sand (Unit 5C) was deposited. This land surface, probably close to the paleo-lake shoreline, was occupied by human groups with a microlithic tool kit, Nderit Ware pottery and domestic animals. The lake subsequently underwent a small scale readvance and the site was submerged. Wave action now thoroughly reworked the artifactual debris (Units 5A and B). Larger fragments of stone, bone, pottery and charcoal were sorted into a series of wave generated forset beds. Mollusc shells were also incorporated into the forsets. The lake underwent a further small scale advance and a second series of oscillating shallow water fine sand and silt horizons (Unit 4) were deposited. Artifacts were not present within these layers. The lake now began a major recession and a suite of coarse gravelly beach sands (Units 2-3) were laid down. Scattered water worn artifacts were present within each level, some of which may have been derived from Unit 5. A final small scale temporary readvance occurred and a fine sandy silt (Unit 1) was deposited. Lastly, an erosional unit, widespread in Areas 102 and 103, of poorly sorted, dirty, coarse sand and gravels with dispersed broken artifacts, mantled the entire sequence and indicated the Holocene lake had further retreated and fluvial/colluvial sedimentation had begun. This unit was present at the outcrop but is not illustrated in Figure 57. The stratigraphic succession at Dongodien is thought to represent sections of the second and third major transgressive-regressive cycles of the Holocene lake.

SAMPLING

An area of discontinuous artifact density along the outcrop's eroded flanks and adjacent flat lying surface to the north was selected for sampling. Because I wanted to intensively sample the site's surrounding area, and time was at a premium, a systematic checker-board strategy was chosen (Redman 1974). The dimensions of the artifact scatter were first determined by a transit and a grid system was formulated. The area to be sampled was divided into seven rows of thirty-two 5 m square units (Fig. 56). Beginning with row one, every other row was sampled. Within each row, 8 squares, or every fourth unit, were collected. Hence:

potential sampling units:	224
units collected:	32
14.3% of the population was sampled	

In addition, six sampling units were collected from the immediate area surrounding Excavation 1. Two further squares from the surface of Excavation 2, as well as the entire surface assemblage of pottery, were also collected.

EXCAVATION

Two areas of considerable artifact density were selected for excavation (Fig. 56). Excavation 1 was the larger excavation and yielded over 99% of the recovered artifacts. A five meter by three meter square was gridded into fifteen 1 m square excavation units. Excavation 1 was excavated to a depth of nearly three meters and all sediment was sieved through a 1:16 inch screen mesh. Individual artifacts and features were plotted within Unit 5C.

Excavation 2 was located 55 m west of the first excavation. Large numbers (N=75) of apparently fire hardened gray to reddish brown, angular sandy blocks were exposed along the margins of the outcrop. With prospects of identifying an archaeological feature, a small excavation measuring three meters by two meters was gridded and excavated to a depth of 75 cms. Ten of the angular blocks, 6-10 cms in size, were recovered from the excavation and the occupation horizon was subsequently correlated with Unit 5 from Excavation 1. Unfortunately, recent erosion had removed most of the blocks and it proved impossible to reconstruct the suspected feature. The original accumulation of the fire hardened blocks may well have represented a cooking area or locality for firing pottery vessels.

CULTURAL REMAINS: SURFACE

An inventory of the sampling units and individually collected pottery sherds is presented in Table 48. The surface assemblage of artifacts was collected before it was established that the Holocene exposures included two separate primary occupations (Dongodien and GaJi 2) as well as artifacts from at least three reworked horizons. The surface collection should be viewed, therefore, as a heterogeneous assemblage that represents a series of occupation episodes, not all of which may have shared similar technological and economic traditions.

Stone Artifacts (Table 49-51)

Microliths were the most common shaped tool followed by *outils écaillés* and various scraper forms.Crescents were well represented among the microliths (Fig. 58). Obsidian accounted for over 70% of the raw material.

A stone bowl fragment manufactured from green metamorphic schist was identified. The artifact appears to have been a shallow bowl with rounded sides. Unfortunately, the provenience of the bowl remains uncertain and it cannot be assigned to the occupations at either Dongodien or GaJi 2. No stone bowls were excavated from either site. Two other stone bowls however, including a complete specimen (GaJj 0/5), were recovered nearby in Area 102.

One dihedral burin was also collected from the surface. The implement represents the only unequivocably identified burin from the numerous late Pleistocene-Holocene sites in the research area. The burin was shaped from a broken obsidian flake fragment. Several

Figure 58. Dongodien: Stone Artifacts. Surface.
(a-i) crescents; (j) double backed percoir; Excavation. (k-n) curved backed flakes; (o-t) double backed. Natural size.

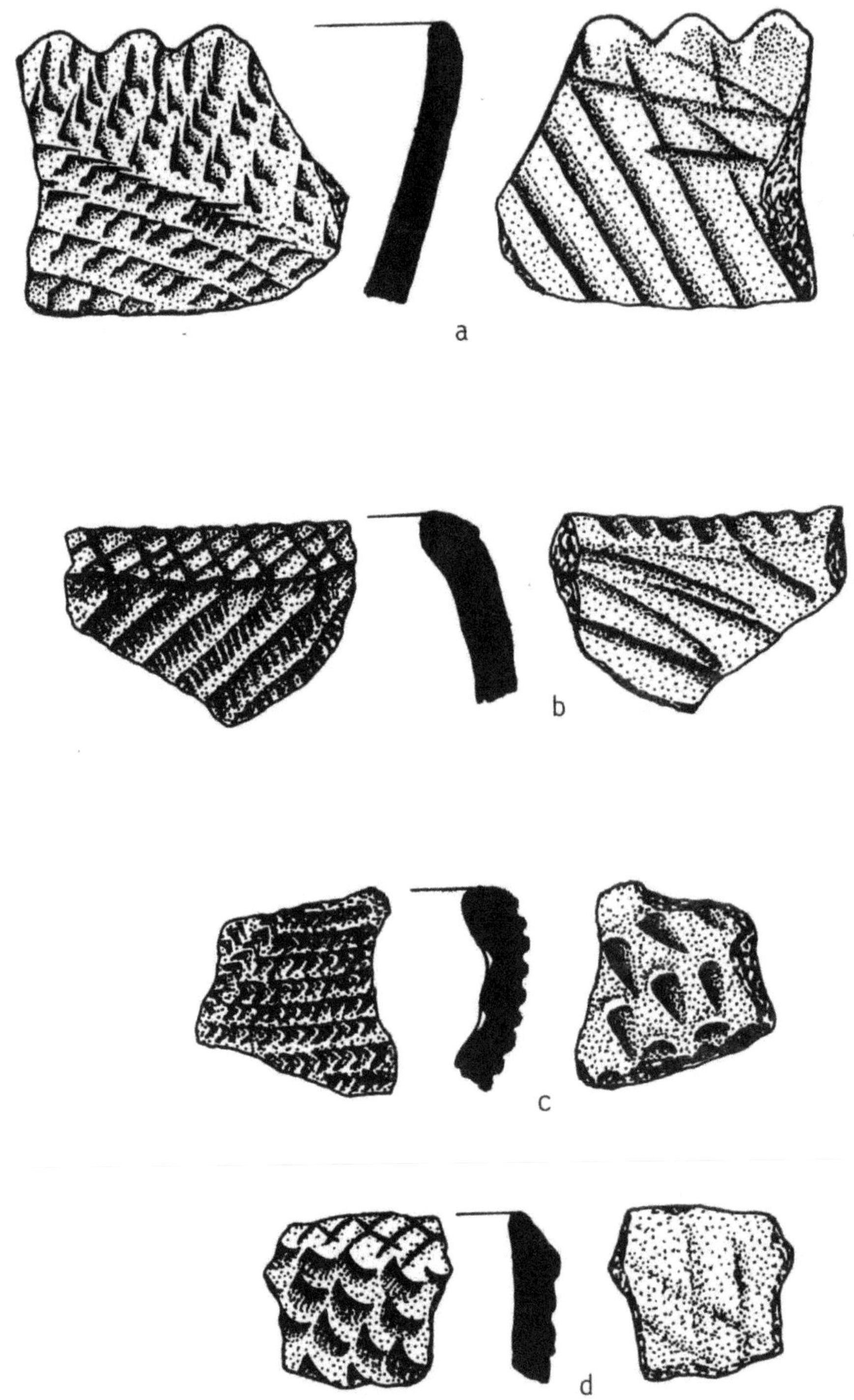

Figure 59. Dongodien. Nderit Ware pottery. Surface. (a-d) decorated rim sherds. Natural size.

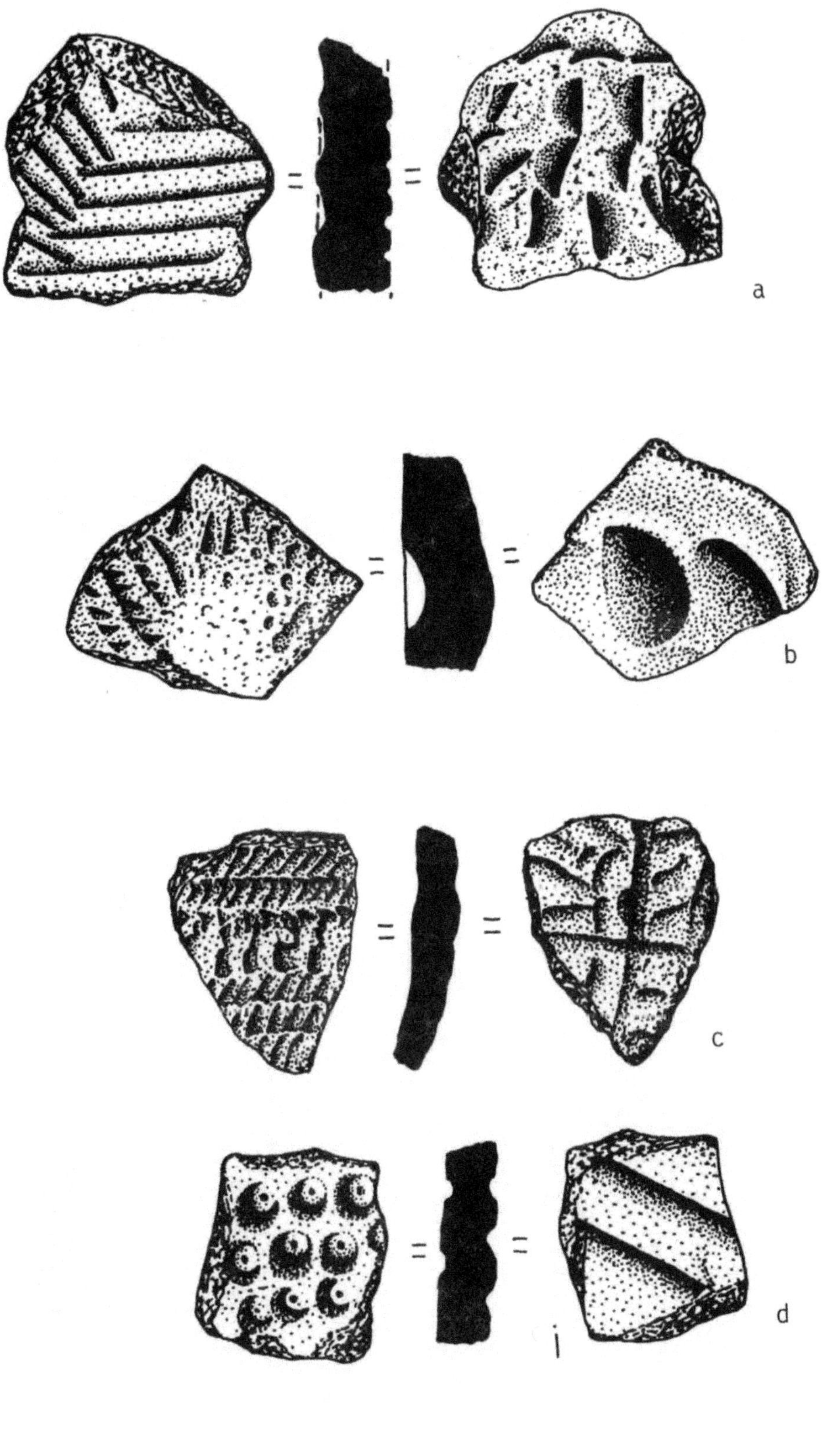

Figure 60. Dongodien. Nderit Ware pottery: Surface. (a-d) decorated body sherds. Natural size.

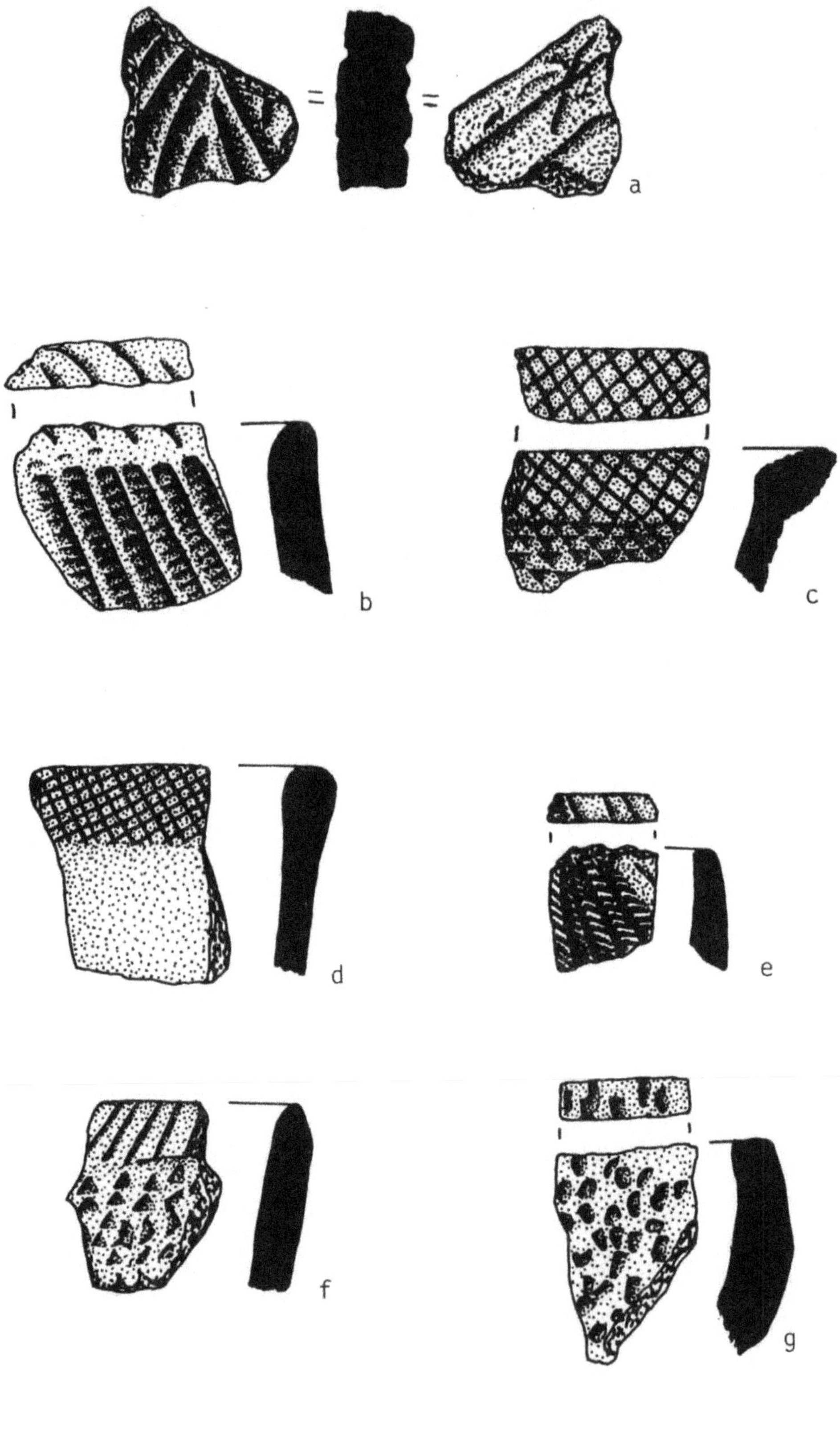

Figure 61. Dongodien. Pottery: Surface.
(a) Nderit Ware decorated body sherd; (b-g) decorated rim sherds. Natural size.

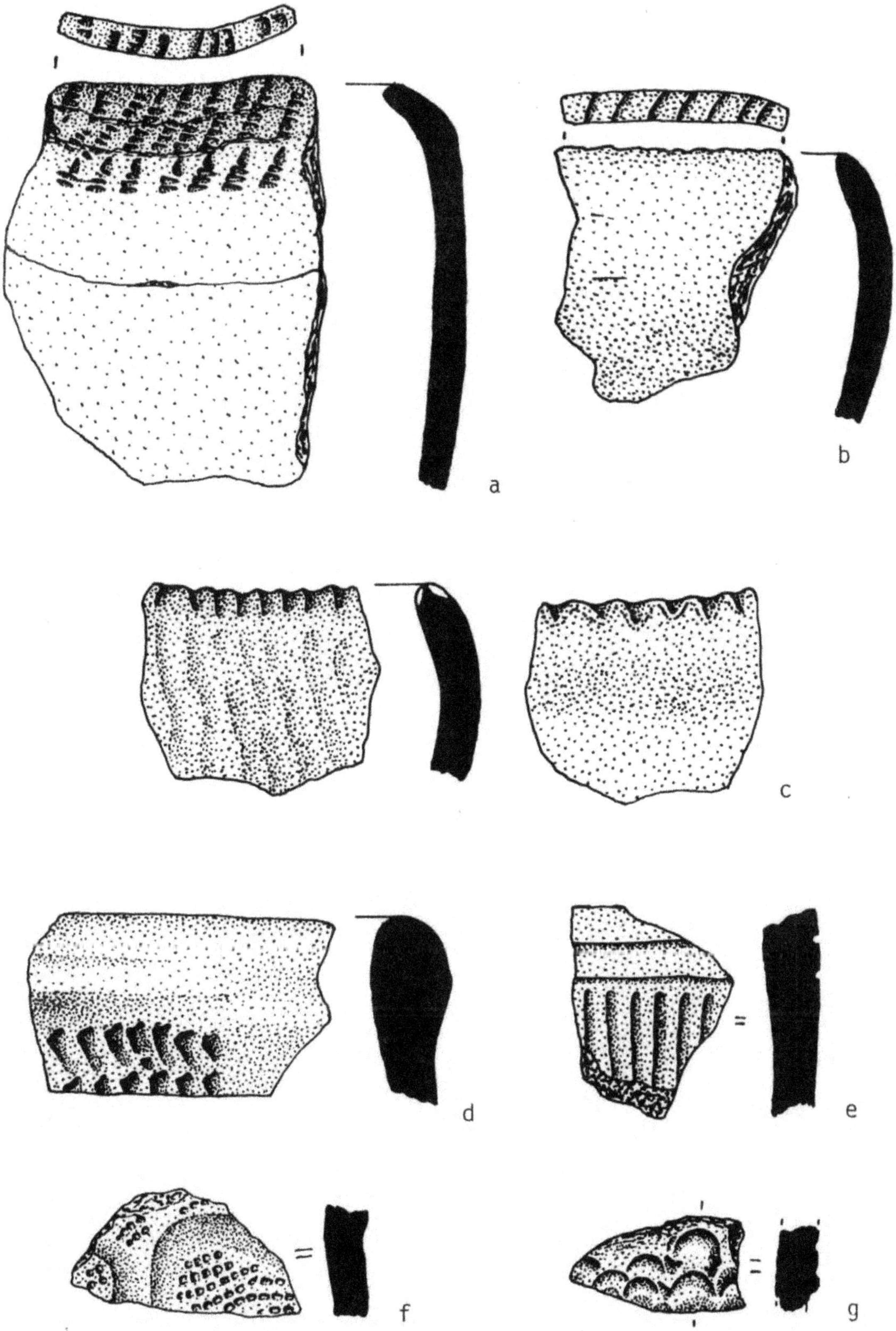

Figure 62. Dongodien. Pottery: Surface.
(a,d) decorated rim sherd; (b-c) rims with decorated lip; (e-g) decorated body. Natural size.

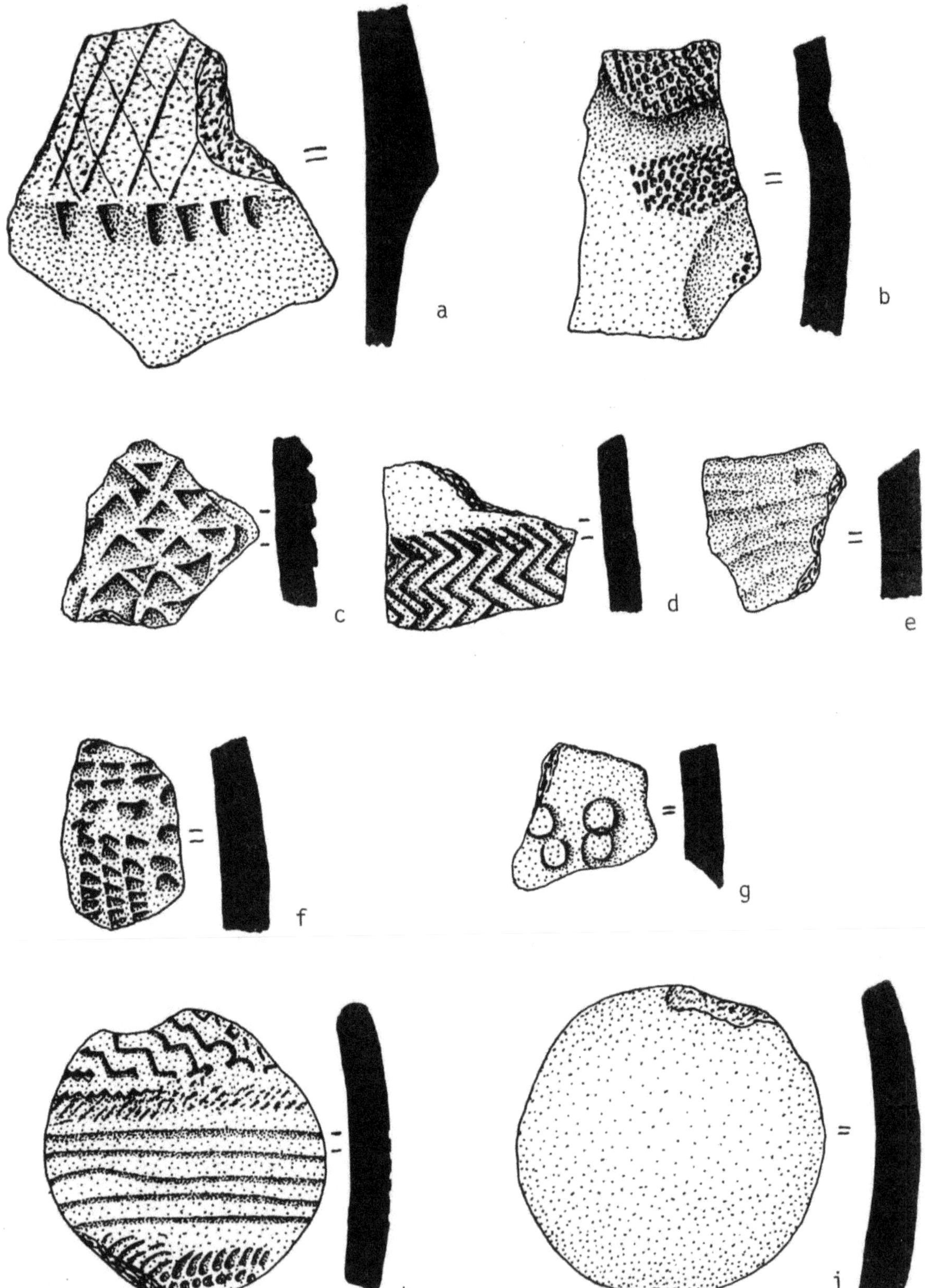

Figure 63. Dongodien. Pottery: Surface.
(a-g) decorated body sherds; (h-i) circular discs. Natural size.

percussion blows at oblique angles produced a dihedral apex. Crushing and small faceting scars along the apex suggest the burin had been utilized. As with the stone bowl fragment, it was not possible to associate the specimen with either of the occupation horizons. Moreover, burins were not excavated from either site.

Pottery (Fig. 59-63)

Over 600 sherds were recovered from the surface (Table 48). The collection of decorated sherds exhibited a considerable diversity of execution techniques and design styles. Both Nderit Ware and sherds with a red slip were identified. The Nderit Ware speçimens displayed a broad variety of external and internal decoration. Several kinds of concentric impressions occurred on the external surface, two of which may have been produced with a Melanoides tuberculata shell (Figs. 59, d; 60, d). The internal surfaces were usually decorated with incised lines but sherds with impressions also occurred. Other examples with external rows of cuniform impressions but without internal decoration were also recovered (Fig. 61, b, e). The remaining surface collection included several discs (Fig. 63, h-i) and rim sherds with cross hatching that resembled pieces from GaJi 2 and the Ileret Stone Bowl Site (Fig. 61, c-d). A single sherd of Ileret pottery was also identified (Fig. 63, d).

Faunal Remains

A small collection of fish and mammal remains were recovered including a fragmentary bovid horn core and several complete bovid teeth. The surface faunal material has not been analyzed.

EXCAVATION

Table 52 presents an inventory of the cultural remains from Excavations 1 and 2.

Stone (Tables 53-55)

Obsidian accounted for over 95% of the raw material at Dongodien. This figure represented the highest raw material percentage for obsidian among the 16 late Pleistocene-Holocene sites analyzed in the study area. The obsidian was usually black, opaque or semi-translucent, with a greenish to brown outer margin. A few specimens were colored greenish gray to dark gray and were totally opaque.

Obsidian debitage consisted of small (>30 mm) to very small (>2 mm) flakes, blades and angular fragments. Cores also tended to be small (15-40 mm). Several casually flaked pieces with remaining cortex suggest the obsidian was collected in the form of small subrounded nodules. A considerable number of angular fragments, small enough to pass through the screen mesh, were retrieved from the back dirt mound. Flakes and blades invariably had crushed platforms and often showed parallel sided dorsal scar ridges (Fig. 64, h, i). A number of alternate ended double and single platform cores also exhibited parallel sided flake scar ridges (Fig. 64, e-g). These observations suggest a flaking pattern in which diminuative flakes and blades were being produced from small obsidian nodules by percussion and perhaps punch techniques. Obsidian may be collected today from the Miocene Suregei-Asilla volcanics (Watkins 1981) and at Shin, a complex of acidic Lower Pliocene lava flows and obsidian intrusions (Vondra and Bowen 1973, 1978). Both sources occur along the northeastern margin of the basin. Recent investigations of obsidian source areas in the Eastern Rift have been initiated by Merrick and Brown. Their preliminary findings, based on the analysis of Ca, Fe and ten other elements by x-ray fluorescence, indicates most, if not all, of the Dongodien obsidian was collected from local sources within the Suregei-Asilla volcanics (Merrick and Brown 1984: 129-152).

Microliths accounted for over 50% of the shaped tools followed by outils écaillés and scrapers. Significantly, neither geometric microliths not crescents were represented. Instead, the majority of microliths consisted of delicately truncated flakes and blades, double backed pieces and broken fragments, some of which may represent waste products of the microburin technique (Fig. 58). Percoirs, including minute double backed examples (Fig. 65, c-e), were also discovered.

A large sample of outils écaillés were recovered during excavation. The obsidian specimens (N=68) ranged in length from 19 mm to 7 mm and had a mean length of 12 mm. As the above figures indicate, the sample consisted predominately of very small specimens. The tools exhibited characteristic crushing on opposite ends and often possessed parallel flake scar ridges. Pieces with crushing on three and four sides also occurred (Fig. 65, s-t).

For several reasons, the majority of the outils écaillés from Dongodien are believed to represent worn out alternate ended or bipolar cores instead of purposefully shaped tools. First, a gradation, based on maximum dimension and thickness, was observed between the outils écaillés and bipolar cores. Second, the presence of parallel scar ridges on many pieces suggests the repeated deliberate removal of blades and flake blades. Third, the small size and relatively distant source area of the raw material may have encouraged intensive flaking of the small nodules by the bipolar core technique. A similar pattern of stone utilization is under investigation from Mumbwa Caves in Zambia (Karla Savage per. comm.). Some of the specimens were extremely small however (Fig. 65, u-v), and their function as bipolar cores may be questioned. Nelson (1973) has discussed a range of possible functions for this tool type. For purposes of this monograph and for comparative analysis with other LSA sites in Eastern Africa, the entire sample has been retained in the shaped tool category.

Scrapers, while not numerous, were represented by a diversity of types (Fig. 64, a-b). Chert was selected over obsidian in the manufacture of scrapers.

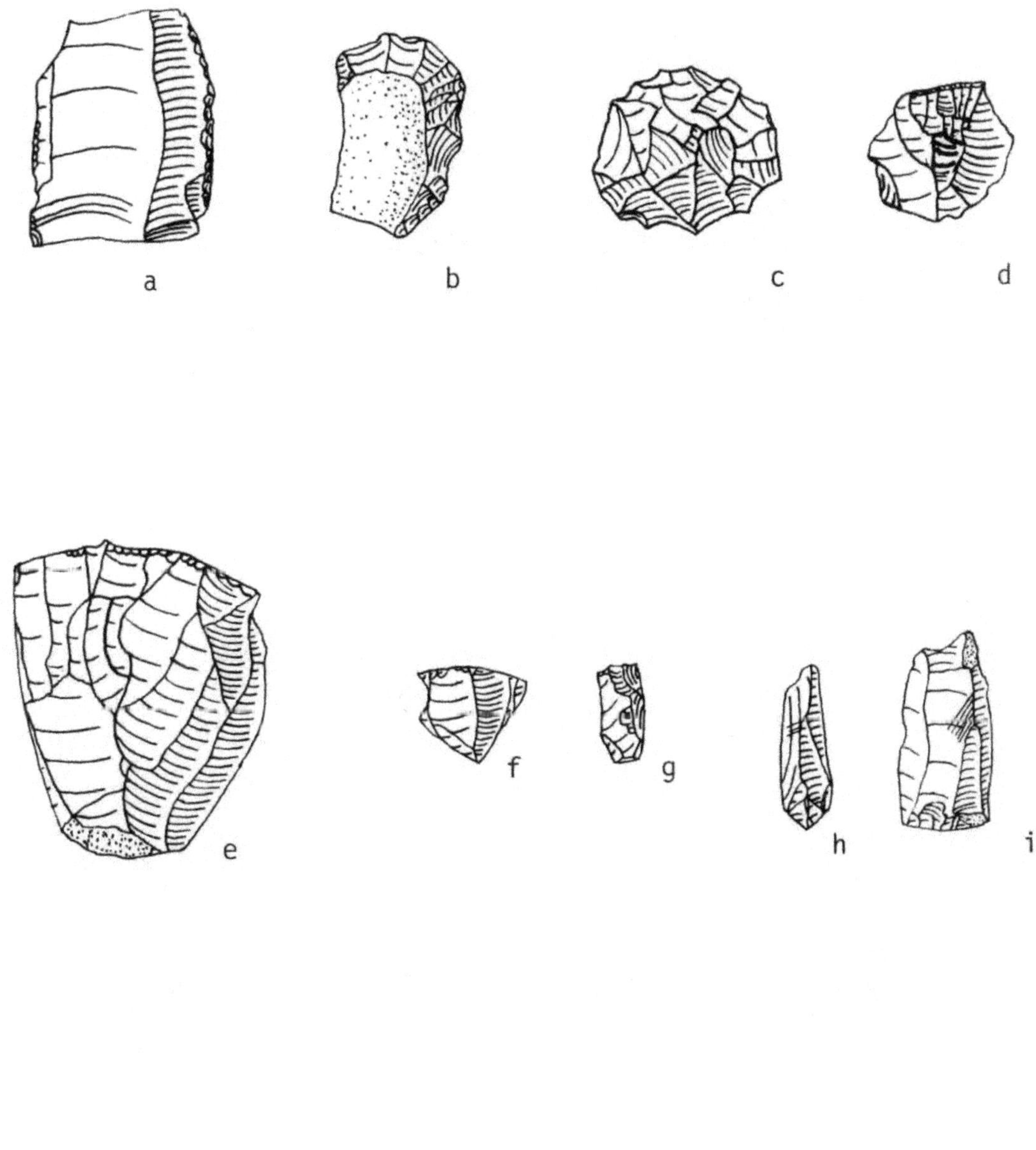

Figure 64. Dongodien. Stone Artifacts: Excavation.
(a-b) scrapers; (c-e) single platform cores; (f-g) double platform cores; (h-i) flake blades. Natural size.

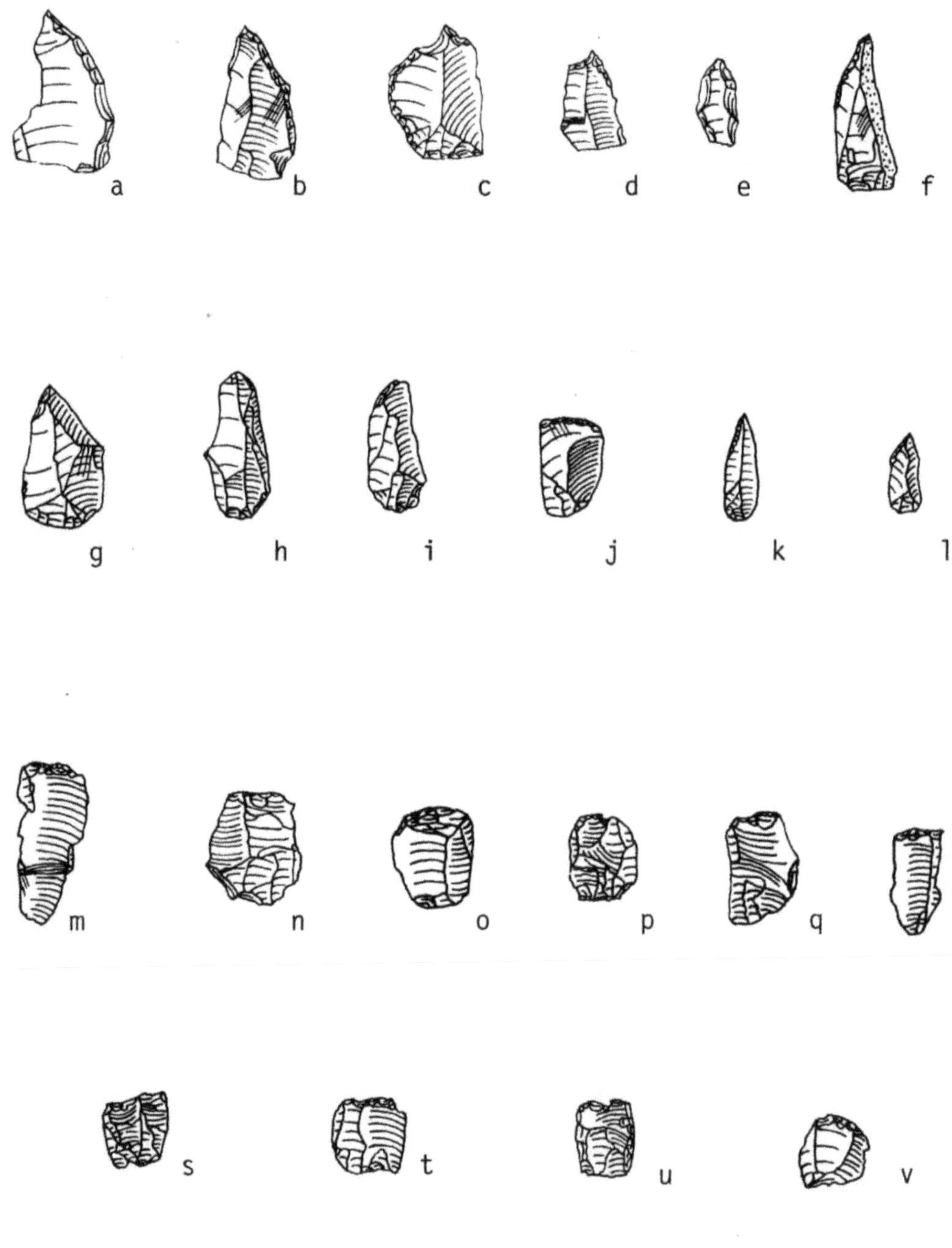

Figure 65. Dongodien. Stone artifacts: Excavation.
(a-b) broken microliths; (c-d) micro-percoir; (e) double backed micro-percoir; (f-l) truncations; (m-z) outils écaillés. Natural size.

POTTERY (Tables 56-57)

Over 350 sherds were recovered from the two excavations including examples of Nderit Ware and a large series of burnished and slipped pieces. An analysis of each pottery group is reported below.

Nderit Ware (N=35)

Controversy has surrounded the stratigraphic association and chronological position of Nderit Ware since the early 1930's when L.S.B. Leakey collected examples of a hitherto unknown type of decorated pottery from Nderit Drift, Stable's Drift and the Makalia Burial Site (L.S.B. Leakey 1931:198-199). Leakey named the pottery 'Gumban A' after a mythical people known as the Gumba who were thought to have inhibited the Nakuru-Elmenteita Basin (Siiriaïnen 1971; Muriuki 1974: 37-42). Wandibba re-examined Leakey's original collections and has proposed the name 'Gumban A' be changed to Nderit Ware. According to Wandibba,

> Nderit Ware... is characterized by vessels that are generally decorated on both the external and internal surfaces. Decoration on the external surface consists of closely set cuneiform impression that, in most cases, cover the whole surface and are arranged in panels. The internal surface is frequently decorated with deep, haphazardly executed grooves or incised lines. In some cases, the interior surface is jabbed or left without decoration. Carinated bowls are the most typical vessel form (in Bower et al. 1977:135).

Leakey believed Nderit Ware sherds were associated with stone bowls and, by inference, with domestic stock. The strati- graphic provenience of the material was unclear however, and the pottery's association with domestic stock and stone vessels has remained uncertain. Since the 1930's surface finds have been reported from southwestern Lake Turkana especially the site of Aipau (Robbins 1972, 1980), Nderit Drift (Bower 1973; Bower et al. 1977), Lukenya Hill (Gramly 1976) and near Rumuruti (Siriianïan 1977). A small set of excavated sherds has also been reported (Robbins 1972; Bower 1973; Bower et al. 1977). Recently, a claim for possible Nderit Ware pottery at Salasun Rockshelter in Central Kenya with a startingly early radiocarbon date on bone apatite of 7255 ± 225 BP has been made (Bower et al. 1977; Bower and Nelson 1978). The excavated sample consisted of only two weathered sherds, neither of which exhibited internal decoration. Although the apparent association of decorated pottery with a radiocarbon age that dates to the end of the eighth millennium BP is most provocative, given the very small sample, poor state of preservation and absence of diagnostic design features, it seems premature to classify the few sherds as Nderit Ware (see also Collett and Robertshaw 1983). Future excavations are planned and a larger pottery inventory should be forthcoming.

At Dongodien, Nderit Ware pottery was directly associated with domestic animals but not with stone bowls. The majority of the excavated sherds closely resemble the decorative features of Leakey's original collection (L.S.B. Leakey 1931:Plates XXII and XXIII). External surfaces were completely covered with rows of closely spaced cuneiform impressions (Fig. 66, e-f). The internal surfaces were decorated or scored with incised or deeply grooved lines. Both external and internal surfaces exhibited other design features including circular and wavy impressions (Fig. 66, c-d). Rim sherds had tapered or flattened lips usually with milling and cross hatching (Fig. 66, a-b). A series (N=7) of externally decorated sherds with cuneiform impressions that lacked internal decoration were also recovered (Fig. 68,b). These pieces are similar to the cuneiform impressed material from Hyrax Hill (M.D. Leakey 1945). Comparable sherds were also collected from the surface at Dongodien.

The assemblage of Nderit Ware sherds was usually dark brown to dark gray in color and was tempered with small angular fragments of quartz. Unfortunately, the fragmentary nature of the collection made it very difficult to reconstruct vessel shape or size. Several rim profiles did suggest however, some of the vessels had open mouths, globular in shape, probably not unlike the reconstructed vessels from Stable's Drift and the Makalia Burial Site. The location of additional sites and locations with Nderit Ware pottery is given in Figure 69.

Burnished and Slipped (N=215)

This decorated pottery group accounted for over 60% of the recovered sherds. External surfaces were burnished and a bright orangish red slip had been applied. A number of internal surfaces also showed traces of burnishing and slipping. Small fragments of haematite were recovered from the excavation and probably were used in the slipping process. Rim sherds were usually tapered and had milled lips (Fig. 67, a-e). Occasionally, both the external and internal surfaces displayed milling (Fig. 67, a). Several rim examples showed cross hatched incised lines (Fig. 67, f). Body sherds exhibited a series of low relief, widely spaced, parallel grooves (Fig. 67, j-k). Additional external or internal decoration was absent. The sample was light brown to bright red in color and had a very sandy paste. Large quartz grains and crushed pottery fragments were used as tempering material. Rim profiles suggest many of the vessels were shaped like small open-mouthed bowls or cups. Many of the sherds classified as undecorated in the total pottery assemblage showed minute traces of red slipping. It is likely that these pieces were formerly completely slipped. Similar pottery has been reported from Shaheniab (Arkel 1953:74-75) in the Sudan. Other comparable pottery to my knowledge has not been reported from Eastern Africa.

Other Decorated Sherds (N=24)

A small collection of decorated sherds other than Nderit Ware and slipped specimens was discovered. Examples of two thin walled rim sherds with impressed designs are illustrated in Figure 67. Figure 68 illustrates two relatively thick body sherds with rows of crescentic impressions and broad grooved lines.

Discs (N=16)

An interesting collection of whole and fragmentary subcircular discs was recovered (Fig. 68, f-i). The sherds had been purposefully shaped by rubbing their edges on an abrasive surface. Most of the pieces had been burnished and slipped. Their function is not known but they may have been used as pottery burnishers.

Undecorated (N=68)

Undecorated sherds accounted for 19% of the excavated collection. Only two rim sherds were identified including an example with a slightly thickened and flattened lip (Fig. 67, i). Body sherds ranged from 5-14 mm in thickness.

FAUNAL REMAINS (Tables 55-56)

Domestic caprines and domestic cattle were identified from the excavated faunal assemblage. Fiona Marshall (1977; 1985 et al.) and Diane Gifford undertook the identification of the mammal, bird and reptile remains while Kathy Stewart (in Marshall et al. 1985) identified the large collection of fish bones.

Nearly 7000 bone fragments were recovered from Excavation 1. Small pieces were numerous as over 2000 elements had a maximum dimension of between 2.5 and 0.5 cm. The collection consisted of 64.7% mammal, 35.3% fish and 0.2% bird and reptile bones. Within the family Bovidae, 3.5% of the faunal remains were attributed to very small birds, 89.8% to small bovids, and 6.7% to medium sized bovids. Only one taxon within each size category was positively identified, *Madoqua* sp., caprini and *Bos* sp. respectively.

The fish bone remains consisted primarily of Nile Perch (*Lates niloticus*), tilapia (*Tilapia nilotica*) and several species of catfish including wahrindi (*Synodontis schall*), *Clarius* sp. and *Bagrus bayad*. Tigerfish (*Hydrocynus* sp.) was also identified. Close to 22% of the sample was identified to element; cranial remains proved the most useful for identification.

For purposes of analysis the family Bovidae was divided into 5 body size groups (Gifford 1980:67). Identifiable bone fragments that could not be assigned to either genus or species were placed into one of the five size groups. This decision resulted in two major methodological problems. First, the sample was extremely fragmentary. Only 59 complete elements were recovered, the majority of which were foot bones. Second, among the group of small size bovids, body parts showed a high degree of similarity. Horn cores were regrettably absent. For the fish remains a scale of 1-5, based on body weights, was used. Catagory 1 included fish of up to one kg while weight catagory 5 held fish up to 40 kg or more. The following section discusses the identified faunal remains by taxa or size group (Marshall et al. 1985).

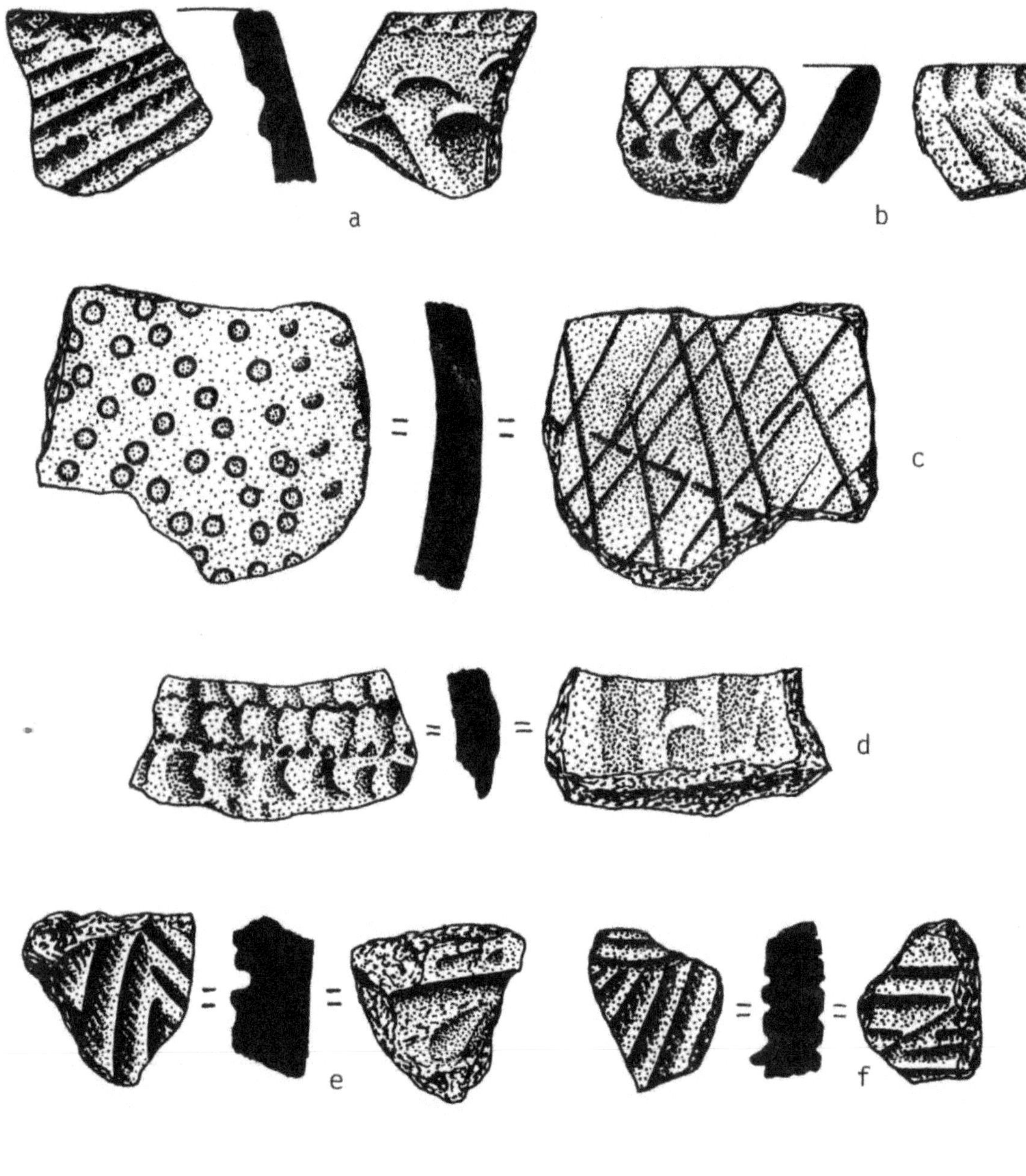

Figure 66. Dongodien. Nderit Ware pottery: Excavation. (a-b) rim sherds; (c-h) body sherds. Natural size.

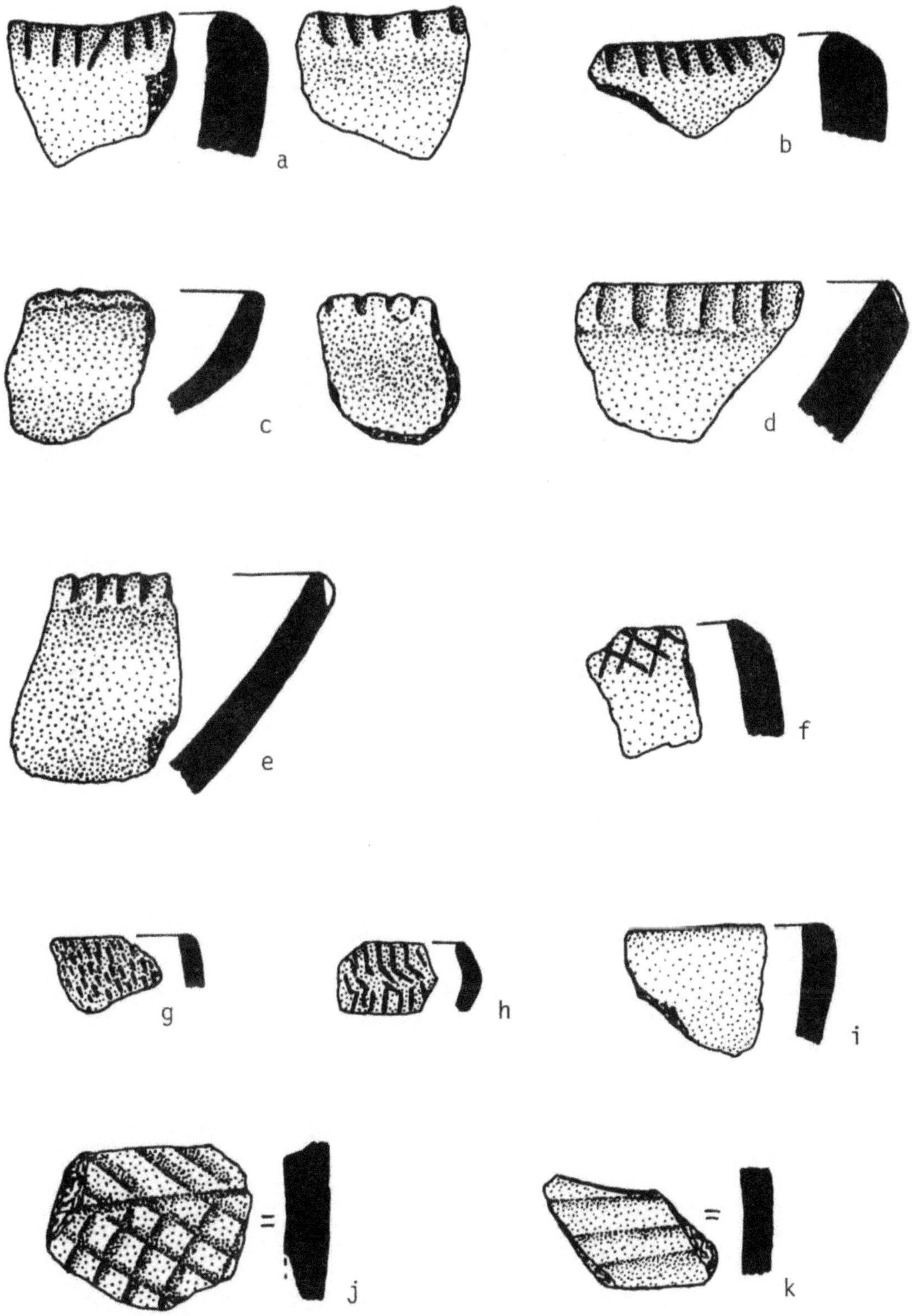

Figure 67. Dongodien. Pottery: Excavation.
(a-e) burnished and slipped rim shreds; (f-h) decorated rims; (i) undecorated rim; (j-k) decorated body. Natural size.

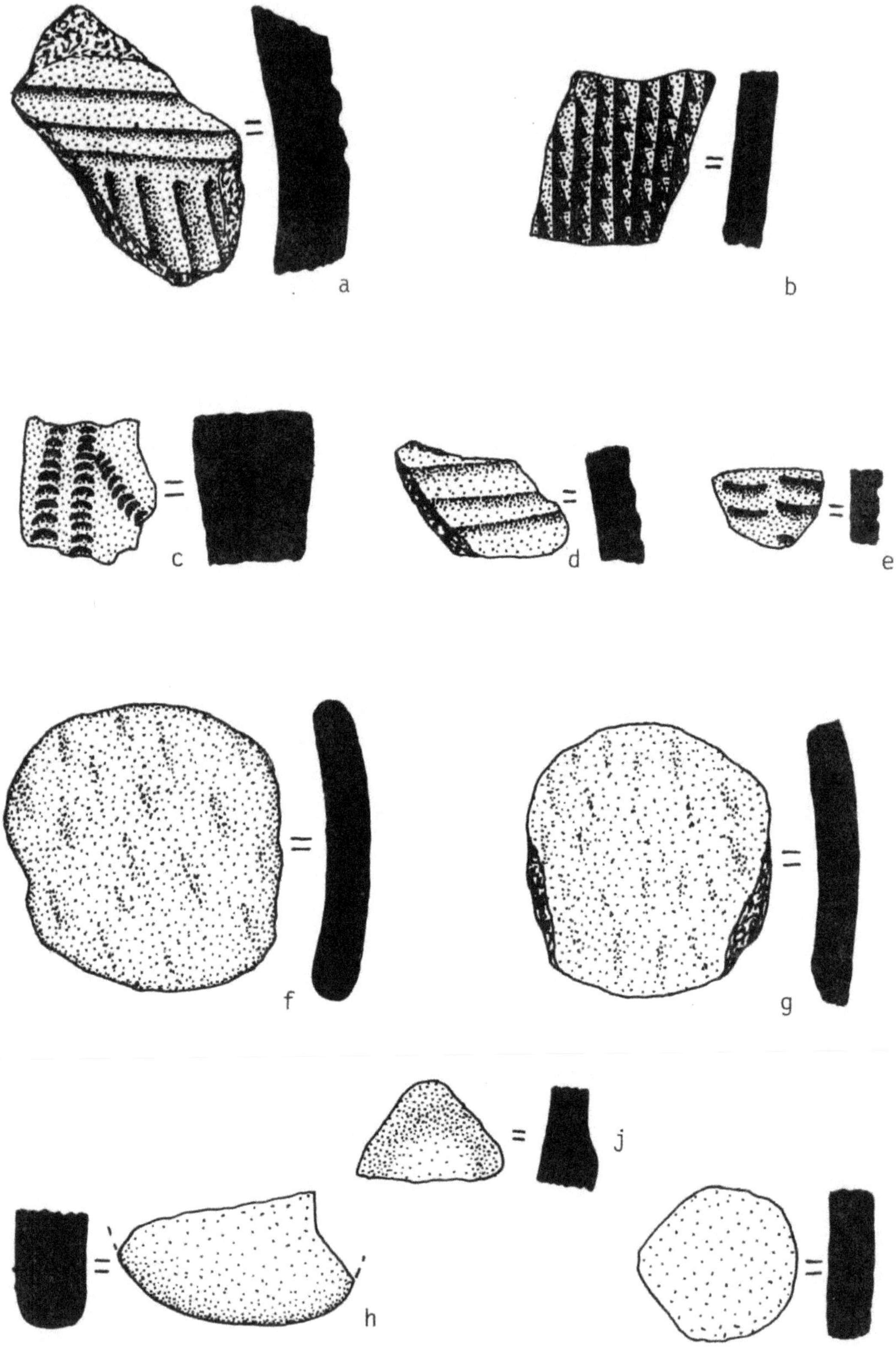

Figure 68. Dongodien. Pottery: Excavation.
(a-e) decorated body sherds; (f-i) discs; (j) possible base. Natural size.

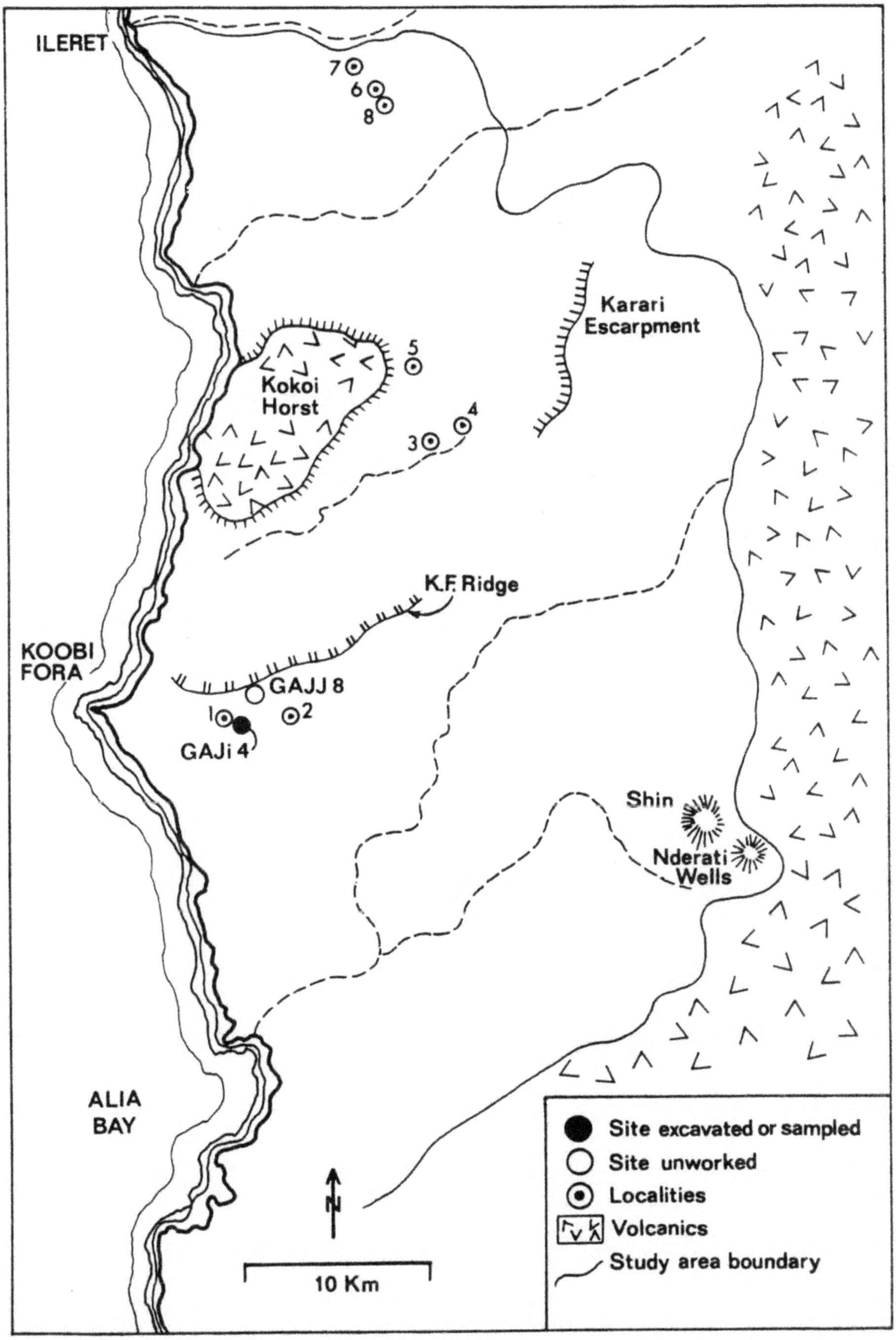

Figure 69. Location of sites and localities with Nderit Ware pottery.

MAMMAL

Bovidae

For the mammal remains, 99.6% of the identifiable bone was attributed to the Bovidae.

Very Small Bovids (N=19)

Dik-Dik (*Madoqua* sp.)

Distinguishing between the sample of fragmentary small bovid elements proved difficult. Nevertheless, a number of whole body parts were assigned to this bovid size group. Several long bones were attributed to *Madoqua* sp.

Small Bovids (N=596)

Nearly 99% of the identifiable bovid remains were assigned to this body size group. Domestic caprini were identified from a combination of dental and post cranial remains. As only caprines were positively identified from size catagory 2, it appears reasonable, though not certain, that many of the bone fragments attributed to this size group can be placed within the caprini. In a preliminary article (Barthelme 1977), the presence of Impala (*Aepyceros melampus*) and Grant's Gazelle (*Gazella granti*) was reported. Further analysis of the faunal collection indicated that neither of the species could be positively identified. The report of wild bovids from this size group should therefore be withdrawn.

Medium Bovids (N=37)

Domestic cattle (*Bos* sp.), were identified on the basis of complete teeth, phalanges and two astragali. Full tooth rows and adult horn cores were not recovered.

Leporidae

Cape Hare (*Lepus capensis*)

A single phalanx was identified to this species.

Hyracoidae

Rock Hyrax (*Procavia* sp.)

The polished and perforated incisor of a male hyrax was recovered. Rock Hyraxes currently can be found at Nderati Wells, along the basin's volcanic perimeter to the east of Koobi Fora.

REPTILE

A single element from a Monitor Lizard (Varanus sp.) and a turtle (Cheloniidae) were identified.

BIRD

Several fragmentary diaphyses and a complete talon of a large raptor, possibly the African Figh Eagle (Haliaetus vocifer), were collected. A small wading bird was also identified from a broken phalanx.

Minimum Numbers

An analysis of the minimum number of individuals was undertaken. Such estimates can be misleading especially if the unidentified portion of the faunal sample is large. In such cases, minimum numbers often reflect an underrepresentation of the total number of individuals originally present (Chaplin 1971; Hill 1980; Klein 1980). In the current study caprines were represented, based on the identification of right upper third molars, by nine individuals. This figure was almost certainly an underestimate. The small size body group contained over 2400 bone fragments, a considerable quantity attributable to only nine animals. Analysis of footbones indicated the presence of both sheep and goat although the relative proportions of each animal were not determined. For cattle, minimum number estimates were based on both teeth and footbones. Two individuals were identified. A single dik-dik, based on longbones, was also identified.

For fish, because only a small proportion of identifiable body parts were useful for taxonomic assignment, minimum number calculations are likely to represent underestimates. Nile Perch were represented by four individuals, while for tilapia and wahirindi two individuals were identified. Other species were represented by one individual.

AGE

A wide range of age catagories was present among the bovids. For the caprini, study of the upper right dentitions indicated that two juveniles, one young adult, three prime adults, one aged and one very old animal were present. For cattle, one adult and one young juvenile were identified. Very small bovids were represented by one young adult.

Utilized Bone and Teeth

A small size bovid femur diaphysis with deeply incised circular grooves was identified (Plate 3). The interior bony matrix had been removed and the remaining shaft was only several millimeters thick. The artifact may have functioned as an instrument for winding twine or perhaps in the manufacture of circular bone objects. A similar broken bone object or 'ring' was collected from the surface. Comparative incised bone artifacts have not been identified from the Turkana

Basin but are known in the Sudan from the site of Shaheinab (Arkell 1951: Plate 27, Fig. 7). The hyrax incisor had a carefully drilled hole at the proximal end and had been polished on each side.

Butchery Patterns

Because of the poor condition of bone preservation, it was difficult to identify cutmarks that may have been made during the processing of animal carcasses. Cutmarks (N=12) were recognized however, on several joint surfaces including the distal humerus-proximal radius and the distal tibia-astragalus of small bovids.

A small quantity (N=173) of mammal remains showed evidence of burning. Burning occurred primarily among fragmentary longbones and rib elements. Several proximal radii as well as two distal humeri of the small sized bovid group were also burned. Such evidence suggests these joints were exposed to fire during cooking (Marshall et al. 1985). Such burning is thought to reflect food preparation as the traces of charcoal, ash and fire cracked rocks in Unit 5C appear to represent former hearth areas.

Data on the burning of fish bone were not recorded as it proved difficult to distinguish burning evidence from manganese staining.

Spatial Distribution

The majority of the faunal remains were concentrated within Unit 5A. Wave action had selectively reworked and winnowed this unit and the high bone densities did not reflect original distribution patterns. Complete bones and teeth were associated with charcoal and ash features in Unit 5C but not in a discernible pattern. Specific activity areas, such as butchery localities, were not recognized.

Faunal Summary

The site of Dongodien represented a shoreline encampment whose occupants possessed domestic cattle, goats and sheep, hunted small wild bovids and engaged in fishing. Some of the nondomestic elements in the bone assemblage, especially fish bone, probably reflect the background density of faunal remains within the Holocene beach deposits. However, the number and size of certain fish species, notably Nile Perch, make it likely that such species represented discarded food remains. The incidence of burned bones and cutmarks indicates cooking and probably butchering of animal carcasses took place at the site.

EXCAVATION FEATURES

Five subcircular patches of charcoal and ash with associated fire reddened stones and consolidated blocks of fire hardened silt were plotted within the western sector of Unit 5C (Fig. 70). The patches had a diameter of 50-100 cm and a thickness of 2-4 cm. The fire reddened stones (N=15) ranged in size from 5-15 cm. The pieces were probably collected from a nearby outcrop of Upper Member bioclastic sandstone and may have functioned as hearth stones. The hardened silty blocks (N=17) measures 6-10 cm in length. Circular pits or shallow depressions filled with ash, sand and faunal debris, like the suspected hearth at GaJi 2, were not identified. No discernible pattern of lithic, pottery or faunal remains was identified within or between the circular features.

DATING

Six C^{14} dates have been received from the site. Four different dating materials, all collected from Unit 5C, were submitted to three separate radiocarbon laboratories. The materials included mollusc shell, fish and mammal bone, and charcoal. A discussion of each material and its associated dates is given below.

Mollusc Shell

The dating of Holocene mollusc shells from the Lake Turkana Basin has been discussed by Butzer (1971, 1980). Older disolved carbonates are apparently assimilated into the biological system of the living mollusc and are thought to yield an age roughly 400 years too old (Butzer et al. 1969). In order to better understand the anomalies involved in dating mollusc shell, an attempt was made to compare age determinations on shell, chiefly *Melanoides turberculata* and *Corbicula africana*, mammal bone and charcoal from the same stratigraphic horizon. The shell date yielded an age of

10,320 ± 150 BP   University of Sydney

This date is thought to reflect the age of the shell sample and not the occupation horizon. Fragmentary mollusc shells from earlier Holocene levels were apparently reworked and incorporated into the archaeological site. A similar situation evidently occurred with the mollusc sample from GaJi 2. The shell date from Unit 5C contrasted sharply with the bone and charcoal determinations and did not reflect the correct age of the occupation horizon.

Bone

The bone sample was divided into fish and mammal components and each material was dated separately. The samples produced dates of:

Fish	4580 ± 170 BP	Geochron Labs
Mammal	3405 ± 130 BP	Geochron Labs

The fish sample yielded an age over 1000 years older than the mammal bone. Several factors may explain this disparity. First, the fish bone may have included, like the mollusc sample, reworked material from earlier Holocene levels. Second, the mammal bone sample may

DONGODIEN: GaJi 4

PLOT OF EXCAVATED CULTURAL REMAINS

KEY

Stone		Body parts	
Stone artifact	⬭	Skull	sk
		Mandible	md
Bone/teeth		Maxillae	mx
Identifiable element: mammal	•	Tooth	th
Fish	o	Scapula	sc
Bird/reptile	θ	Rib	rb
Non-identifiable: mammal	/	Vertebrae	vt
		Humerus	hu
Pottery		Radius	ra
Nderit Ware	▲	Ulna	ul
Other	Δ	Innominate	in
Concentrations of charcoal and ash		Femur	fe
		Tibia	tb
Bovid size groups		Metacarpal	mc
Small	s	Metapodial	mp
Medium	m	Astagalus	as
Medium-large	l	Navicular-cuboid	nc
Other		Phalange	ph
Bird	BD	Odontoid	od
Reptile	RP	Scaphoid	sh

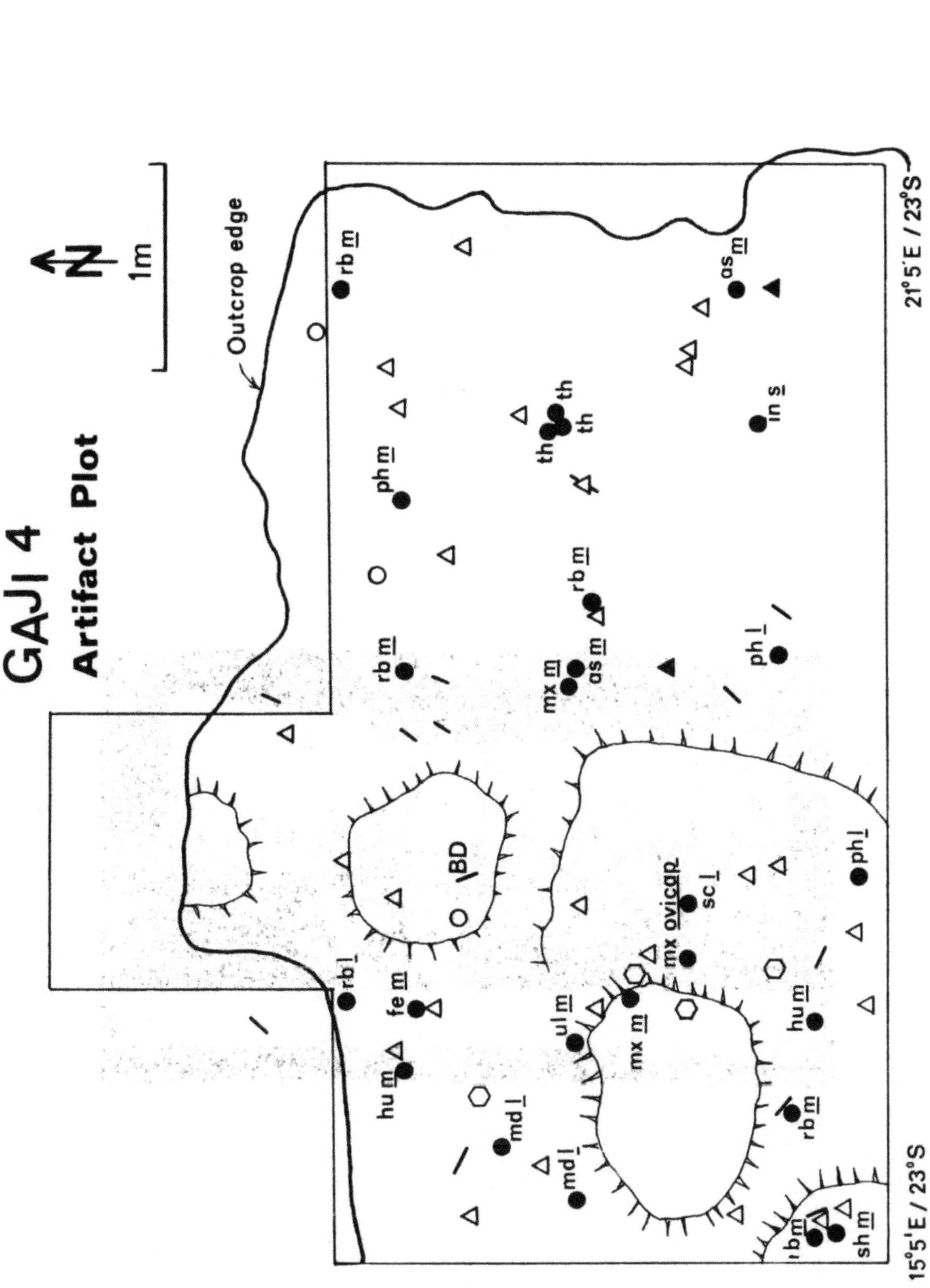

Figure 70. Dongodien. Plot of excavated cultural remains.

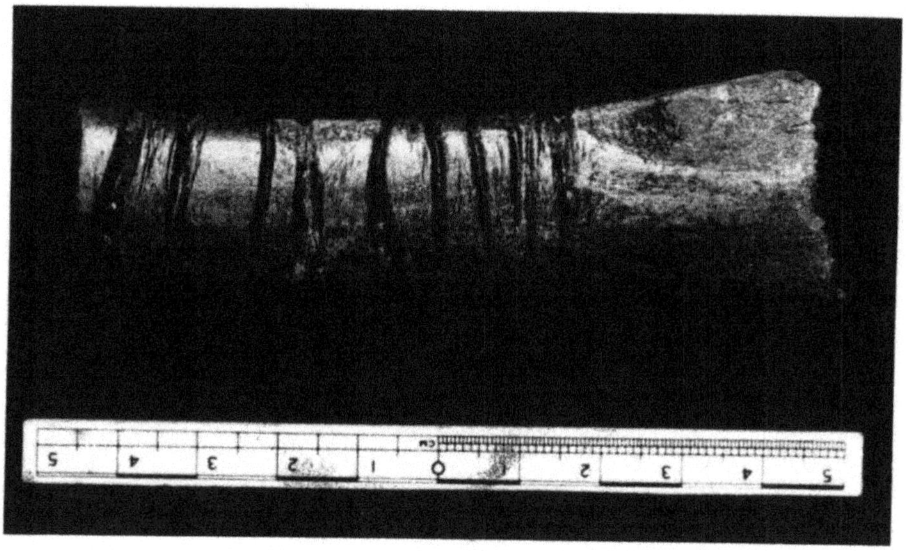

Plate 3. Dongodien: Bone artifact with circular incisions.

have been subjected to geochemical alteration processes (Bower and Nelson 1978) that did not affect the fish bone. Third, one or both of the dates may reflect laboratory error. Currently, the age discrepancy is not fully understood. A future study of the methodology and results of dating bone by the radiocarbon technique is planned by the author. However, the bone dates do suggest an age of between 3500-4500 BP for the site. The bone apatite dates from Dongodien clearly demonstrate that single bone determinations from Lake Turkana's Holocene deposits should be interpreted cautiously and with healthy skepticism. Whenever possible, bone dates should be corroborated with charcoal dates from the same stratigraphic horizon.

Charcoal

Abundant charcoal fragments, 5-12 mm in length, occurred in Unit 5C. A large sample of charcoal fragments and associated fine sand was divided into equal portions and sent to different radiocarbon laboratories. The sample yielded dates of:

3945 ± 135 BP	charcoal	University of Sydney
4100 ± 125 BP	humic acid residue from the charcoal sample	University of Sydney
3890 ± 60 BP	charcoal	University of Pennsylvania

The independently dated charcoal samples, as well as the humic acid residue, produced comparable results. The standard deviations of the charcoal samples overlap and are thus viewed as statistically equivalent. The age of the occupation horizon is therefore interpreted as 3900-4100 BP.

PRELIMINARY INTERPRETATION

Excavations at Dongodien established the direct association of Nderit Ware pottery with domestic animals that dated to 4000 BP. This date represents, along with the sites of GaJi 2, the Ileret Stone Bowl Site and North Horr, one of the earliest excavated occurrences of domestic animals in Eastern Africa (Owen et al. 1982; Collett and Robertshaw 1983).

The site was situated close to the shoreline when Lake Turkana stood nearly 50 m higher than today. The site's inhabitants continued to utilize the rich fish resources of the lake but use of bone harpoons had evidently ceased and the hunting of wild game animals had substantially declined. Portions of the occupation horizon were subsequently reworked by small scale advances of the lake. Probably shortly thereafter, a major regression of the Holocene lake began.

The site's subsistence activities are open to several different interpretations. The recovery of domestic animals, fish refuse and small wild bovids suggests the site may have represented a home base for stock herding peoples who practiced a diversified subsistence economy. Alternatively, the site have been occupied by cattle herding groups who had lost much of their large stock or by hunter-gatherer bands who may have acquired and butchered numbers of the domestic animals (Marshall et al. 1985). These questions are clearly important and larger faunal samples from expanded excavations are needed to help provide the necessary answers.

The association of a microlithic tool kit based on the production of small obsidian flakes and blades, new techniques and styles of pottery decoration, the absence of bone harpoons, and the introduction of domestic animals strongly suggests the northeastern sector of the Lake Turkana Basin underwent a significant cultural shift by at least 4000 years ago.

Table 48

GaJi 4

INVENTORY OF CULTURAL REMAINS BY SAMPLING UNIT

Unit	1	2	3	4	5	6	7	8	9	10	11	12	13	14	15	16	17	18	19	20	21	22	23	24	25	26	27
Stone:																											
Microliths	1	1	-	-	-	1	-	-	-	-	-	1	1	1	-	-	1	-	1	-	1	-	-	-	-	-	-
Scrapers	-	-	-	-	1	-	-	-	-	-	-	-	-	-	-	-	-	-	-	-	-	-	-	-	1	1	-
Outils écaillés	-	-	-	-	1	-	-	1	-	-	-	-	-	-	-	-	-	-	1	-	-	-	-	-	-	-	-
Burins	-	-	-	-	-	-	-	-	-	-	-	-	-	1	-	1	-	-	-	-	-	-	-	-	-	-	-
Core/chopper	-	-	-	-	-	-	-	-	-	-	-	-	-	-	-	-	-	-	-	-	-	-	-	-	-	-	-
"MSA" point	-	-	-	-	-	-	-	-	-	-	-	-	-	-	-	-	1	-	-	-	-	-	-	-	-	-	-
Cas. trimm.	-	-	-	-	-	1	-	-	1	-	-	-	-	-	-	-	-	-	-	-	-	1	-	-	-	1	1
Cas. retouch	-	-	-	-	-	-	-	-	-	-	-	-	-	-	-	-	-	-	1	-	-	-	-	-	-	-	-
Whole flakes	-	1	1	1	5	5	2	-	-	-	2	-	3	2	-	2	-	-	8	1	5	1	1	-	-	-	4
Broken flakes	-	1	-	1	2	-	1	-	-	2	-	-	-	5	1	1	-	-	7	4	1	-	-	-	1	4	-
Angular frags.	-	-	-	1	4	1	2	-	1	-	-	-	2	-	-	-	1	-	9	2	3	-	1	-	-	2	2
Cores	-	-	-	-	-	2	-	-	-	2	-	-	-	-	-	-	1	-	-	1	2	-	-	-	-	-	-
Hammerstones	-	-	-	-	-	-	-	-	-	-	-	-	-	-	-	-	-	-	-	1	-	-	-	-	-	-	-
Grindstone	-	-	-	-	-	-	-	-	-	-	-	-	-	-	-	-	-	-	-	-	-	-	-	-	-	-	-
Stone Bowl	-	-	-	-	-	-	-	-	-	-	-	-	-	-	-	-	-	-	-	-	-	-	-	-	-	-	-
Pottery:																											
Nderit Ware	1	2	-	-	-	-	-	-	-	-	-	-	-	-	-	-	-	-	2	1	-	-	-	-	-	-	-
Burnished/slipped	-	5	-	-	1	-	1	1	-	-	5	-	-	1	-	-	-	-	10	-	-	-	-	-	-	-	1
Other decorated	-	-	-	-	8	-	2	2	-	-	-	-	2	3	-	1	-	-	9	3	1	-	-	1	-	-	-
Disc	-	1	-	-	-	-	-	-	-	-	-	-	-	-	-	-	-	-	-	-	-	-	-	-	-	-	-
Undecorated	-	3	-	1	11	7	7	4	-	1	4	2	7	3	2	8	2	2	12	9	10	-	-	3	-	-	-
Bone/teeth:																											
Bone frags. (NID):																											
Mammal	-	2	21	6	-	36	2	2	5	-	1	12	-	4	1	3	1	-	3	2	7	6	-	-	-	1	3
Fish	-	2	14	2	35	3	12	1	-	11	3	14	4	3	15	16	14	5	8	-	9	5	-	-	-	-	13
Turtle	-	-	-	-	4	-	-	-	-	-	-	-	-	-	-	-	-	-	-	-	-	-	-	-	-	-	-
Bird	-	-	-	-	-	-	-	-	-	-	-	-	-	-	-	-	-	-	-	-	-	-	-	-	-	-	-
Bones w/art. sur.	-	-	9	4	-	7	-	-	1	1	1	-	-	1	1	-	-	-	-	-	-	-	-	-	-	-	-
Horn core	-	-	-	-	-	-	-	-	-	-	-	-	-	-	-	-	-	-	-	-	-	-	-	-	-	-	-
Teeth:																											
Mammal	-	-	-	2	6	-	-	-	-	-	1	-	-	-	-	-	-	-	-	1	1	-	-	-	-	-	2
Fish	-	-	-	-	1	-	-	-	-	-	-	-	-	-	-	-	-	-	-	-	-	-	-	-	1	-	-
Other:																											
Circular bone 'ring'	-	-	-	-	-	-	-	-	-	-	-	-	-	-	-	-	-	-	-	-	-	-	-	-	-	-	-
Angular silty blocks	-	-	-	-	-	-	-	-	-	-	-	-	-	-	-	-	-	-	-	-	-	-	-	-	-	-	-

Table 48 Continued

GaJi 4

INVENTORY OF CULTURAL REMAINS BY SAMPLING UNIT

Unit	28	29	30	31	32	Totals: Units I-32	1974 Collection	1976 Collection	1976 Plotted	6 Units on Excavation Areas	Total Surface Collection (N)
Stone:											
Microliths	1	-	-	-	-	10	2	21	7	4	44
Scrapers	-	-	1	-	-	4	1	3	1	-	9
Outils écaillés	1	-	-	-	-	4	-	12	-	1	17
Burins	-	-	-	-	-	2	-	-	-	-	2
Core/chopper	-	-	-	-	-	-	-	-	1	-	1
"MSA" point	-	-	-	-	-	1	-	-	-	-	1
Cas. trimm.	-	1	-	1	-	7	-	-	-	-	7
Cas. retouch	-	-	-	-	-	1	-	-	-	-	1
Whole flakes	2	9	2	-	4	61	3	87	-	12	163
Broken flakes	2	8	2	-	6	51	-	58	-	12	121
Angular frags.	6	6	4	2	2	51	-	72	-	9	132
Cores	1	-	-	-	-	9	3	6	1	1	20
Hammerstones	-	-	-	-	-	1	-	-	-	-	1
Grindstone	-	-	-	-	-	-	-	-	1	-	1
Stone Bowl	-	-	-	-	-	-	-	-	1	-	1
Pottery:											
Nderit Ware	-	-	-	-	-	6	-	10	4	2	22
Burnished/slipped	-	2	-	-	-	27	-	43	-	6	78
Other decorated	-	1	-	1	-	34	3	59	7	-	103
Disc	-	-	-	-	-	1	1	8	-	-	10
Undecorated	5	3	2	-	-	108	5	270	-	10	393
Bone/teeth:											
Bone frags. (NID):											
Mammal	8	1	2	-	-	129	-	-	-	63	192
Fish	3	1	1	-	-	191	-	-	-	20	211
Turtle	-	-	-	-	-	4	-	-	-	-	4
Bird	-	-	-	-	-	-	-	-	1	-	1
Bones w/art. sur.	-	-	-	-	-	25	-	-	2	20	47
Horn core	-	-	-	-	-	-	-	-	1	-	1
Teeth:											
Mammal	1	-	2	-	-	15	-	-	11	7	33
Fish	-	-	-	-	-	2	-	-	-	-	2
Other:											
Circular bone 'ring'	-	-	-	-	-	-	-	-	1	-	1
Angular silty blocks	-	-	-	-	-	-	2	-	-	2	4

Table 49
GaJi 4: Surface

FREQUENCY OF TRIMMED PIECES

	Obsidian	Chert	Quartz	Lava	Total (N)	Total (%)
Microliths:					(44)	(54.3)
Crescents	9	2	-	-	11	13.6
Curved backed flakes	1	3	-	-	4	4.9
Straight backed flakes	1	-	-	-	1	1.2
Truncations	15	-	-	-	15	18.7
Double backed micro-percoir	1	-	-	-	1	1.2
Double backed: broken	4	1	-	-	5	6.2
Fragmentary	6	1	-	-	7	8.6
Scrapers:					(9)	(11.1)
Side	1	3	-	-	4	4.9
Double side	-	-	-	1	1	1.2
End	1	-	-	-	1	1.2
Notch	-	2	-	-	2	2.5
Semi-circular	-	1	-	-	1	1.2
Outils écaillés	13	4	-	-	17	21.2
Burin:						
Technical	-	-	1	-	1	1.2
Dihedral	1	-	-	-	1	1.2
Core/chopper	-	-	-	1	1	1.2
Cas. trimm.	-	7	-	-	7	8.6
Cas. retouch	-	1	-	-	1	1.2
Raw material (N)	53	25	1	2	81	100.0
(%)	65.4	30.9	1.2	2.5		

Table 50

GaJi 4: Surface

FREQUENCY OF CORE TYPES

	Chert	Lava	Quartz	Obsidian	Total (N)	Total (%)
Single platform	4	1	-	1	6	30.0
Double: alternate	-	1	-	1	2	10.0
other	2	-	-	2	4	20.0
Multi	2	-	-	2	4	20.0
Disc	-	-	-	-	-	-
Casual	-	-	-	-	-	-
Fragmentary	-	-	-	4	4	20.0
Raw material (N)	8	2	-	10	20	100.0
(%)	40.0	10.0	-	50.0		

Table 51

GaJi 4: Surface

FREQUENCY OF MAJOR ARTIFACT CLASSES

	Chert	Lava	Quartz	Obsidian	Total (N)	Total (%)
Trimmed pieces	25	2	1	53	81	15.7
Cores	8	2	-	10	20	3.8
UTFF	85	13	-	318	416	80.5
Raw material (N)	118	17	1	381	517	100.0
(%)	22.8	3.3	0.2	73.7		

Table 52

GaJi 4: Excavation

INVENTORY OF CULTURAL REMAINS

	Unit 2	Unit 3	Excavation 1 Unit 5							Excavation 2 Unit 5			Unit 5 Total (N)
			5A	5B	5C	Mixed Levels	Plotted 5A-5C	Eroding on Outcrop	Total (N)	5A-C	Eroding on Outcrop	Total (N)	
Stone:													
Microliths	-	-	15	4	56	4	-	-	79	6	-	6	85
Scrapers	-	-	-	1	4	1	-	-	6	-	1	1	7
Outils écaillés	1	-	8	9	30	6	1	3	57	-	-	-	57
Cas. trimm.	-	-	-	1	2	-	-	-	3	-	1	1	4
Cas. retouch	-	-	-	1	1	-	-	-	2	-	-	-	2
Whole flakes	4	2	67	51	382	30	2	-	532	-	3	3	535
Broken flakes	2	9	88	77	428	42	-	-	635	-	3	3	638
Angular frags.	2	-	105	54	293	46	1	-	499	-	3	3	502
Cores	-	-	-	3	11	2	1	-	17	-	1	1	18
Pottery:													
Nderit Ware	-	-	6	7	5	2	4	3	27	1	-	1	28
Burnished/slipped	1	-	14	11	164	9	16	-	215	1	-	1	216
Other decorated	2	1	1	5	14	2	6	2	33	-	1	1	34
Disc	-	-	2	1	6	-	7	-	16	-	-	-	16
Undecorated	5	5	18	13	20	4	3	2	70	-	8	8	78
Bone/teeth:													
Bone frags. (NID):													
Mammal	40	27	591	294	638	158	48	-	1729	1	17	18	1747
Fish	96	71	2268	102	759	291	17	-	3437	4	31	34	3471
Bones w/art. sur.	-	1	133	27	86	18	38	10	312	1	3	4	316
Horn core	-	-	-	-	-	-	-	-	-	-	1	1	1
Teeth:													
Mammal	1	-	55	23	77	19	8	-	182	-	1	1	183
Fish	-	-	8	-	-	2	-	-	10	-	1	1	11
Other:													
Bone artifact	-	-	-	-	-	-	-	1	1	-	-	-	1
Hyrax canine	-	-	-	-	-	-	1	-	1	-	-	-	1
Ostrich e.s. beads	-	-	1	3	8	1	-	1	14	-	-	-	14
Angular silty blocks	1	-	53	16	13	7	4	-	93	10	-	10	103
Fire-cracked rocks	-	1	1	1	5	-	10	-	17	-	-	-	17
Haematite	-	-	11	2	61	1	-	-	75	-	-	-	75
Calcite crystal	-	-	-	-	2	-	-	-	2	-	-	-	2
Ostrich e.s. frags.	1	2	5	1	13	3	-	-	22	-	-	-	22

Table 53

GaJi 4: Excavations 1 and 2

FREQUENCY OF TRIMMED PIECES

	Obsidian	Chert	Total (N)	Total (%)
Microliths:			(85)	(54.8)
Crescents	-	-	-	-
Curved backed flakes	4	-	4	2.6
Double backed flakes	12	-	12	7.7
Truncations	29	-	29	18.7
Micro-percoir	1	3	4	2.6
Micro-percoir with notch	2	-	2	1.3
Miscellaneous	4	-	4	2.6
Fragmentary	27	3	30	19.4
Scrapers:			(7)	(4.5)
Side	1	2	3	1.7
Side and end	1	-	1	0.6
Notch	-	1	1	0.6
Double notch	1	-	1	0.6
Core scraper	-	1	1	0.6
Outils écaillés	56	1	57	36.8
Cas. trimm.	2	2	4	2.6
Cas. retouch	2	-	2	1.3
Raw material (N)	142	13	155	100.0

Table 54

GaJi 4: Excavation

FREQUENCY OF CORE TYPES

	Chert	Lava	Quartz	Obsidian	Total (N)	Total (%)
Single platform	-	-	-	4	4	22.0
Double: alternate	-	-	-	7	7	38.0
other	-	-	-	4	4	22.0
Multi	1	-	-	3	4	22.0
Disc	-	-	-	-	-	-
Casual	-	-	-	1	1	6.0
Fragmentary	-	-	-	1	1	6.0
Raw material (N)	1	-	-	17	18	100.0
(%)	6.0	-	-	94.0		

Table 55

GaJi 4: Excavation

FREQUENCY OF MAJOR ARTIFACT CLASSES

	Chert	Lava	Quartz	Obsidian	Total (N)	Total (%)
Trimmed pieces	13	-	-	142	155	8.4
Cores	1	-	-	17	18	1.0
UTFF	67	5	3	1600	1675	90.6
Raw material (N)	81	5	3	1759	1848	100.0
(%)	4.3	0.3	0.2	95.2		

Table 56

GaJi 4: Surface and Excavation

FREQUENCY OF POTTERY SHERDS

	Surface (604)		Excavation (358)	
	(N)	(%)	(N)	(%)
Rim sherds	57	9.4	25	7.0
Body sherds	537	88.9	317	88.5
Disc	10	2.7	16	4.5
Decorated	203	33.6	278	77.7
Undecorated	401	66.4	80	22.3

Table 57

GaJi 4: Surface and Excavation

FREQUENCY OF DECORATED SHERDS

	Surface (203)		Excavation (278)	
	(N)	(%)	(N)	(%)
Nderit Ware	22	10.9	28	10.0
Burnished/slipped	78	38.4	216	77.7
Other	103	50.7	34	22.3

Table 58 A

Dongodien: Excavation

INVENTORY OF IDENTIFIABLE FAUNAL REMAINS

	N	%
MAMMAL		
Bovids	(652)	(54.0)
Very small	19	1.6
Small (caprine size)	596	49.3
Medium (cattle size)	37	3.1
Hyrax	1	0.1
Leporid	1	0.1
BIRD	3	0.2
REPTILE	3	0.2
FISH	550	45.6
Assemblage Total	1210	100.0

INVENTORY OF FAUNAL REMAINS Table 58 B

	N	%
MAMMAL	4446	64.7
FISH	2437	35.3
REPTILE	4	>0.1
BIRD	9	>0.1
Assemblage Total	6896	100.0

Table 59

Dongodien: Excavation

INVENTORY OF IDENTIFIABLE BODY PARTS (BOVIDS)

	Madoqua sp.	Very Small Bovid	Caprinae	Small Bovid (ovicaprid size)	*Bos* sp.	Medium Bovid (cattle size)
Horncore	---	---	---	---	---	---
Cranium	---	---	---	1	---	---
Maxilla	---	---	---	---	---	---
Mandible	---	---	---	2	---	---
Upper Ch. Th.	---	---	24	8	2	---
Lower Ch. Th.	---	---	17	6	---	---
Incisors	---	---	15	2	---	---
Vertebrae	---	3	1	23	---	---
Rib	---	3	---	33	---	5
Sternum	---	---	---	---	---	---
Scapula	---	---	2	8	---	2
Humerus	---	---	4	9	---	2
Radio-Ulna	1	1	11	12	---	1
Scaphoid	---	---	2	2	1	---
Lunate	---	---	1	---	---	---
Cuneiform	---	---	2	3	---	---
Magnum	---	---	3	---	---	---
Unciform	---	---	---	2	---	---
Pisiform	---	---	1	---	---	---
Metacarpal	---	---	1	10	---	---
Pelvis	---	---	1	18	---	---
Femur	1	---	2	9	---	1
Tibia	---	---	---	6	---	1
Calcanium	---	---	---	2	---	---
Astragalus	---	---	4	2	2	---
NVC	---	---	2	1	---	---
Med. Cun.	---	---	2	---	---	---
Lat. Cun.	---	---	---	4	---	---
Metatarsal	4	---	1	4	---	---
Met. Indet.	---	1	---	26	---	2
Phalanges	2	---	11	20	4	---
Patella	---	---	---	---	---	---
Assemblage Total (N)	8	8	107	213	9	14

INTRODUCTION

The Ileret Stone Bowl site (FwJj 5) was located in Area 10 approximately 1/2 km south of the Il-Erriet River (Fig. 50). Abundant stone tools, pottery and faunal remains, including domestic caprines and cattle, were excavated from a level of coarse sand and gravels. Four ceramic vessels were partially reconstructed and over 2000 pottery sherds were recovered. A new pottery taxon, Ileret type, was identified and described. Fragments of six stone bowls were discovered in the site's general vicinity (Fig. 71). They are thought to be associated with the domestic animals and large pottery inventory. The site was sampled and a small excavation was undertaken. A single radiocarbon determination of 4000 ±140 BP dated the main archaeological horizon.

STRATIGRAPHY

The Galana Boi deposits at FwJj 5 lay upon exposed Upper Member units of the Koobi Fora Formation and had a cumulative thickness of just over 2 1/2 m (Fig. 72). The lower Holocene levels were composed of fine to coarse clean gravelly sands. Several horizons had rich accumulations of unbroken mollusc shells, chiefly Melanoides tuberculata and Cleopatra, and fish bone. An isolated uniserial bone harpoon was collected along the northern margin of the outcrop. The association of molluscs, fish bone, bone harpoons and clean gravelly sands indicates these lower Galana Boi units were former beach sediments.

The upper Holocene level (Fig. 73, bed A) was composed of coarse dirty sands, broken mollusc shell and large angular rock fragments. Recent erosion had exposed considerable numbers of stone artifacts and pottery sherds along the flanks of this horizon. One such area, located on a low lying sand ridge, was chosen for excavation. Details of the excavation stratigraphy are reported below.

Lithology	Description
A	light gray to pale orange semi-consolidated medium sand; scattered broken mollusc shell and angular pebbles; silt particles adhering to sand grains; few artifacts; 35-55 cms.
B	Upper level: pale gray, poorly sorted, consolidated coarse sand; bedding structures not present; scattered angular rock fragments, rounded cobbles and broken mollusc shell; main archaeological horizon; 60-80 cms.
	Lower level: medium to dark gray, poorly sorted, very consolidated coarse sand; sand coated clay pellets at base; calcium carbonate cement; abundant rock fragments and broken mollusc shell (Melanoides, Corbicula, Mutela and Pila ovata); broken pumice pebbles; numerous artifacts, many of which were broken and heavily water abraided; 20-40 cms.

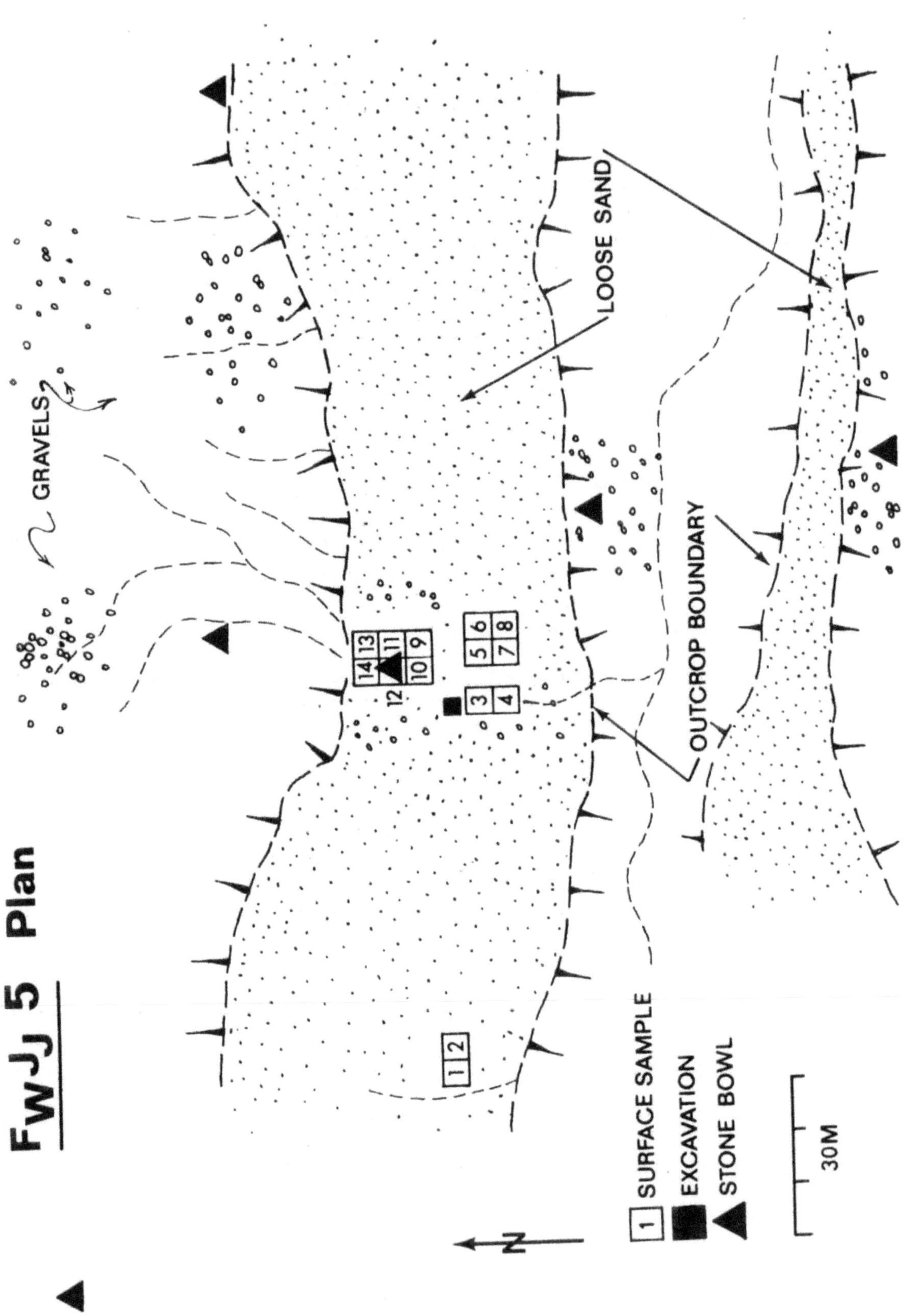

Figure 71. Ileret Stone Bowl Site (FwJj 5). Sketch plan of the Holocene outcrop showing location of sampling units and stone bowls.

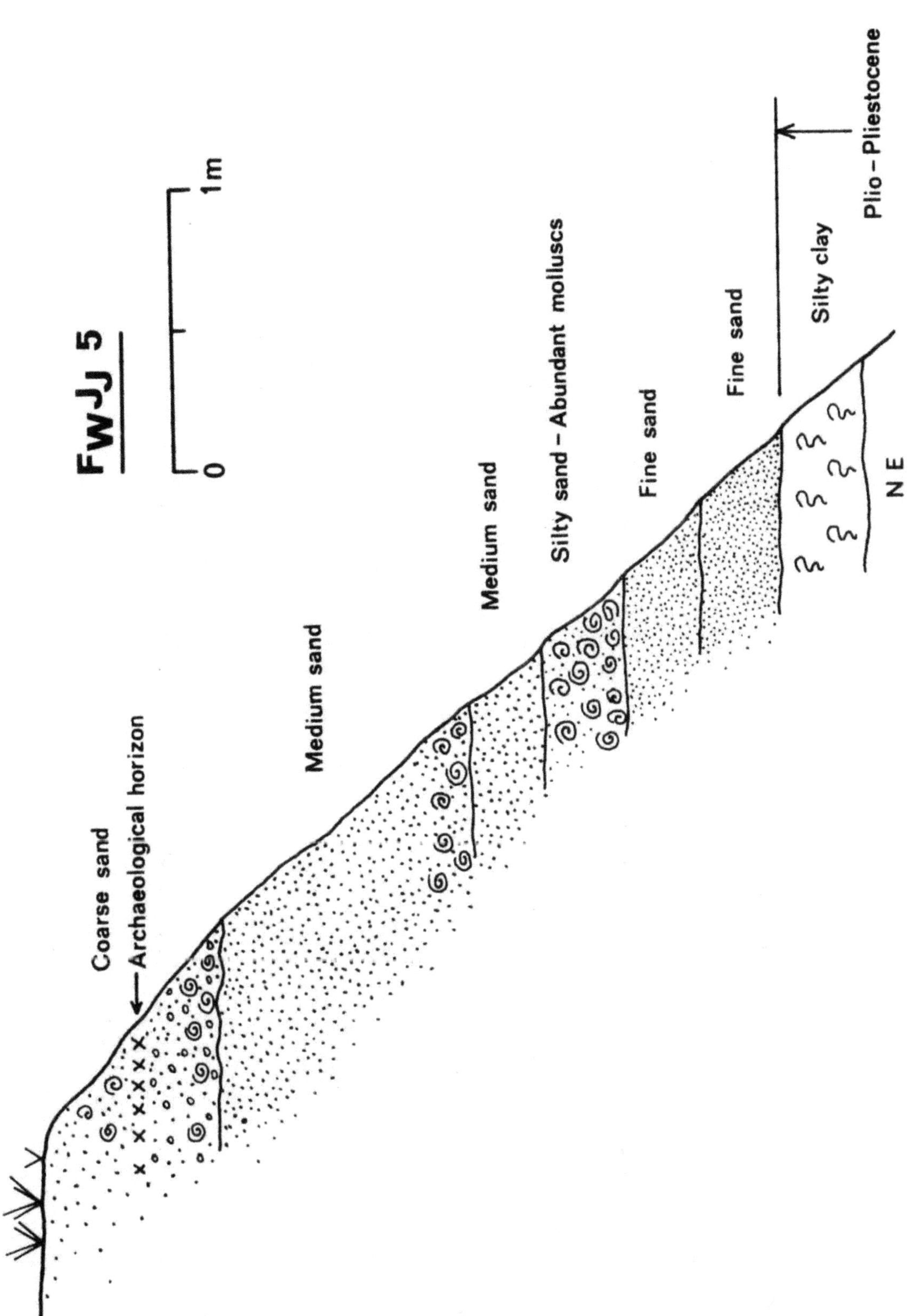

Figure 72. Ileret Stone Bowl Site. Profile of geological trench showing Plio-Pleistocene and Holocene beds.

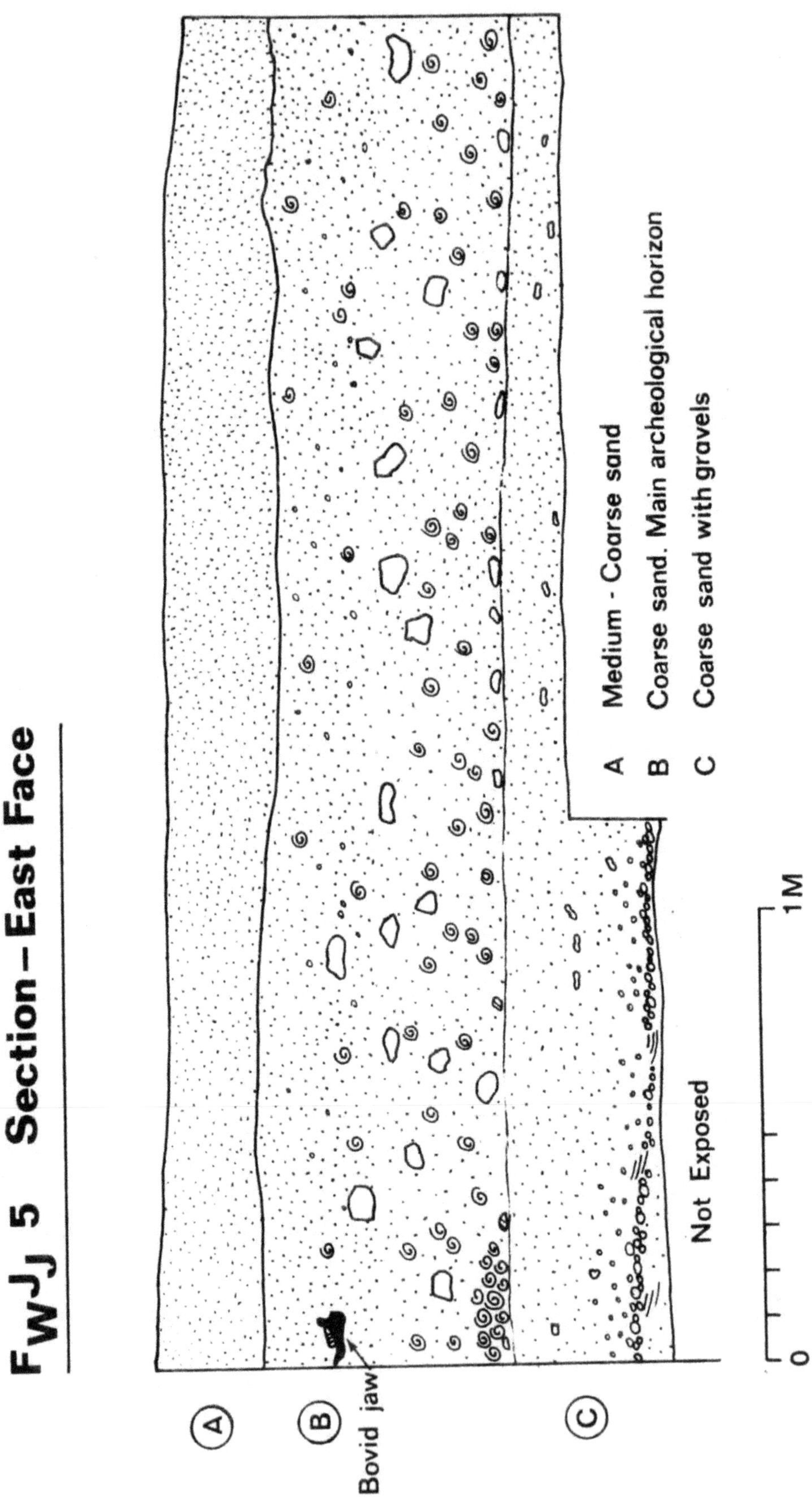

Figure 73. Ileret Stone Bowl Site. Excavation profile - east face.

C light gray consolidated coarse sand; calcium carbonate concretions; concentrations of broken Pila ovata shells; abundant fish bones; a few water worn artifacts; 60-80 cms.

The coarse gravelly sands in Unit C (Fig. 73) are thought to represent an early stage of deposition of reworked beach and channel deposits after regression of the middle Holocene paleo-lake shoreline. The subsequent accumulation of a 20-40 cm horizon of coarse sand, large rock fragments and water worn cobbles probably represents an episode of vigorous channel flow and the possible formation of a channel bar. An analogous setting can be found today along the distributory channels of the Il-Erriet River. Scattered water worn artifacts, similar to those in the succeeding unit, occurred within the upper part of Unit C. The site was subsequently occupied, following deposition of a distinct cobble horizon in Unit B, by early pastoralist groups. The presence of small flaking debris and numerous rejoined pottery sherds suggests some areas of the occupation horizon were not significantly disturbed by subsequent burial. Within the upper level of Unit B, the sands began to fine upward. The topmost sandy deposits were probably of mixed eolian and fluvial origin. Only a few dispersed artifacts were recovered from the upper part of the section.

The overall stratigraphic relationships of the Holocene deposits in Area 10 proved exceedingly complex. An inital reconstruction of the sequence of Plio-Pleistocene and Holocene beds is presented in Figure 74. It should be stressed, however, that further geological trenching and field mapping are required to confirm the following reconstruction.

The lowermost beds in the area consisted of Plio-Pleistocene silty clays (unit A) and sandy (tuffaceous?) silts (unit B). A widespread erosion surface (unit C) mantled these beds. In certain areas, Middle Stone Age bifacial points and levallois cores were recovered from the erosional gravels. These artifacts were often broken, water abraided, and partially desilicified. A thick sequence of characteristic Galana Boi littoral sands (unit D) overlay the erosion surface. In places, the MSA erosional gravels had been reworked and incorporated within the base of the littoral sands. Occasionally, these mixed units contained MSA points and bone harpoons. A distinctive surface of coarse muddy sand and silty clay (unit E) occurred to the southeast of the beach sands. This surface is thought to be laterally equivalent to unit D and to represent back beach mudflats or seasonally flooded marshes. Similar silty clay surfaces lateral to beach deposits were identified in Areas 116, 117 and 127. A second erosion surface mantled the beach and laterally equivalent deposits. Channel incision of the semi-consolidated beach and backbeach units, probably by the proto Il-Erriet River, had apparently set in as the lake began a major regression. Coarse channel sands, gravels, large rock fragments and reworked artifacts were deposited over the landscape (unit F). Following this episode of fluvial activity, human groups with domestic stock, pottery and stone bowls occupied the area. Their abandoned settlement was subsequently buried by a combination of channel, floodplain and eolian deposits.

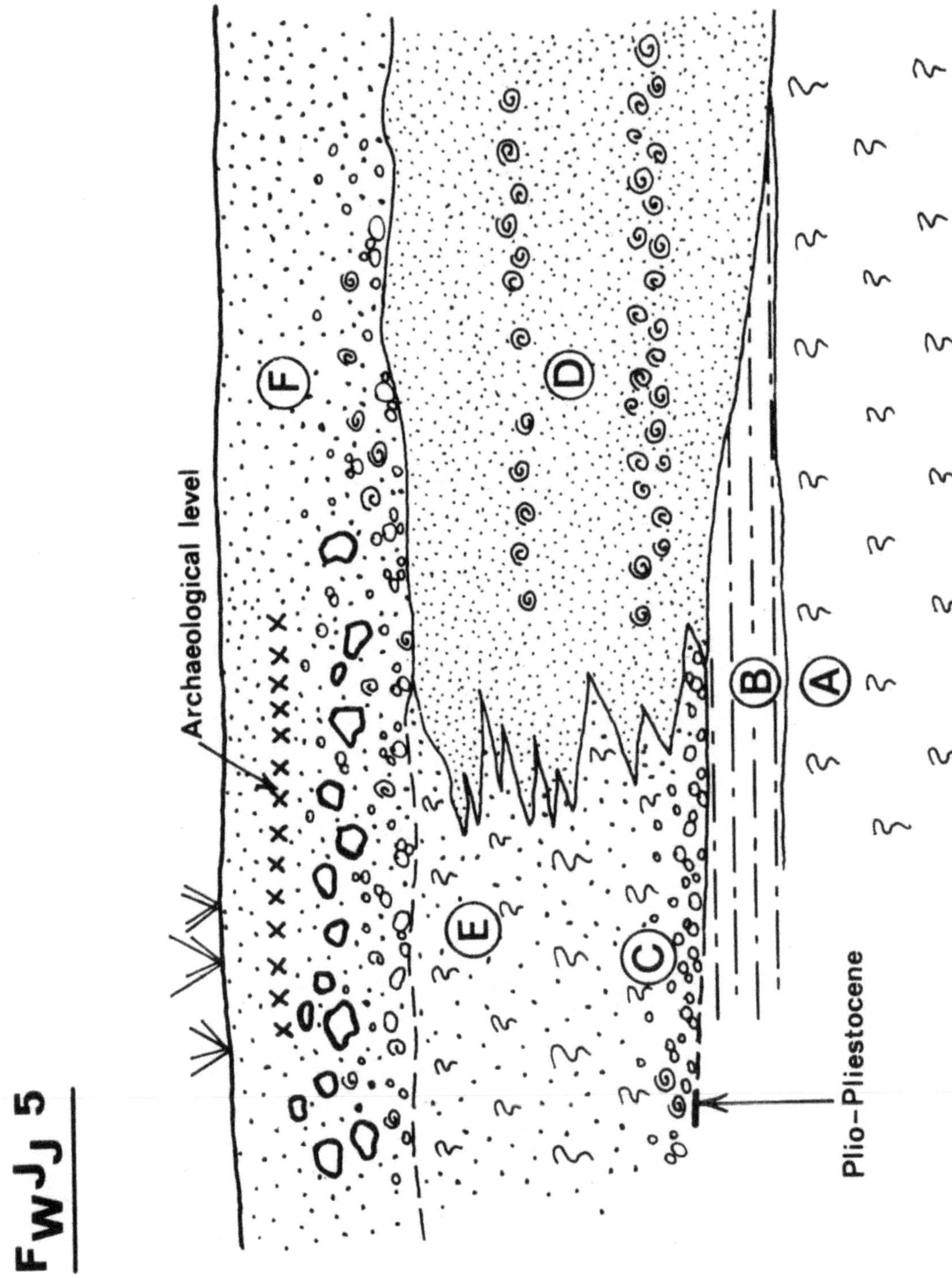

Figure 74. Ileret Stone Bowl Site. Schematic representation of the succession of Holocene and Plio-Pleistocene beds in Area 10.

SAMPLING

Along the crest of a low lying sand ridge, recent erosion had exposed high density patches of artifacts. Three concentrated patches were chosen for sampling (Fig. 71). A total of fourteen units were collected.

CULTURAL REMAIN: SURFACE

An inventory of the surface finds by collecting unit is presented in Table 60.

Stone Artifacts (Tables 61-63: Fig. 75)

Microliths were the most common shaped stone artifact followed by scrapers and *outils écaillés*. Among the microliths, double backed specimens were well represented. A diverse set of scraper forms was recovered including side, notched and composite examples. The notched specimens often displayed distinct crushing and hinge fractures. Several diminuative *outils écaillés* manufactured from obsidian resembled the excavated finds from GaJi 2 and Dongodien. Chert accounted for 85% of the raw material followed by quartz, obsidian and lava. Unlike the chert debitage, obsidian flakes and flakeblades showed crushed platforms and parallel sided dorsal scar ridges.

Pottery

Nearly 600 pottery sherds were recovered from the surface. The collection exhibited an especially wide range of rim and body decorative features. The most prominent decorative style consisted of overlapping crescentic impressions that often occurred in rows (Fig. 76, d). This pattern of decoration has been designated Ileret type. Nearly 60% of the decorated sherds could be attributed to this design motif. Other rim sherds had thickened lips (Fig. 76, a) and flattened, cross-hatched lips with raised knobs (Fig. 76, b-c). One group of body sherds displayed a series of closely spaced geometric impressions that were circumscribed by rectangular and rounded panels (Fig 76, e-f). These pieces were similar to the excavated, partially reconstructed, pot in Figure 80. The fragmentary nature of the sherds made it very difficult to reconstruct either vessel size or shape from the surface collection. The wide variety of decorative designs and rim profiles does suggest a considerable diversity of pottery vessels was represented, possibly from several distinct occupation episodes.

Stone Bowls

Six fragmentary stone bowls were collected within 100 m of the excavation area, including one example from surface collecting unit 12 (Fig 71). A seventh bowl fragment was discovered in a small sandy blowout approximately 350 m to the north. A thorough search of the surrounding countryside (500 m radius) failed to turn up additional specimens. While stone bowls were not recovered from the excavation,

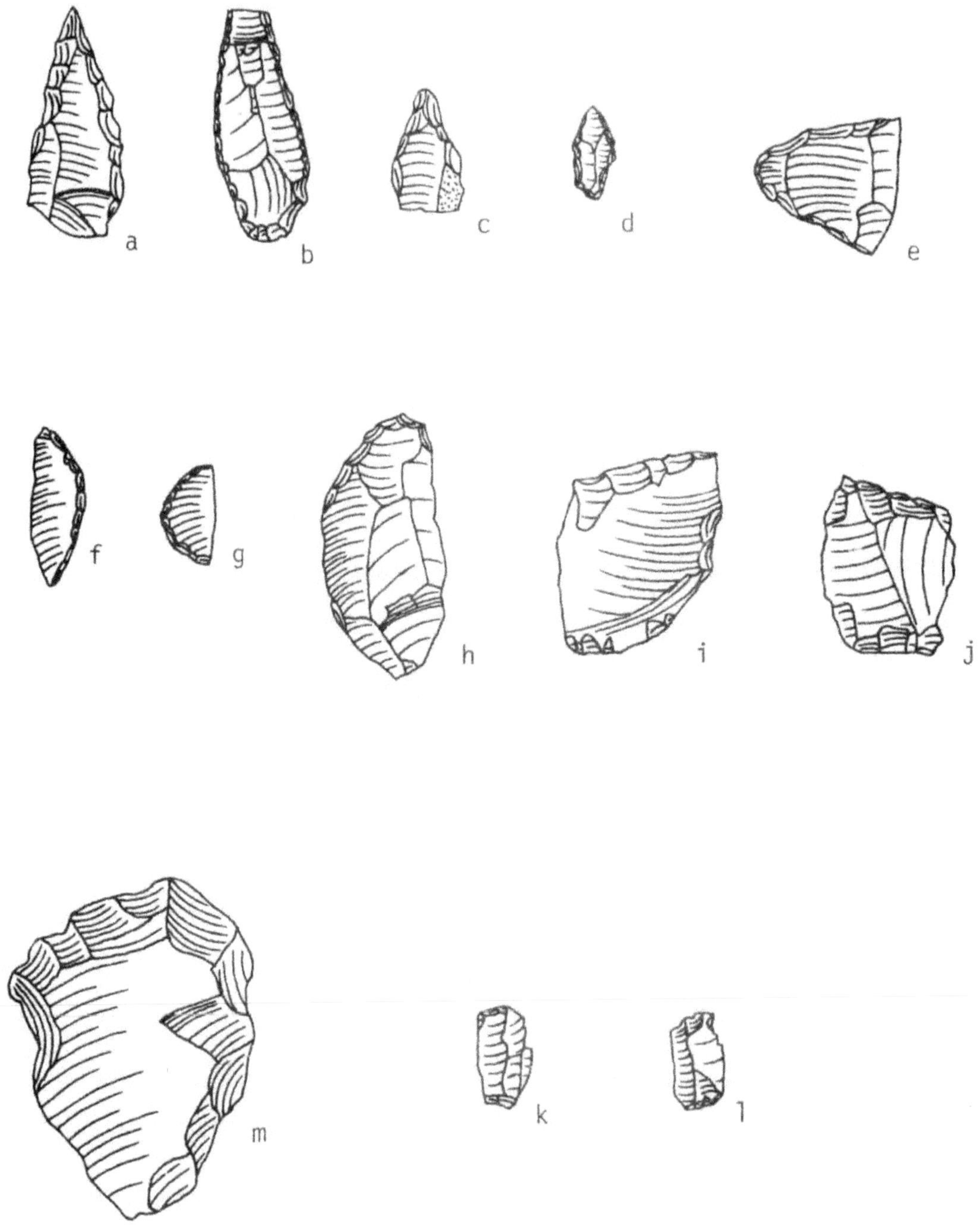

Figure 75. Ileret Stone Bowl Site. Stone artifacts: Surface. (a-b) double backed percoirs; (c-d) micro-percoir; Excavation. (e-g) crescents; (h) curved backed flake; (i-l) *outils écaillés*; (m) scraper. Natural size.

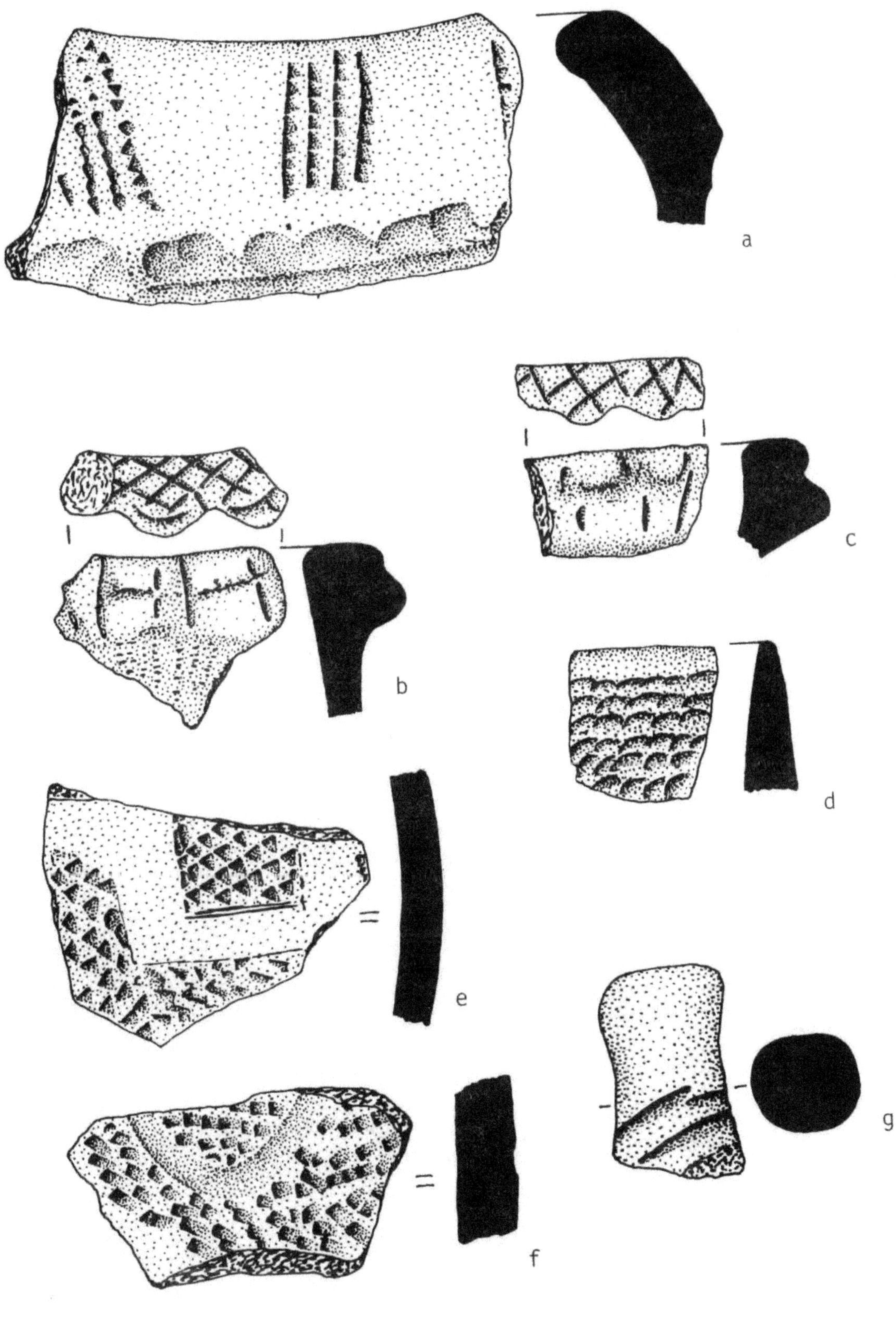

Figure 76. Ileret Stone Bowl Site. Pottery: Surface. (a-c) decorated rim sherds; (d) Ileret type rim; (e-f) decorated body; (g) lug. Natural size.

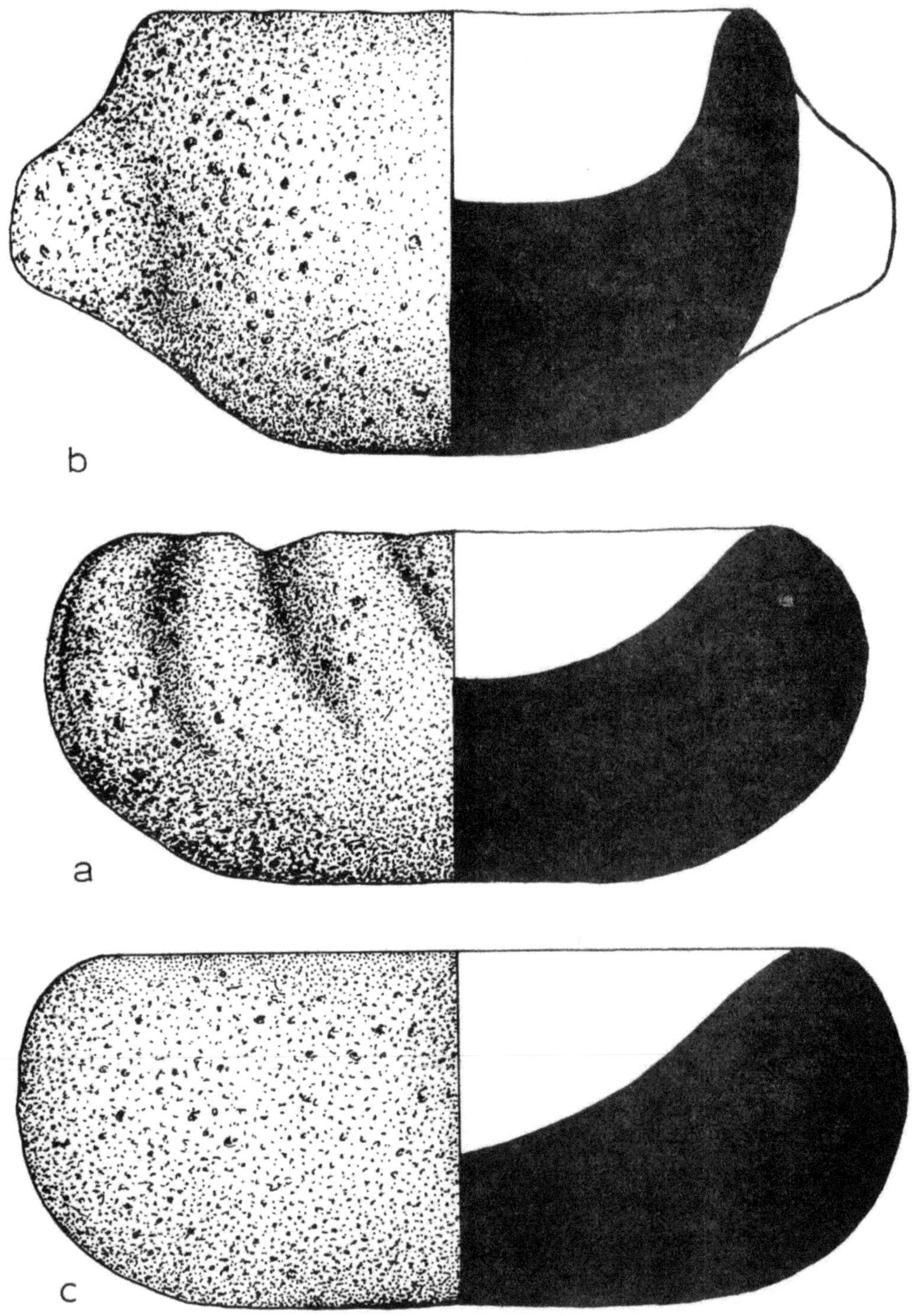

Figure 77. Ileret Stone Bowl Site. Surface.
(a) bowl with "handle", vesicular basalt; (b) bowl with beaded rim, basalt; (c) bowl with flattened base, basalt. 3:4 size.

the close proximity of the surface finds strongly suggests that they were associated with the excavated pottery and domestic animals.

Three stone bowl fragments are illustrated (Fig. 77). One example had an unusual small lateral projection, perhaps a handle, near its base. The vessel may have had a second projection on the opposite side. To the writer's knowledge, the specimen represents a hitherto unknown East African stone bowl configuration, although Desmond Clark (pers. comm.) has recently collected a stone bowl from Lake Besaka, Ethiopia, that is reminiscent of the Ileret specimen.

A second bowl fragment had a milled or beaded rim. The third example was a well crafted bowl with a smooth rim and flattened base. The remaining four bowl fragments were relatively steep sided and had either flattened or rounded bases. Six of the vessels were manufactured from basalt or fine grained lava. The seventh bowl was manufactured from a coarse grained quartzite. No traces of red ochre or evidence of oxidation staining were observed on the specimens.

CULTURAL REMAINS: EXCAVATION

One area of artifact concentration along the surface of the ridge crest was chosen for excavation. During the 1975 field season, a trial three meter square was excavated to the top of Unit C (Fig. 73). The excavation area was expanded to a three by four and one half meter square and deepened to the base of Unit C during the 1976 field season. Over 250 artifacts and large rock fragments were individually plotted in Unit B. Excavation proceeded, depending on the density of finds, in a series of 5, 10, and 20 cm spits.

An inventory of the excavated finds by stratigraphic unit is presented in Table 64. The great majority of artifacts were recovered from Unit B. For purposes of analysis, artifacts from each stratigraphic unit have been combined.

Stone Artifacts (Tables 65-67)

Microliths were the most common shaped stone artifact followed by *outils écaillés* and a diverse series of scrapers (Fig. 75). Small obsidian *outils écaillés*, similar to those collected from the site's surface, were numerous. Core/choppers were not identified. Debitage consisted primarily of single platform chert flakes struck from a variety of core types. Obsidian debitage usually exhibited parallel dorsal scar patterns and multiple striking platforms. Chert accounted for 74% of the raw material followed by obsidian and quartz. Several well patinated and desilicified chert flakes were recovered from the base of Units B and C. These finds are thought to represent reworked artifacts from either the underlying beach sediments or the MSA erosion surface.

Pottery (Tables 68-69; Figs. 78-82)

A total of 1706 sherds were recovered from the excavation. The collection included four partially reconstructed vessels and a large assemblage of sherds that shared a distinctive set of decorative motifs. An inventory of the excavated pottery is presented in Table 68.

Nearly 75% of the decorated sherds can be assigned to a single group, henceforth termed Ileret type. This group is characterized by a distinctive series of impressed and incised herringbone design motifs and straight or slightly concave sided vessel forms. The occurrence of noticeably different design techniques, styles and vessel shapes should be demonstrated in repeated association before they can be accepted as a unified ware. Consequently, the presence of new and, as yet incompletely defined, multiple design modes at FwJj 5 strongly argues that the taxon ware not be employed. Instead, the largest and most distinctive segment of the pottery assemblage has been designated as a type. Key attributes of Ileret type are described below.

ILERET TYPE

Design

The basic decorative motif consisted of incised or impressed herringbone patterns. Motifs were grouped by pattern irrespective of technique of manufacture. One group of sherds exhibited crescentic or lozenge shape impressions that occurred in closely spaced, often overlapping, horizontal and vertical rows(Fig 78, a-c). The majority of sherds displayed a series of zig-zag incised lines that formed a herringbone pattern (Fig. 78 e,i,k-l). Decoration was consistently placed just below the rim. No rim sherds with flattened decorated lips were identified.

Surface Finish

No evidence of external or internal burnishing was recognized.

Temper

A coarse sub-angular quartz grit was used as temper. Particle size was usually less than 1 mm although occasional fragments of 2-3 mm were found. Evidence for the use of broken sherds or large rock fragments was lacking.

Thickness

A sample of 35 body sherds yielded measurements of between 5-8 mm. The mean thickness was 6.2 mm.

Paste

A moderately porous silty clay was used as paste. Analysis of the clay type has not been undertaken.

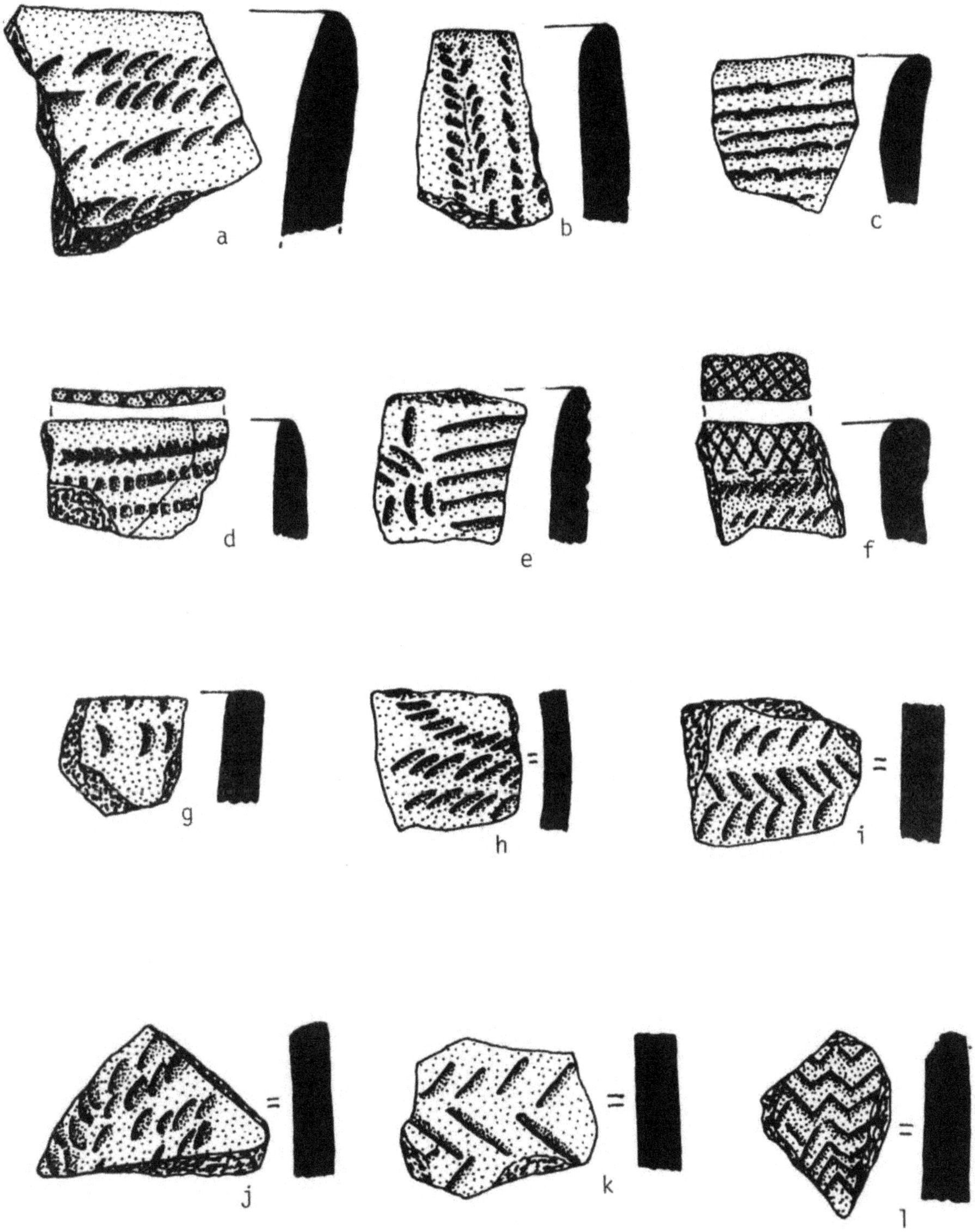

Figure 78. Ileret Stone Bowl Site. Ileret type pottery: Excavation. (a-g) decorated rim sherds; (h-l) decorated body. Natural size.

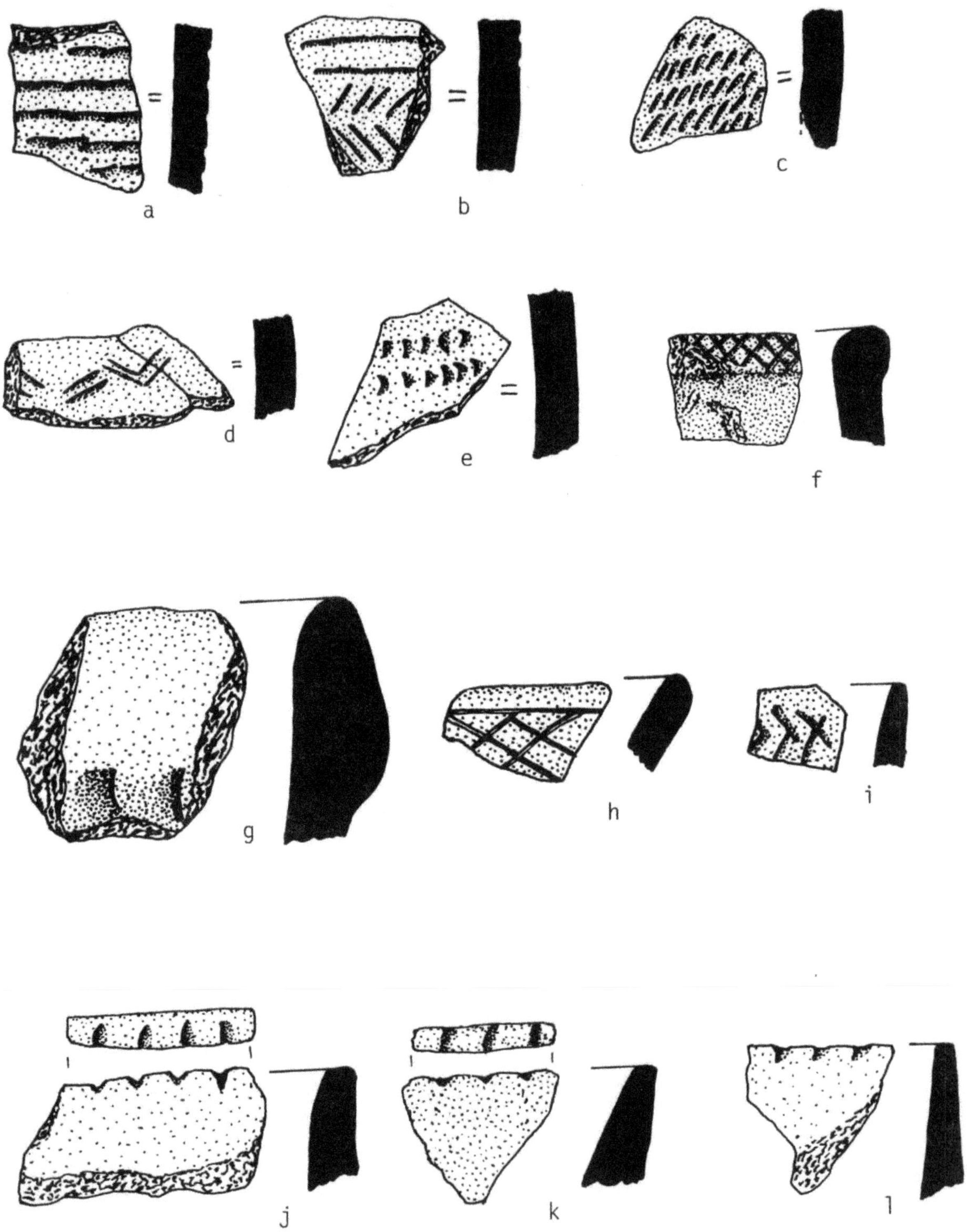

Figure 79. Ileret Stone Bowl Site. Pottery: Excavation.
(a-e) Ileret type body sherds; (f-i) decorated rim; (j-l) decorated lip.
Natural size.

Slip

While sherds with red slipping occurred among the total pottery assemblage, this feature was lacking on Ileret type.

Vessel Form

Unfortunately, it was not possible to reconstruct the shape of Ileret type vessels. However, the illustrated tapered and flattened rim sherds (Fig 78, a-g) suggest the existence of both straight sided and slightly concave pots.

Distribution

The distribution of Ileret type appears to be primarily concentrated in northwestern Kenya, within and near the Turkana Basin. Besided the numerous finds reported along the northeastern margin of the Basin (this monograph), other sherds have been collected at Eliye Springs, to the southwest of the lake (Jacobs and Soper 1972). Phillipson (1977b: 71-73) has reported sherds with similar affinities from North Horr, 150 km east of Lake Turkana. There, sherds are thought to be associated with stone bowls and possibly with domestic animals. Sherds with comparable decoration have also been collected from Nyero and Magosi Rock shelters, both in northeastern Uganda (Charles Nelson pers. comm.). As well, sherds resembling Ileret type have been collected from Lesayu, a site located along the lower Omo valley in southwestern Ethiopia (British Institute collections, Kenya).

Besides the apparent distribution center in northwestern Kenya, and probably including Karamoja and southwestern Ethiopia as well, other sherds resembling Ileret type have been reported from the important Pastoral Neolithic site of Narosura in the Central Rift of Kenya. There, pottery has been described as having distinctive bands of comb stamping and incision below the rim. Decorative motifs occur as closely spaced vertical, oblique or cross hatched designs and occasionally as pendant triangles (Odner 1972). Several Ileret type sherds show similar characteristics (Fig 79, b,f) although the greater resemblance is with comb stamped sherds(Fig. 79, h; Fig. 80, j; Fig. 81).

Other Pottery

Four non-Ileret type vessels were partially reconstructed. Figure 81 a illustrates a vessel with a tapered undecorated rim. Set within a wide (7 cm) horizontal boundary, decoration consisted of small grooved panels that circumscribed triangular and rectangular figures. Closely spaced rows of very small comb stamped geometric representations occurred inside the larger figures. The vessel had been burnished and a reddish purple slip (5R 3/4 - 5R 2/6) had been applied. Over 100 similarly decorated and/or slipped sherds were identified but could not be rejoined to the vessel. A second pot with similar design features was also reconstructed. However, this vessel was neither burnished nor slipped. Sherds with similar comb stamped motifs are known from North Horr (Phillipson 1977b:73; Stiles 1982) and from

Central Kenya at Narosura (Odner 1972:65) and possibly Crescent Island (Onyango-Abuje 1977:155).

A third partially reconstructed pot had a flattened cross-hatched lip with deep external stamping separated by wide (5-8 mm) undulating panels (Fig. 81, b). The rim profile suggests the vessel was originally a narrow mouth bowl.

A globular shaped vessel with a slightly constricted neck is illustrated in Figure 82. Decoration consisted of two horizontally stamped rows of sub-circular representations followed by a sinuous narrow wavy band. Small vertical incisions lay beneath the band. The vessel had a crumbly texture and was tempered with mica.

The majority of the non-Ileret type decorated rim sherds were tapered and had either stamped or incised cross-hatched designs (Fig. 79, h-i). A number of these specimens resembled rim sherds from Narosura (Odner 1972:64d, 65d-h). Other decorated rim sherds had thickened everted lips, sometimes with a flattened cross-hatched top (Fig. 79, f). These examples resembled rim sherds from GaJi 2. Rim sherds with milled and flattened lips were well represented (Fig. 79,j-l; Fig 80,a-b). Undecorated rim sherds occurred in a variety of lip profiles (Fig. 80 c-e).

A representative series of non-Ileret type decorated body sherds is illustrated in Figure 80. Comb stamping and incision techniques were employed (Fig. 80, f-k). A single internally scored sherd was identified from the excavation (Fig. 80, m). The exterior surface consisted of cross-hatched incised lines and did not resemble the large collection of Nderit Ware sherds from Dongodien.

An unusual set of small fragmentary rim sherds was also excavated. The specimens were decorated with rows of very small stamped geometric motifs and shallow incisions (Fig. 80, l). Their rim profiles suggest a shallow bowl form. The reconstructed diameter of one specimen was only 3-4 cm wide. Two of the rim sherds have also been burnished and red slipped. Figure 80 n, illustrates a broken circular sherd with rounded edges. Similar sherds were recovered from Dongodien. They may have functioned as pot burnishers.

Summary

The pottery from the Ileret Stone Bowl site consisted of a highly varied and often intricately decorated series of sherds. Well over 2000 specimens were collected from surface and excavated samples. Impression, comb stamping and incision were the predominate modes of decoration. Four vessels were partially reconstructed including two that were burnished and slipped. Internally scored pottery, similar to Nderit Ware, was not recovered. A new pottery taxon, Ileret type, was proposed to describe a distinctive herringbone design motif. Ileret type appears to have a distribution restricted to northwestern Kenya and possibly to northeastern Uganda, southwestern Ethiopia and central Kenya. Several non Ileret type reconstructed vessels also

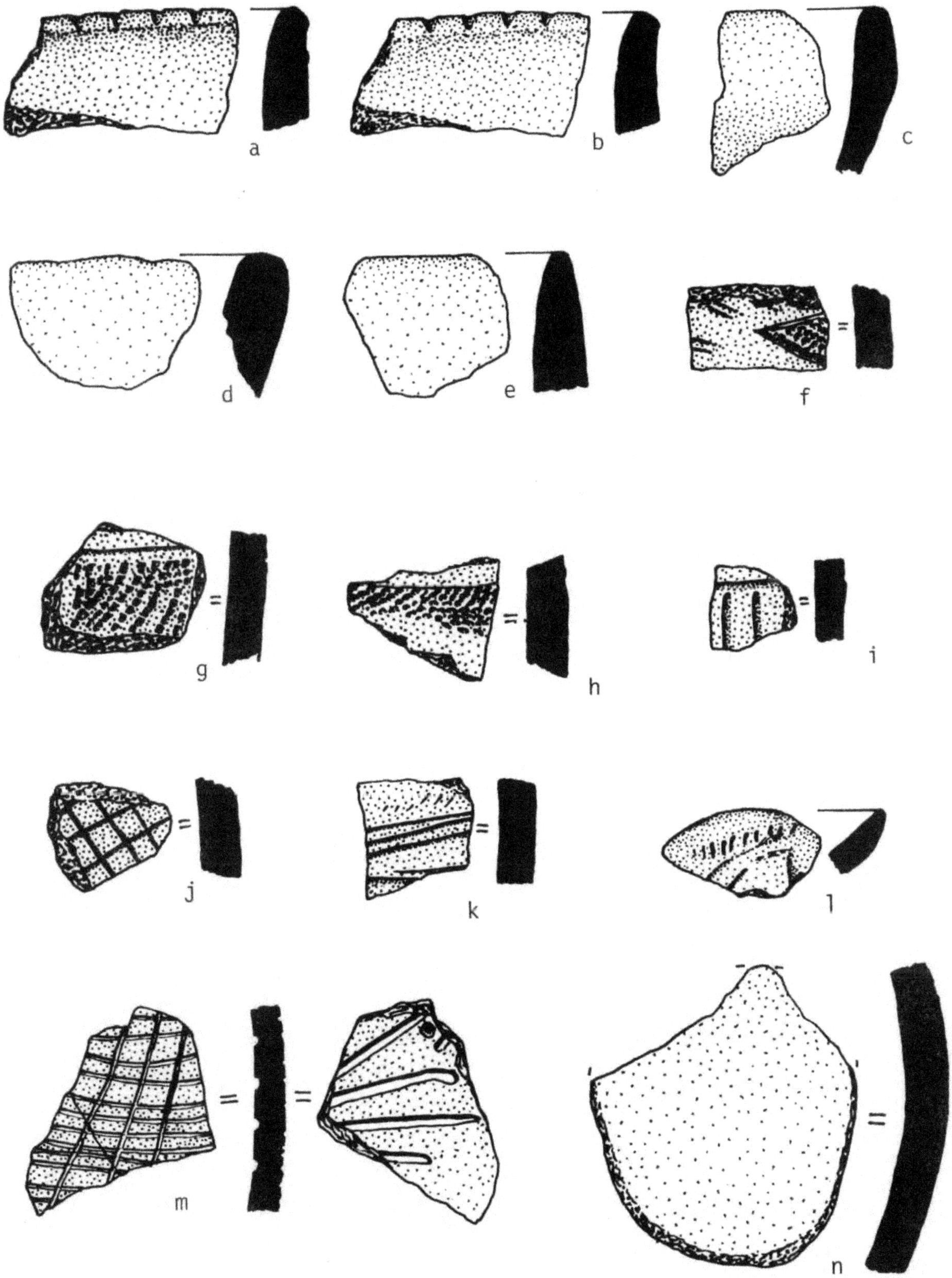

Figure 80. Ileret Stone Bowl Site. Pottery: Excavation.
(a-b) decorated lips; (c-e) undecorated rim; (f-k) decorated body; (l) rim; (m) internal scoring; (n) disc. Natural size.

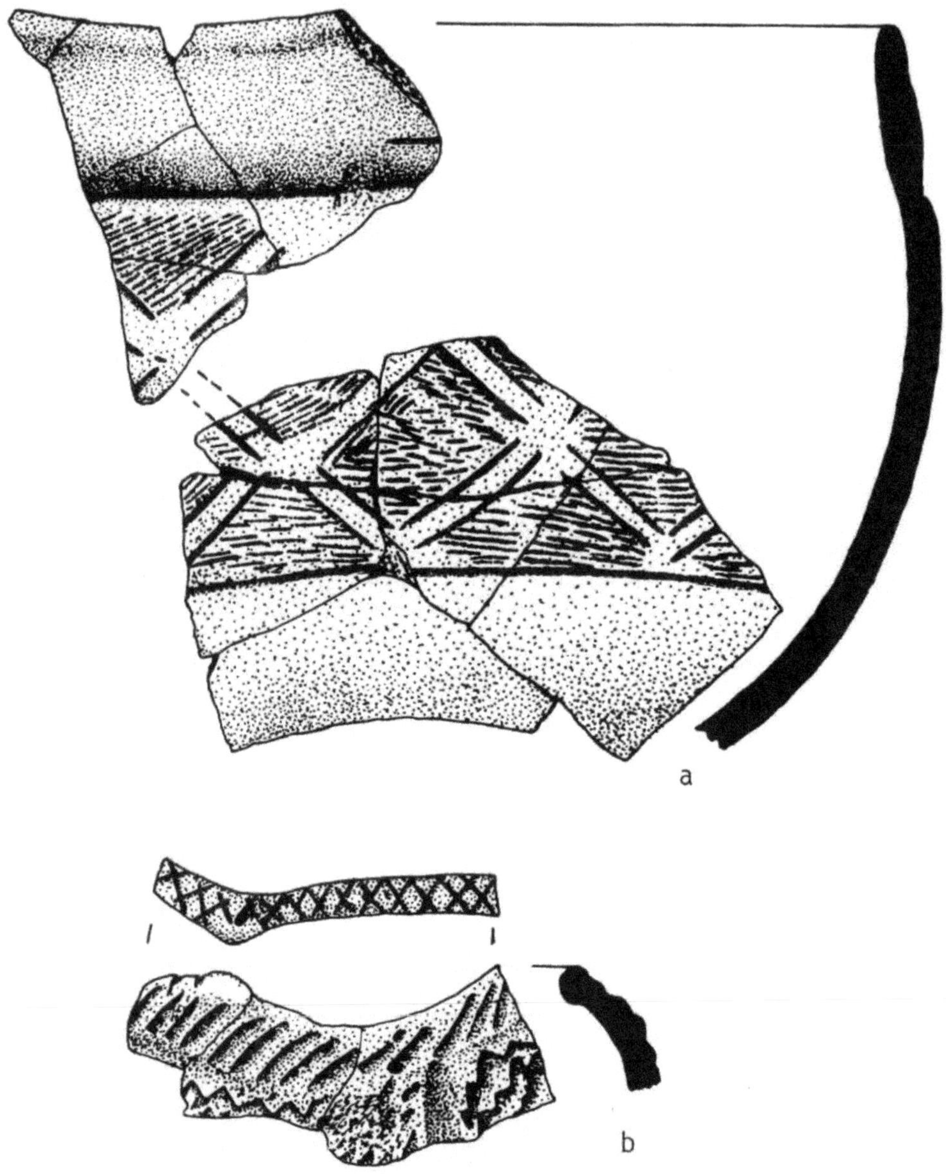

Figure 81. Ileret Stone Bowl Site.
(a) partially reconstructed vessel with comb stamping and incised lines; (b) rim sherds of partially reconstructed pot. 3:4 size.

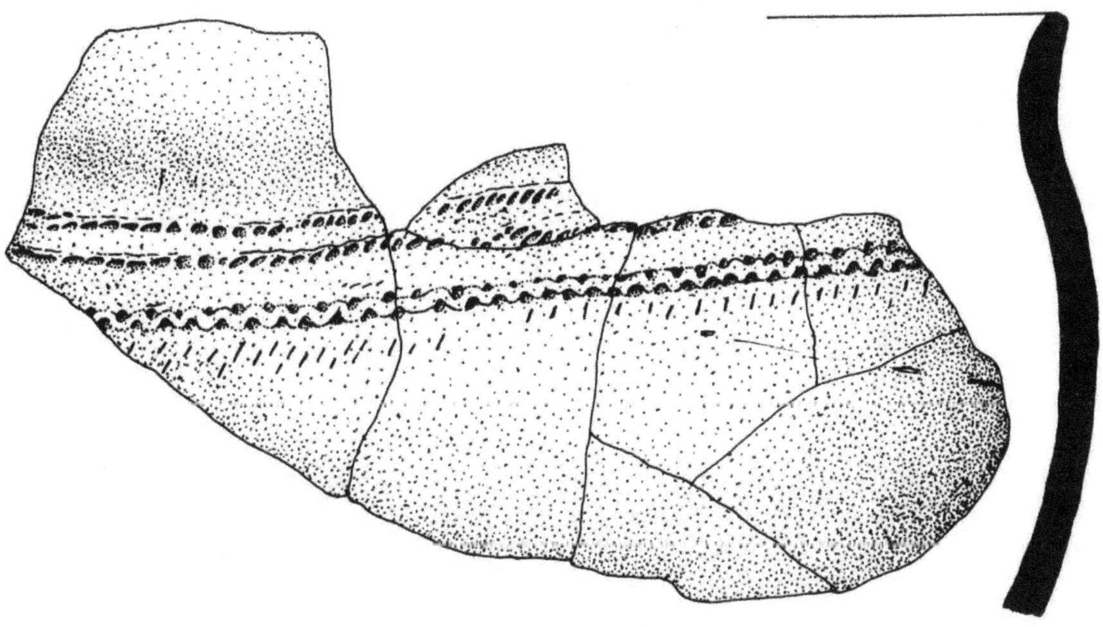

Figure 82. Ileret Stone Bowl Site.
Partially reconstructed vessel with stamped decoration. 3:4 size.

show similarities with the Pastoral Neolithic settlement at Narosura in central Kenya.

FAUNAL REMAINS

The excavated faunal remains were fragmentary and poorly mineralized. Identifications were, as a result, determined entirely from dental remains. Identifiable body parts other than teeth were grouped according to body size. John K. Mengitch of the Kenyan National Museum analyzed the mammal bones. His identifications are summarized in Tables 70 and 71. Domestic caprines and cattle were identified.

A fragmentary maxillae with teeth of a sub-adult cow was identified from Unit B. Most of the dental and postcranial remains however, came from adult animals. The remains of wild bovids and other game animals were not present. Fish bone, based on raw counts only, comprised nearly 40% of the total assemblage. The bulk of fish bone was recovered below the main archaeological horizon from the base of Units B and C. Portions of the fish bone assemblage may represent either discarded food refuse of reworked remains from the underlying Galana Boi beach sediments. The collection of fish bone remains to be analyzed in detail.

Human Remains

Six well preserved cranial fragments were collected at the erosional interface of units D and F approximately 350 m northeast of the excavation area(Fig. 74). Five of the fragments were discovered immediately downslope below the interface contact while the remaining piece was partially imbedded within the erosional gravels. Three of the six cranial fragments were rejoined.

Karen Bell reports (in lit.) that the cranial sutures were not fused and the individual may have attained an age of approximately 15 years. It was not possible to determine the sex of the individual nor was the cause of death evident.

The fragmentary human remains lay stratigraphically below the main occupation horizon. While the fragments were most problably late Pleistocene to Holocene in age, it is not currently possible to assign the cranial fragments to the main level of occupation.

DATING

A sample of fragmentary mammal bone from the archaeological horizon in Unit B was submitted for radiocarbon analysis. Measurement of the apatite fraction (collagen and bone gelatin were not present in sufficient quantity) yielded an age of

4000 ± 140 BP Geochron Labs

This date is in general agreement with the C^{14} determinations from North Horr. There, Ileret type pottery, in apparent association with stone bowls and possibly with domestic stock, was dated to between 3000 and 5000 BP (Phillipson 1977b:71-73). It must be noted however, that single radiocarbon dates, especially on bone apatite, should be viewed with extreme caution. A growing body of evidence has questioned the accuracy of radiocarbon dates run on bone apatite (Michels 1973; Schoeninger and De Niro 1982; Collett and Robertshaw 1983).

Originally, the mammal bone sample from FwJj 5 was to have been divided into equal weights and sent to separate radiocarbon laboratories for bone collagen analysis. The series of radiocarbon results from Dongodien had clearly demonstrated the danger of relying on single bone apatite results. Unfortunately, the sample proved too small to divide equally. Future dating samples, mammal bone and ostrich eggshell, will be submitted.

PRELIMINARY INTERPRETATION

The Ileret Stone Bowl site was occupied by Neolithic pastoralists who tended domestic cattle and caprines and who manufactured stone bowls, microlithic stone tools and various pottery wares. Several pottery designs suggest connections with sites in the Central Rift of Kenya. Site location was along the margins of a major river drainage, probably the proto Il-Erriet, during retreat of the second Holocene lake transgression. The site consisted of one or more occupation surfaces of unknown duration spread across over 30,000 square meters. An age, based on one radiocarbon date, of 3500-4000 years BP is tentatively suggested for the occupation. Confirmatory dates are needed.

Table 60

Ileret Stone Bowl Site

INVENTORY OF CULTURAL REMAINS BY SAMPLING UNITS

Unit	1	2	3	4	5	6	7	8	9	10	11	12	13	14	Within 100 M Radius of the Site	Total (N)
Stone:																
Microliths	3	3	3	2	5	5	3	4	3	3	3	9	3	5	-	54
Scrapers	7	1	6	4	-	8	4	-	2	2	1	3	1	2	-	41
Outils écaillés	1	-	-	1	3	2	1	-	1	4	-	3	2	4	-	22
Bec	-	-	-	-	-	1	1	-	1	1	1	2	1	-	-	8
Cas. trimm.	2	1	-	1	1	-	1	1	-	-	1	3	1	2	-	14
Cas. retouch	-	-	-	-	-	-	2	-	-	-	-	-	-	2	-	4
Whole flakes	21	8	21	20	15	23	12	10	9	13	8	34	9	23	-	226
Broken flakes	31	25	16	16	30	39	16	15	17	21	16	36	18	30	-	326
Angular frags.	39	47	77	32	65	83	49	32	45	66	35	144	35	98	-	847
Cores	2	4	2	5	-	10	4	2	1	-	2	13	-	7	-	52
Pottery:																
Rim: decorated	-	-	1	-	-	2	-	1	1	-	-	3	-	4	-	12
undecorated	-	-	1	-	-	-	-	-	3	4	-	1	-	4	-	13
Body: decorated	1	3	16	6	9	15	14	9	5	14	2	20	4	8	-	126
undecorated	3	20	47	17	23	52	28	35	11	19	11	67	7	33	-	373
Red slip	-	1	-	-	-	4	3	6	1	3	1	25	-	8	-	52
Internally scored	-	-	-	-	-	-	-	1	-	-	-	-	-	-	-	1
Handles	-	-	-	-	-	-	-	2	-	-	-	-	-	-	-	2
Bone/teeth:																
Bone frags. (NID)																
Mammal	-	-	32	6	3	10	7	4	3	-	1	101	6	28	-	201
Fish	-	-	2	2	-	-	-	-	-	-	-	-	-	-	-	4
Bones w/art. sur.	-	-	-	-	-	1	-	-	-	-	1	6	-	-	-	8
Teeth: mammal	-	-	3	-	-	-	-	-	-	-	1	20	3	16	-	43
Stone Bowls:	-	-	-	-	-	-	-	-	-	-	-	1	-	-	5	6
Raw counts (N)																
Stone	106	88	125	81	119	171	93	64	79	110	67	247	70	173	5	(1596)
Bone/teeth	-	-	37	8	3	11	7	4	3	-	3	127	9	44	-	(256)

Table 61

Ileret Stone Bowl Site: Surface

FREQUENCY OF TRIMMED ARTIFACTS

	Chert	Quartz	Obsidian	Total (N)	Total (%)
Microliths:				(52)	(40.0)
Crescents	4	-	-	4	2.8
Curved backed flakes	3	-	1	4	2.8
Straight backed flakes	1	-	-	1	0.7
Truncations	1	-	-	1	0.7
Double backed	2	-	1	3	2.1
Double backed micro-percoir	11	-	1	12	8.4
Miscellaneous	5	-	-	5	3.5
Fragmentary	22	-	2	22	15.4
Scrapers:				(41)	(28.7)
Side	4	-	-	4	2.8
Double side	5	-	-	5	3.5
Three sided	1	-	-	1	0.7
Side-end	1	-	-	1	0.7
Side-bec	1	-	-	1	0.7
Double side-bec	1	-	-	1	0.7
End	4	-	-	4	2.8
Convergent	5	-	-	5	3.5
Notch	10	-	-	10	7.0
Notch-nosed	2	-	1	3	2.1
Other	6	-	-	6	4.2
Outils écaillés	20	1	1	22	15.4
Bec	8	-	-	8	5.6
Cas. trimm.	14	-	-	14	9.8
Cas. retouch	4	-	-	4	2.8
Raw material (N)	135	1	7	143	100.0

Table 62

Ileret Stone Bowl Site: Surface

FREQUENCY OF CORE TYPES

	Chert	Lava	Quartz	Obsidian	Total (N)	Total (%)
Single platform	6	-	-	1	7	13.5
Double: alternate	12	-	-	-	12	23.1
other	2	-	-	-	2	3.8
Multi	10	-	-	-	10	19.2
Disc	2	-	-	-	2	3.8
Casual	6	-	-	-	6	11.5
Fragmentary	12	-	1	-	13	25.0
Raw material (N)	50	-	1	1	52	100.0
(%)	96.2	-	1.9	1.9		

Table 63

Ileret Stone Bowl Site: Surface

FREQUENCY OF MAJOR ARTIFACT CLASSES

	Chert	Lava	Quartz	Obsidian	Total (N)	Total (%)
Trimmed pieces	135	-	1	7	143	9.0
Cores	50	-	1	1	52	3.0
UTFF	1171	48	100	82	1401	88.0
Raw material (N)	1356	48	102	90	1596	100.0
(%)	85.0	3.0	6.4	5.6		

Table 64

Ileret Stone Bowl Site: Excavation

INVENTORY OF CULTURAL REMAINS

	Unit A	Unit B	Plotted B	Unit C	Total (N)
Stone:					
Microliths	-	32	-	2	34
Scrapers	1	9	1	1	13
Outils écaillés	2	22	-	1	25
Bec	-	1	1	-	2
Cas. trimm.	-	4	-	-	4
Cas. retouch	-	2	-	-	2
Whole flakes	13	188	6	10	217
Broken flakes	12	494	12	14	532
Angular frags.	32	547	12	24	615
Cores	-	18	14	1	33
Pottery:					
Rim: decorated	-	21	3	-	24
undecorated	-	33	3	1	36
Body: decorated	16	347	20	22	384
undecorated	42	1045	52	16	1155
Red slip: rim	-	3	-	-	3
body: decorated	-	14	-	-	18
undecorated	4	71	9	-	84
Internally scored	-	1	-	-	1
Disc	-	4	1	-	5
Bone/teeth:					
Bone frags. (NID):					
Mammal	1	391	8	20	420
Fish	5	202	-	112	319
Turtle	-	-	-	-	-
Crocodile	-	-	-	-	-
Bones w/art. sur.	-	9	1	2	12
Teeth:					
Mammal	-	55	4	-	59
Fish	-	5	-	-	5
Other:					
Beads: ostrich e.s.	-	1	-	-	1
Frags: ostrich e.s.	-	35	-	-	35
Unmodified stone	-	1	109	-	110
Haematite fragment	-	1	-	-	1
Calcite crystal	-	2	-	-	2

Table 65

Ileret Stone Bowl Site: Excavation

FREQUENCY OF TRIMMED ARTIFACTS

	Chert	Quartz	Obsidian	Total (N)	Total (%)
Microliths:				(34)	(42.5)
Crescents	5	-	1	6	7.5
Curved backed flakes	2	-	-	2	2.5
Truncations	-	-	2	2	2.5
Miscellaneous	3	-	-	3	3.8
Fragmentary	17	-	4	21	26.3
Scrapers:				(13)	(16.3)
Side	3	-	-	3	3.8
Double side	2	-	-	2	2.5
Side-end	1	1	-	2	2.5
End	1	-	-	1	1.3
Convergent	2	-	-	2	2.5
Notch	2	-	-	2	2.5
Multiple	1	-	-	1	1.3
Outils écaillés	14	4	7	25	31.3
Bec	2	-	-	2	2.5
Cas. trimm.	4	-	-	4	5.0
Cas. retouch	2	-	-	2	2.5
Raw material (N)	61	5	14	80	100.0

Table 66

Ileret Stone Bowl Site: Excavation

FREQUENCY OF CORE TYPES

	Chert	Lava	Quartz	Obsidian	Total (N)	Total (%)
Single platform	4	-	-	-	4	12.1
Double: alternate	3	-	-	1	4	12.1
other	1	-	-	-	1	3.0
Multi	6	1	-	-	7	21.3
Disc	-	-	-	-	-	-
Casual	5	-	2	-	7	21.3
Fragmentary	9	-	1	-	10	30.2
Raw material (N)	28	1	3	1	33	100.0
(%)	84.5	0.3	0.9	0.3		

Table 67

Ileret Stone Bowl Site: Excavation

FREQUENCY OF MAJOR ARTIFACT CLASSES

	Chert	Lava	Quartz	Obsidian	Total (N)	Total (%)
Trimmed pieces	61	-	5	14	80	5.4
Cores	28	1	3	1	33	2.2
UTFF	997	39	137	191	1364	92.4
Raw material (N)	1086	40	145	206	1477	100.0
(%)	73.6	2.7	9.8	13.9		

Table 68

Ileret Stone Bowl Site

INVENTORY OF POTTERY SHERDS

	Surface (579)		Excavation (1706)	
	(N)	(%)	(N)	(%)
Rim sherds	25	4.3	63	3.7
Body sherds	552	95.4	1637	96.0
Handles	2	0.3	-	-
Disc	-	-	5	0.3
Decorated	139	23.9	409	24.0
Slipped	52	9.0	101	5.9
Undecorated	388	67.1	1196	71.1

Table 69

Ileret Stone Bowl Site

INVENTORY OF DECORATED SHERDS BY DESIGN FEATURES

	Excavation (338)			Surface (126)		
	Decorated Body	Red Slip	(N)	Decorated Body	Red Slip*	(N)
Ileret Type	246	-	246	66	-	66
Non Ileret Type:	(74)	(18)	(92)	(60)	-	(60)
Comb stamped	39	12	51	21	-	21
Comb stamped and paneled	12	-	12	13	-	13
Incised	12	4	16	11	-	11
Punctuations	3	2	5	4	-	4
Internally scored	1	-	1	1	-	1
NID	19	-	19	10	-	10

* This attribute was not determined for the surface collection

Table 70

Ileret Stone Bowl Site: Excavation

INVENTORY OF FAUNAL REMAINS

	N	%
MAMMAL		
Bovids	(26)	(3.2)
Caprine	3	0.4
Small bovid (caprine size)	5	0.6
Cattle	2	0.2
Medium bovid (cattle size)	1	0.1
Size not determined	15	1.8
Carnivores		
Small carnivore	1	0.1
Non-identifiable fragments	463	57.1
REPTILE		
Crocodile	1	0.1
Turtle	1	0.1
FISH	319	39.4
Assemblage totals	811	100.0

Table 71

Ileret Stone Bowl Site

INVENTORY OF IDENTIFIABLE BODY PARTS: MAMMAL

Body Part	Caprine	Small Bovid	Cattle	Medium Bovid		Small Carnivore
		(Caprine size)		(Cattle size)	Bovid (size not determined)	
Mandible	-	-	-	-	-	-
Maxillae	-	-	1	-	-	-
Isolated teeth	3	-	1	-	15	1
Femur	-	1	-	-	-	-
Metapodial	-	1	-	-	-	-
Calcanium	-	1	-	-	-	-
Astagalus	-	1	-	-	-	-
Cuneiform	-	-	-	1	-	-
Phalange	-	1	-	-	-	-
Assemblage total (N)	3	5	2	1	15	1

CHAPTER V

OTHER SITE AND LOCALITY REPORTS

Chapter V presents four site reports that describe ceramic and aceramic Later Stone Age occupations. An account of the excavation of two human burials, as well as short descriptions of unsampled archaeological localities, is also included.

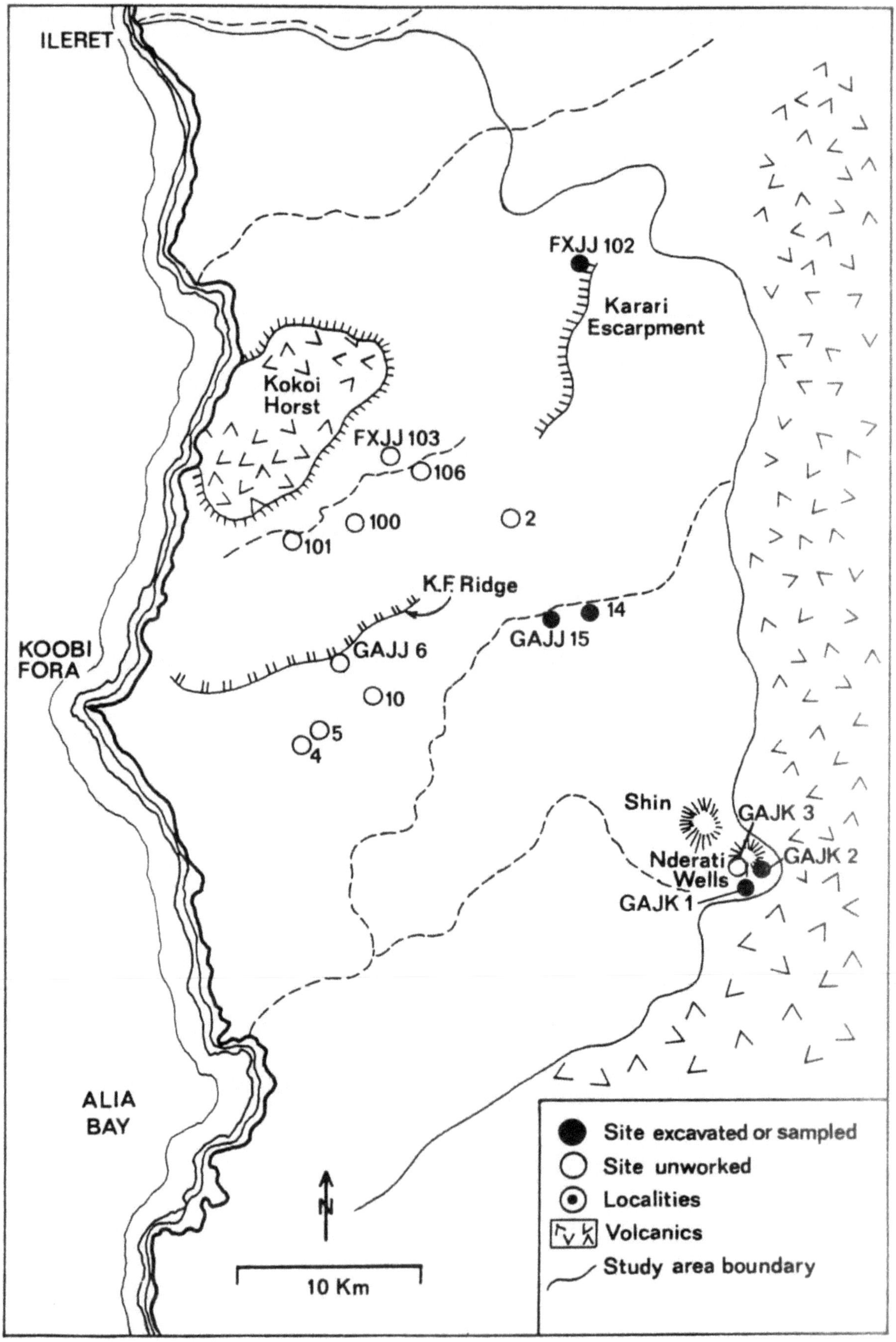

Figure 83. Location of selected sites referred to in Chapter V.

INTRODUCTION

The site of GaJj 14 was located to the northeast of Area 127 along the modern floodplain of the Bakate River (Fig. 83). Microlithic artifacts, pottery and faunal remains were sampled along the eroded margins of a low lying outcrop of coarse gravelly sands.

STRATIGRAPHY

A twisting exposure of coarse sands and gravels lay 20-80 m to the south of the Bakate River. The exposure was approximately 150 m in length and just under one meter in height. Fine grained eolian sands mantled the gravelly horizons. A low but persistent density of cultural remains was continuously distributed along the flanks of the exposure. Concentrated patches of artifacts were present in several areas. The exposure is thought to represent a former river terrace of coarse overbank sediments that was associated with the proto-Bakate River.

SAMPLING

Two areas of artifact concentration were preferentially chosen for sampling (Fig. 84). A total of three 5 m square units were collected.

CULTURAL REMAINS

An inventory of the collected artifacts by sampling unit is presented in Table 72.

Stone

The stone artifact collection had a characteristic Later Stone Age aspect. Microliths, scrapers and *outils écaillés* were identified (Fig. 85). The debitage consisted primarily of micro-blades and flakes manufactured from small nodules of chert. Similar chert nodules may be collected today from gravel screes along the Bakate River. Quantities of lava chippage as well as smaller amounts of quartz and obsidian were also recovered. Several large discolored sandstone blocks, perhaps the remains of a hearth or pottery firing area, were identified near sampling unit 2. No traces of charcoal were present.

Pottery

A small collection (N=70) of fragmentary and weathered pottery sherds with impressed and incised line decorations were identified. One flattened and thickened rim sherd with cross hatching (Fig. 85, j) resembled examples from the Ileret Stone Bowl site. A second rim sherd showed external and internal decoration (Fig. 85, g). Burnished and slipped sherds were not identified. In general, the pottery sample showed design similarities with the group of excavated Pastoral Neolithic sites. However, the sample was not large enough to associate with either a single pottery type or broader, more distinctive ware.

Faunal Remains

A small collection of fragmentary and poorly mineralized mammal bone was recovered. A careful search along the exposure did not yield fish bones. Regrettably, the mammal remains were not identifiable to taxa.

PRELIMINARY INTERPRETATION

The collection of microliths and microlithic debitage indicates the site represents a Later Stone Age occupation that was established along the flood plain of the proto-Bakate River. It was not possible to determine age, spatial distribution of artifacts nor number of occupation episodes. The decorated pottery sherds showed general similarities with several excavated Pastoral Neolithic sites.

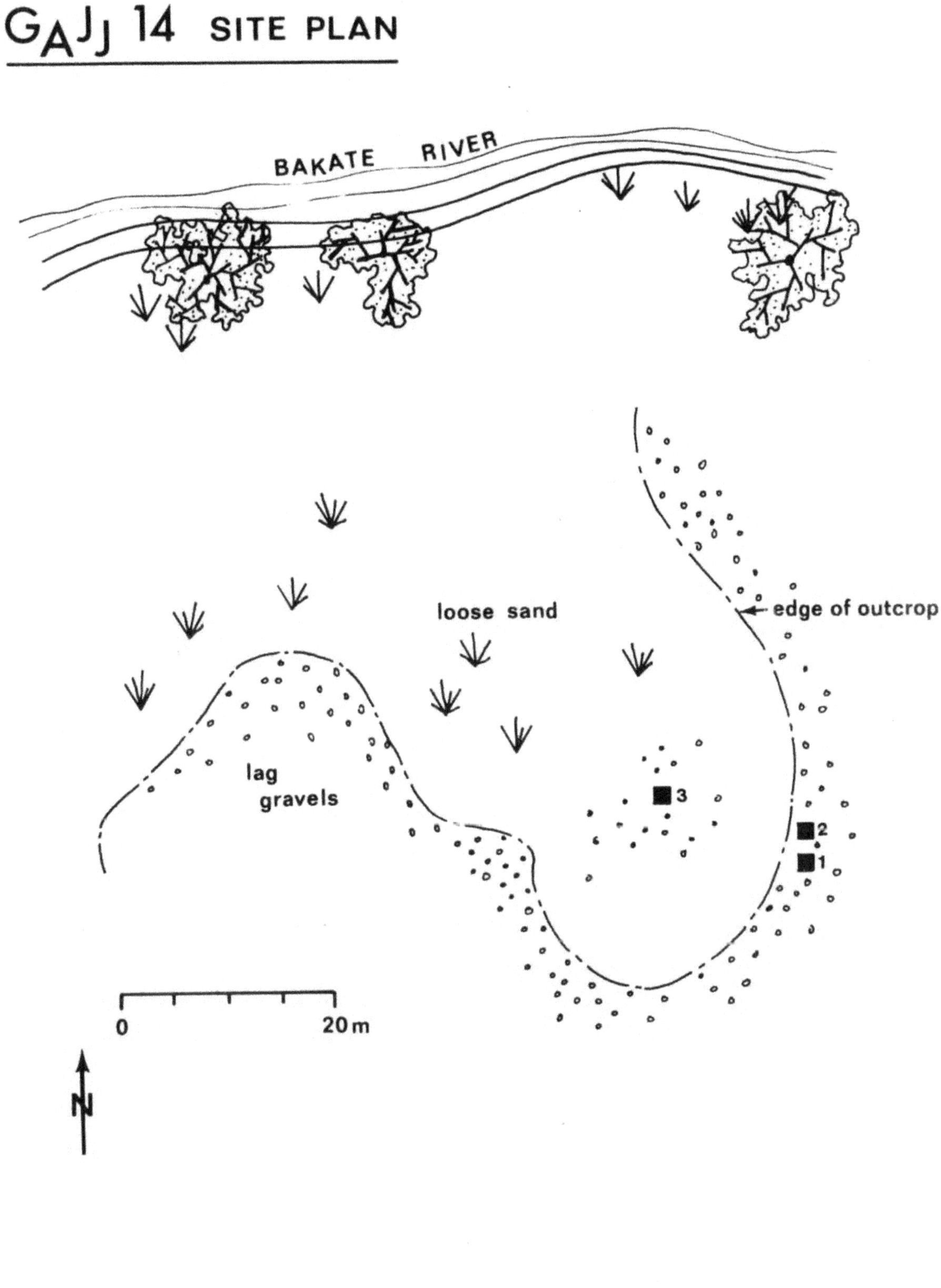

Figure 84. GaJj 14. Sketch of site plan showing location of the sampling units.

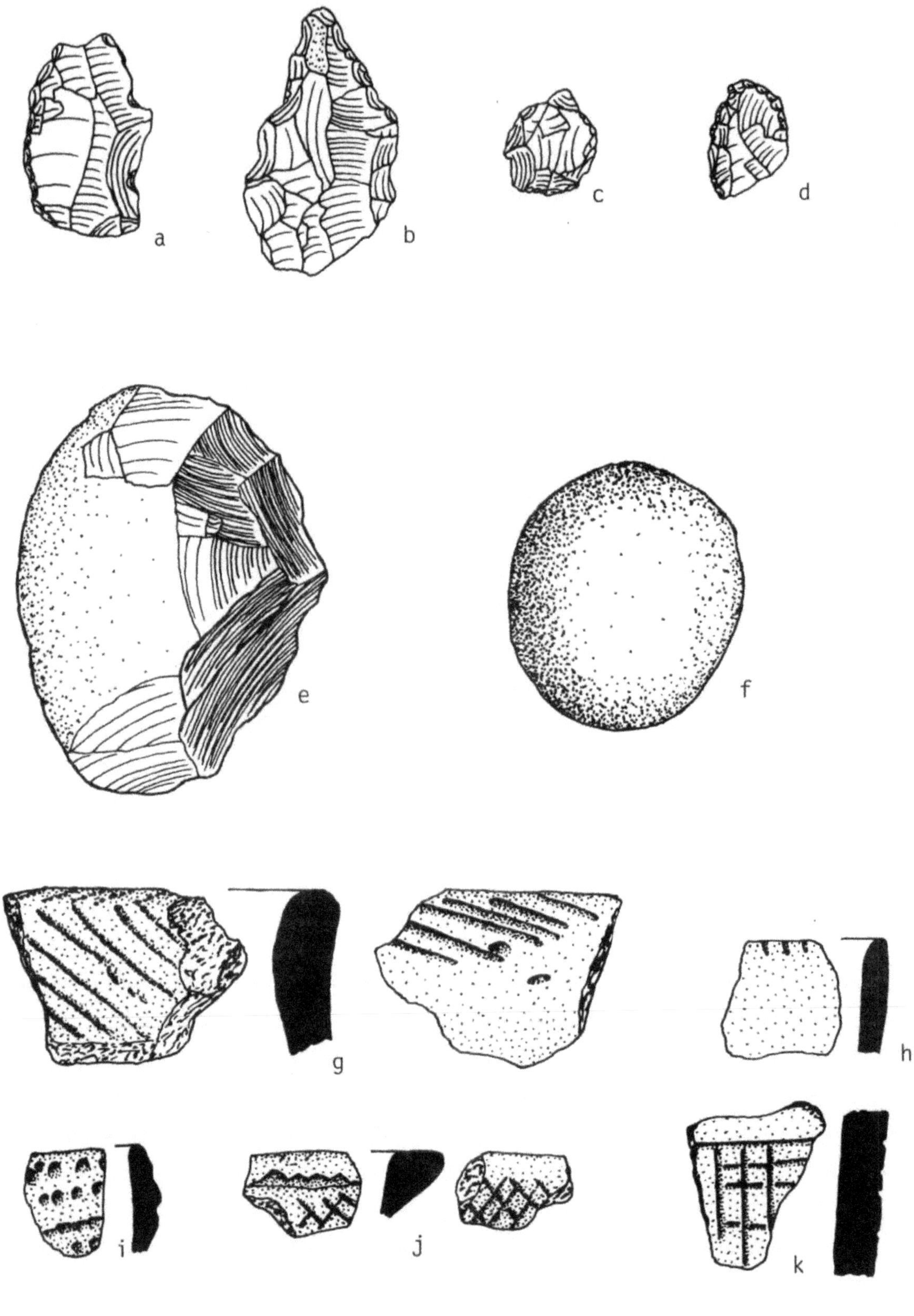

Figure 85. GaJj 14. Stone artifacts and pottery sherds.
(a) curved backed flakes; (b-e) scrapers; (f) hammerstone; (g-j) decorated rim sherds; (k) decorated body. Natural size.

Table 72

GaJj 14

INVENTORY OF CULTURAL REMAINS BY SAMPLING UNIT

	1	2	3	Total (N)
Stone:				
Microliths	3	5	3	11
Scrapers	3	3	2	8
Outils écaillés	1	-	-	1
Cas. trimm.	2	1	1	4
Cas. retouch	2	-	-	2
Whole flakes	14	17	5	36
Broken flakes	31	45	14	90
Angular fragments	31	84	16	131
Cores	13	17	1	31
Hammerstone	1	-	-	1
Pottery:				
Rim: decorated	3	-	-	3
undecorated	-	-	-	-
Body: decorated	9	1	-	10
undecorated	11	15	31	57
Bone:				
Bone frags. (NID): Mammal	5	13	13	31
Bones w/art. sur.	-	2	-	2
Ostrich e.s. frags.	4	3	3	10
Raw counts (N)				
Stone	101	172	42	315
Bone	5	15	13	33

Table 73

GaJj 14

FREQUENCY OF MAJOR ARTIFACT CLASSES

	Chert	Lava	Quartz	Obsidian	Total (N)	Total (%)
Trimmed pieces	22	1	1	2	26	8.3
Cores	29	1	1	-	31	9.9
UTFF	186	40	17	14	257	81.8
Raw material (N)	237	42	19	16	314	100.0
(%)	75.5	13.4	6.1	5.1		

INTRODUCTION

The site of FxJj 102 was situated in Area 129 along the western flank of the Karari Escarpment (Fig. 83). Abundant Later Stone Age lithics and a small scatter of pottery sherds were collected from the surface of a coarse brown sand. The site was sampled and a single altimeter reading indicated the occupation lay 175 m above the June 1976 level of Lake Turkana.

STRATIGRAPHY

Cultural materials were exposed by erosion on top of a coarse, well sorted reddish brown sand (10 R 4/6 - 10 R 3/4) that uncomformably overlaid Upper Member beds of the Koobi Fora Formation. The sands, of mixed fluvial and eolian origin, were variable in thickness but did not exceed one meter. Numerous stone artifacts and unmodified basalt cobbles were clustered within an area coincident with a network of small erosional gullies. The majority of the artifacts were concentrated within the gully area but a low density scatter also occurred over a larger region approximately 200 m north-south by 100 m east-west. The Ileret Road ran directly through the site (Fig. 86) and pottery sherds were collected just across the road on the eastern side. The site's upland location would have offered panoramic views of Lake Turkana as well as the surrounding volcanic highlands.

A small cluster of levallois flakes, cores and several bifacial MSA points occurred along the outcrop's western margin. Several well mineralized faunal remains, including a worn proboscidian tooth fragment, were associated with the stone artifacts. A Middle Stone Age occupation apparently underlies the well exposed scatter of Later Stone Age material. Intermixing of the two occupation sufaces has occurred along the edge of the escarpment where erosion has removed the upper coarse sand.

SAMPLING

A modified version of the systematic checkerboard sampling technique was chosen for the site (Redman 1974). An area of artifact concentration, 25 m by 20 m, was chosen for sampling. Five rows of sampling units were gridded and within each row, two units were marked for recovery. A total of 10 units, each a 5 m square, was collected.

CULTURAL REMAINS

An inventory of the artifacts by sampling unit is presented in Table 74.

Stone (Tables 75-77; Fig. 87)

Nearly 3000 stone artifacts were collected from the sampling units. A wide variety of scraper types, including notched and composite forms, accounted for over 40% of the trimmed pieces (Fig. 87,g-j). Crescents were the most common of the microlithic tools. Outils écaillés, some comparatively large, were manufactured from chert and quartz cobbles (Fig. 87, k-n). Eight quartz hammerstones, a bifacial lava chopper, double pitted anvil stone and a broken grindstone were also recovered. The debitage consisted primarily of single platform side struck flakes made from water worn cobbles of chert and quartz. Few examples of blades or flake blades were identified. A small number of patinated and desilicified flakes were also collected. These pieces are thought to be derived from the underlying MSA occupation and were not included in the final inventory.

Pottery

A small group of pottery sherds was discovered 30 m east of sampling unit 2 (Fig. 88). The sherds were directly associated with debitage similar to that collected from the sampling area and were most probably associated with the Later Stone Age occupation. The sherds exhibited a sandy paste and large angular grains of quartz were used as temper. Two rim sherds had tapered and bevelled lips (Fig. 88, a-b). Decorated body sherds included triangular and crescentic impressions. Several additional body sherds showed parallel grooved lines. The bevelled rim sherd and rows of crescentic impressions resembled sherds from the Ileret Stone Bowl site. However, the collection was too small and fragmentary to be assigned to a specific type or ware.

Faunal Remains

The faunal collection consisted of only a few highly fragmentary and unidentifiable elements.

PRELIMINARY INTERPRETATION

The site apparently represents, like GaJj 14, a Later Stone Age encampment with pottery that was situated between the former high Holocene lake shoreline and the eastern volcanic highlands. Some of the recovered sherds show affinities with the nearby Pastoral Neolithic site of FwJj 5. The inhabitants may have herded domestic stock, but direct evidence was not collected. The high percentage of scraper forms was unusual among the Later Stone Age sites in the research area and may signify that specialized activities, such as animal butchery or woodworking, were taking place (Binford 1979). While the site may well represent the remains of a former encampment of early pastoralists, the absence of radiocarbon dates, domestic animals and only fragmentary undiagnostic pottery precludes positive assignment to the Pastoral Neolithic group.

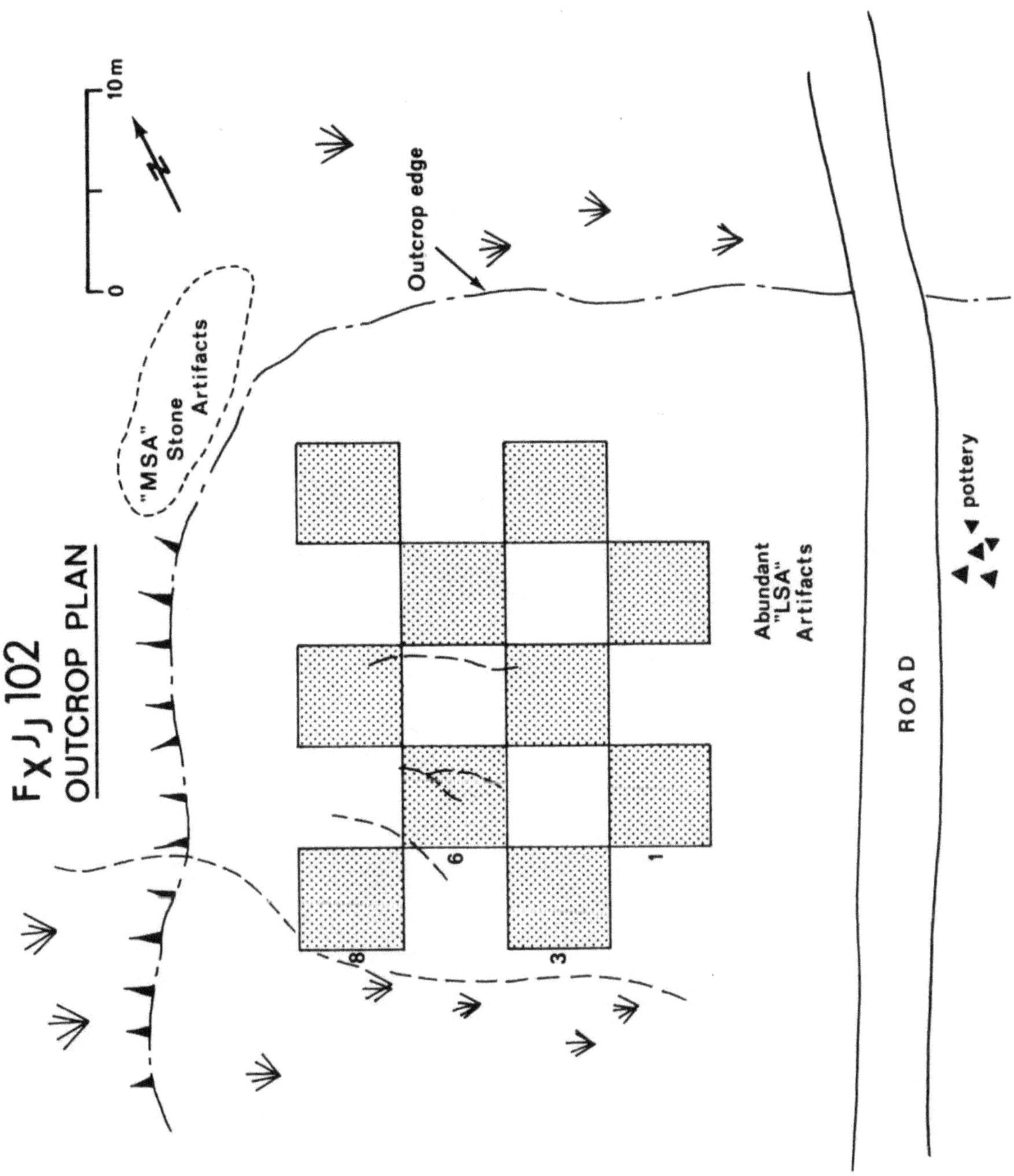

Figure 86. FxJj 102. Sketch plan showing location of sampling units.

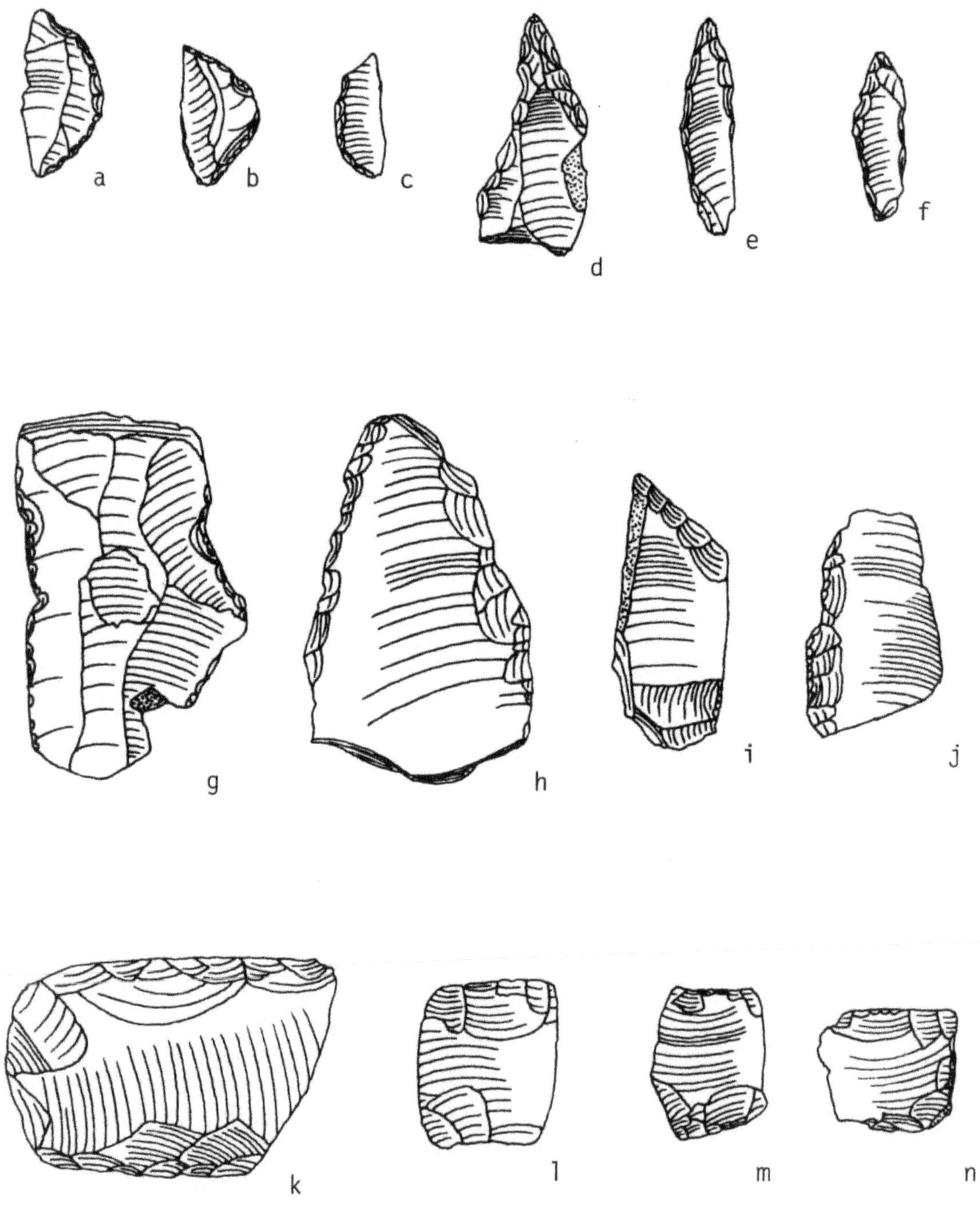

Figure 87. FxJj 102. Stone Artifacts.
(a-c) crescents; (d-f) awls; (g-j) scrapers; (k-m) <u>outils écaillés</u>.
Natural size.

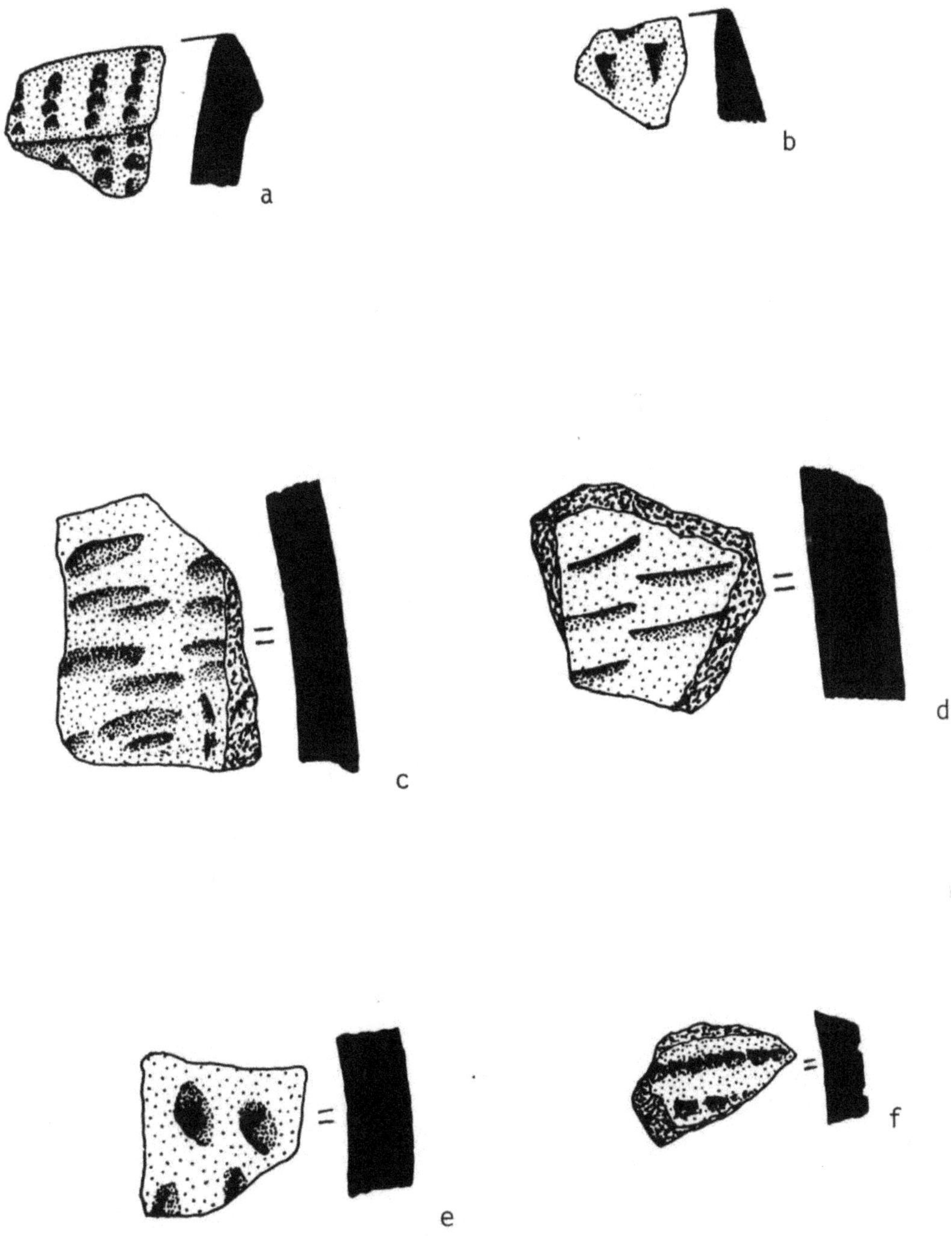

Figure 88. FxJj 102. Pottery.
(a-b) decorated rim sherds; (c-f) decorated body. Natural size.

Table 74

FxJj 102

INVENTORY OF CULTURAL REMAINS BY SAMPLING UNIT

Unit	1	2	3	4	5	6	7	8	9	10	Outside of Units	Assemblage Total (N)
Stone:												
Microliths	-	-	-	-	3	3	3	6	1	7	-	23
Scrapers	-	1	3	1	8	15	9	26	14	8	-	85
Outils écaillés	-	-	-	2	6	11	9	18	2	4	-	49
Awls	-	-	-	-	-	2	-	-	4	-	-	6
Becs	1	-	-	1	-	4	1	5	1	-	-	13
Core/chopper	-	-	-	-	-	1	-	-	-	-	-	1
Cas. trimm.	-	-	1	1	1	2	1	9	3	1	-	19
Cas. retouch	-	-	1	-	1	1	1	8	-	3	-	15
Hammerstones	-	-	1	-	-	-	-	6	-	1	-	8
Grindstone	-	-	1	-	-	-	-	-	-	-	-	1
Anvil	-	-	-	-	1	-	-	-	-	-	-	1
Whole flakes	8	1	12	5	55	106	39	154	83	54	-	517
Broken flakes	16	3	19	14	61	168	80	234	173	151	-	919
Angular frags.	23	8	23	20	79	214	72	244	133	108	-	924
Cores	11	3	32	33	43	48	58	83	64	24	-	399
Pottery:												
Rim: decorated	-	-	-	-	-	-	-	-	-	-	1	1
Body: decorated	-	-	-	-	-	-	-	-	-	-	17	17
undecorated	-	-	-	-	-	-	-	-	-	-	2	2
Bone/Teeth:												
Bone frags. (NID) mammal	-	-	-	2	1	-	1	-	-	-	-	4
Bones w/artic. sur.	-	-	-	-	-	-	1	-	-	-	-	1
Teeth: mammal	-	-	-	-	-	-	1	-	-	-	-	1
Ostrich e.s. frags.	-	-	-	-	1	-	-	-	-	-	-	1
Raw counts (N) Stone:	59	16	93	77	258	575	270	793	478	361	-	2970
Bone/teeth:	-	-	-	2	1	-	3	-	-	-	-	6

Table 75

FxJj 102

FREQUENCY OF TRIMMED PIECES

	Chert	Lava	Quartz	Obsidian	Total (N)	Total (%)
Microliths:					(23)	(10.9)
Crescents	9	-	-	-	9	4.3
Truncations	3	-	-	-	3	1.4
Miscellaneous	4	-	-	-	4	1.9
Fragmentary	7	-	-	-	7	3.3
Scrapers:					(85)	(40.3)
Side	8	-	-	-	8	3.8
Double side	2	-	-	-	2	0.9
End: simple	7	-	-	-	7	3.3
oblique	2	-	-	-	2	0.9
concave	1	1	-	-	2	0.9
steep	2	-	-	-	2	0.9
Semi-circular:	3	-	-	-	3	1.4
concave	3	-	-	-	3	1.4
double concave	1	-	-	-	1	0.5
Notch: single	34	-	1	-	35	16.6
double	1	1	-	-	2	0.9
multiple	7	-	-	-	7	3.3
Nosed	1	-	-	-	1	0.5
Composite	9	-	-	-	9	4.3
Miscellaneous	1	-	-	-	1	0.5
Outils écaillés	36	-	11	2	49	23.2
Awls	6	-	-	-	6	2.8
Bec	13	-	-	-	13	6.2
Core/chopper	-	1	-	-	1	0.5
Cas. trimm.	19	-	-	-	19	9.0
Cas. retouch	15	-	-	-	15	7.1
Raw material (N)	194	3	12	2	211	100.0

Table 76

FxJj 102

FREQUENCY OF CORE TYPES

	Chert	Lava	Quartz	Obsidian	Total (N)	Total (%)
Single platform	39	6	1	-	46	11.5
Double: alternate	34	1	23	-	58	14.5
other	38	-	5	-	53	13.3
Multi	85	3	6	-	94	23.6
Disc	25	3	3	-	31	7.8
Casual	74	3	8	-	85	21.3
Fragmentary	27	4	1	-	32	8.0
Raw material (N)	332	20	47	-	399	100.0
(%)	83.2	5.0	11.8	-		

Table 77

FxJj 102

FREQUENCY OF MAJOR ARTIFACT CLASSES

	Chert	Lava	Quartz	Obsidian	Total (N)	Total (%)
Trimmed pieces	194	3	12	2	211	7.1
Cores	332	20	47	-	399	13.4
UTFF	1799	175	383	3	2360	79.5
Raw material (N)	2325	198	442	5	2970	100.0
(%)	78.3	6.7	14.9	0.1		

INTRODUCTION

The site of GaJj 15 was located to the northeast of Area 127 along the crest of an escarpment overlooking the Bakate River (Fig. 83). Abundant Later Stone Age lithic artifacts and a low density scatter of bone fragments were concentrated across the escarpment's surface. Pottery sherds were not present. Three stone structures were situated near the northwest perimeter of the ridge top. The area of artifact concentration was sampled.

STRATIGRAPHY

The occupation area was located on the crest of an undisected, elevated remnant of earlier Plio-Pleistocene sedimentary deposits. Lava cobbles mantled the surface of the escarpment. The scatter of stone artifacts measured approximately 5,000 square meters and lay directly upon the pavement of volcanic cobbles. Two brief excavation attempts did not yield artifacts beneath the cobble horizon.

SAMPLING

A linear transect sampling technique was used at the site. The transect, 65 m long and 5 m wide, was located within a concentrated area of stone artifacts (Fig. 89). Five sampling units, beginning arbitrarily with the first unit, were collected for every third meter square.

CULTURAL REMAINS

An inventory by sampling unit of the recovered artifacts is presented in Table 78.

Stone (Tables 79-81)

Over 2500 stone artifacts were collected from the sampling units. Microliths accounted for the majority of the trimmed pieces followed by *outils écaillés* and scrapers. Crescents were well represented (Fig. 90, f-m). A diverse series of scraper types, including notched and semicircular forms, were recovered (Fig. 91, c-d). Many *outils écaillés* showed parallel sided scar ridges and the collection formed a morphological continuum with bipolar cores (Fig. 91, j-p). The debitage consisted primarily of blades and flake-blades struck from small nodules of chert. The distinctive small size of the debitage, selection of chert nodules, and high percentage of microliths is comparable to the lithic assemblages from Nderati (GaJk 2) and Buluk (FwJk 2) Wells.

Three stone structures or cairns, occurred near the escarpment's northern perimeter. Each structure was approximately 3 m wide and 50 cms in height. One cairn was enclosed by a circle, 7 m in diameter, of large angular sandstone blocks. Numerous quartz pebbles and

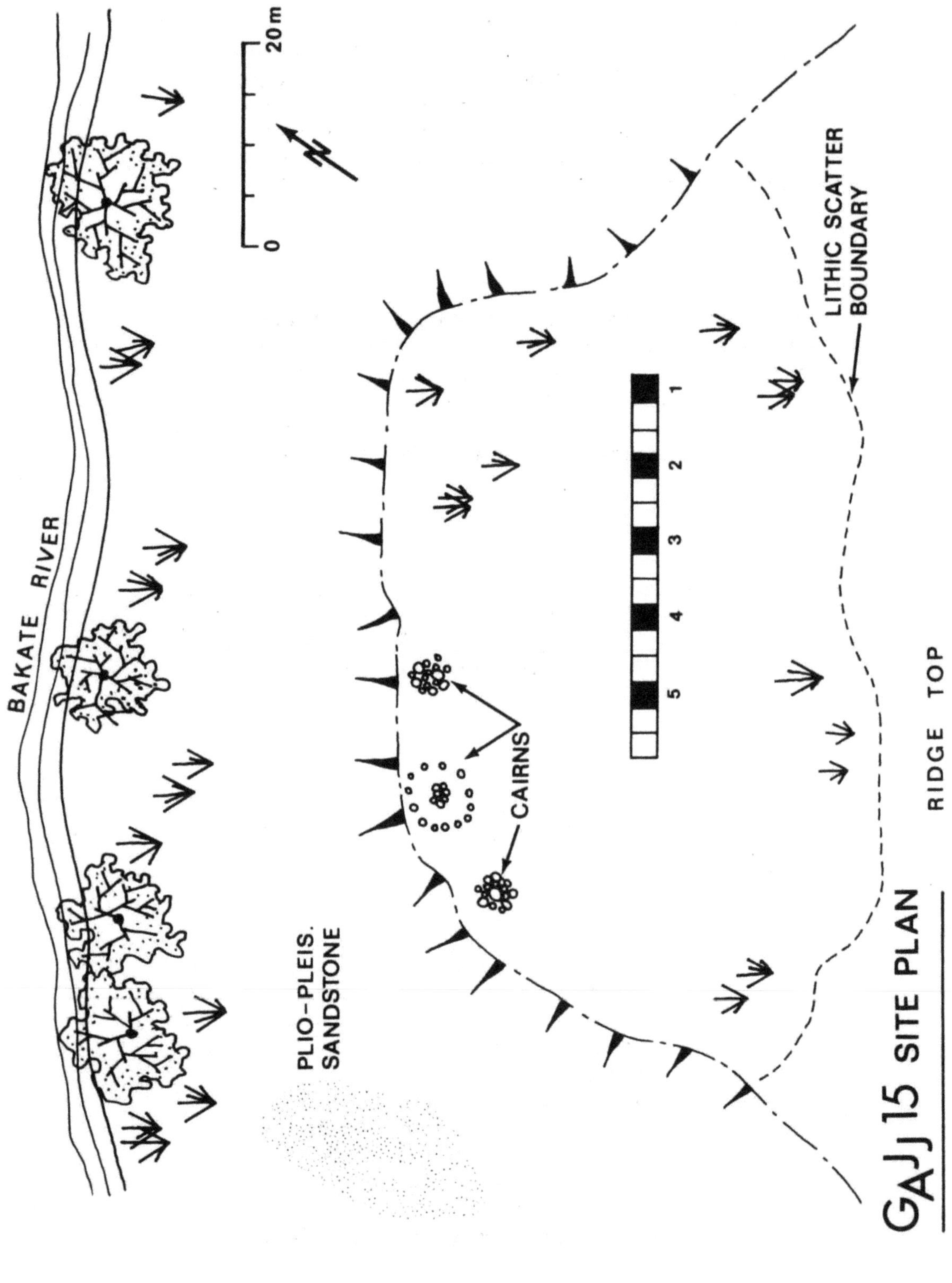

Figure 89. GaJj 15. Sketch plan showing location of sampling units.

chert debitage were concentrated within the circle. The placement of the brightly colored stones was apparently intentional and produced a striking visual effect.

The age and cultural affinity of the stone structures remains unclear. Soper and Jacobs (1972), based on historical and ethnographic evidence, tentatively concluded that a Cushitic speaking people, perhaps groups of Rendille or Borana, were responsible for many of the stone structures in the vicinity of Koobi Fora. They report:

> . . . in the event of the death of a "ritual expert," the Borana often construct a circular wall of stones around the plain cairn of stones which covers the grave; and . . . that white, translucent stones, such as quartz or silicate pebbles, are commonly thrown on burial cairns by passing Borana as a mark of respect and honour. (p. 17 based on information provided by Paul Baxter).

A thorough search of the immediate area surrounding each stone structure did not yield either human bones or associated grave goods. After much deliberation, I decided not to excavate any of the structures. Such an undertaking would have required detailed analysis of the recent history and ethnography of the peoples in the East Turkana area and therefore was outside the scope of the research project.

Faunal Remains

A small collection of well mineralized mammal bone and tooth fragments were recovered. Unfortunately, the fragments could not be identified to either taxa or body part.

PRELIMINARY INTERPRETATION

The site represents a Later Stone Age occupation that lacked bone harpoons and pottery and was situated near the eastern margin of the sedimentary basin. The lithic assemblage, especially the manufacture of microliths from small chert nodules, revealed marked similarities with the lithic collections from several oases (Nderati and Buluk Wells, Karsa Waterhole) within the eastern volcanic highlands. These sites suggest the presence of an aceramic late Pleistocene-Holocene cultural tradition that may have preceded or coexisted with early fishing and/or pastoralist groups. Three associated stone cairns were probably of recent origin, possibly constructed by contemporary Cushitic speaking peoples.

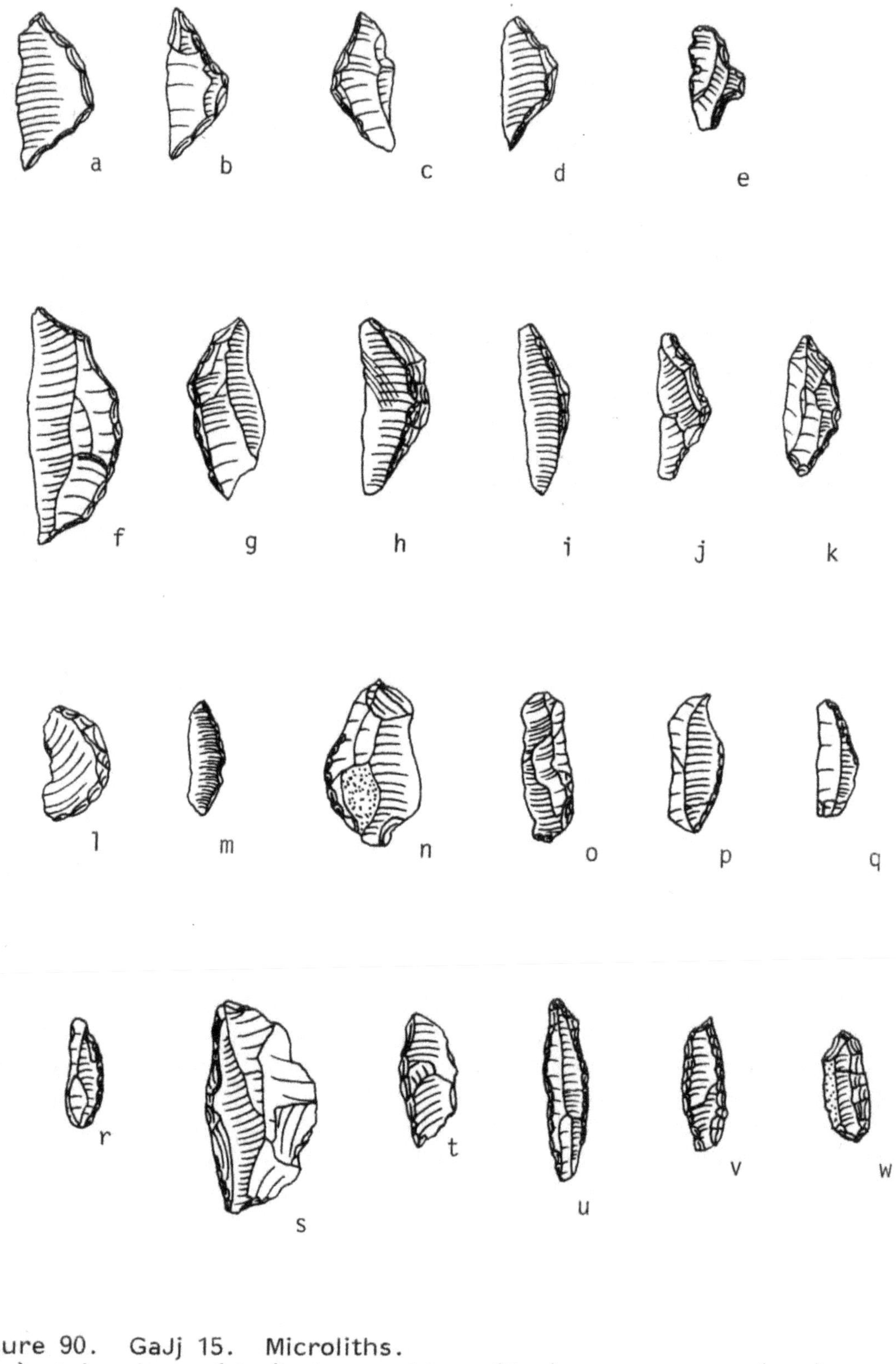

Figure 90. GaJj 15. Microliths.
(a-c) triangles; (d-e) trapezoids; (f-m) crescents; (n-r) curved backed flakes; (s-t) straight backed; (v-w) double backed. Natural size.

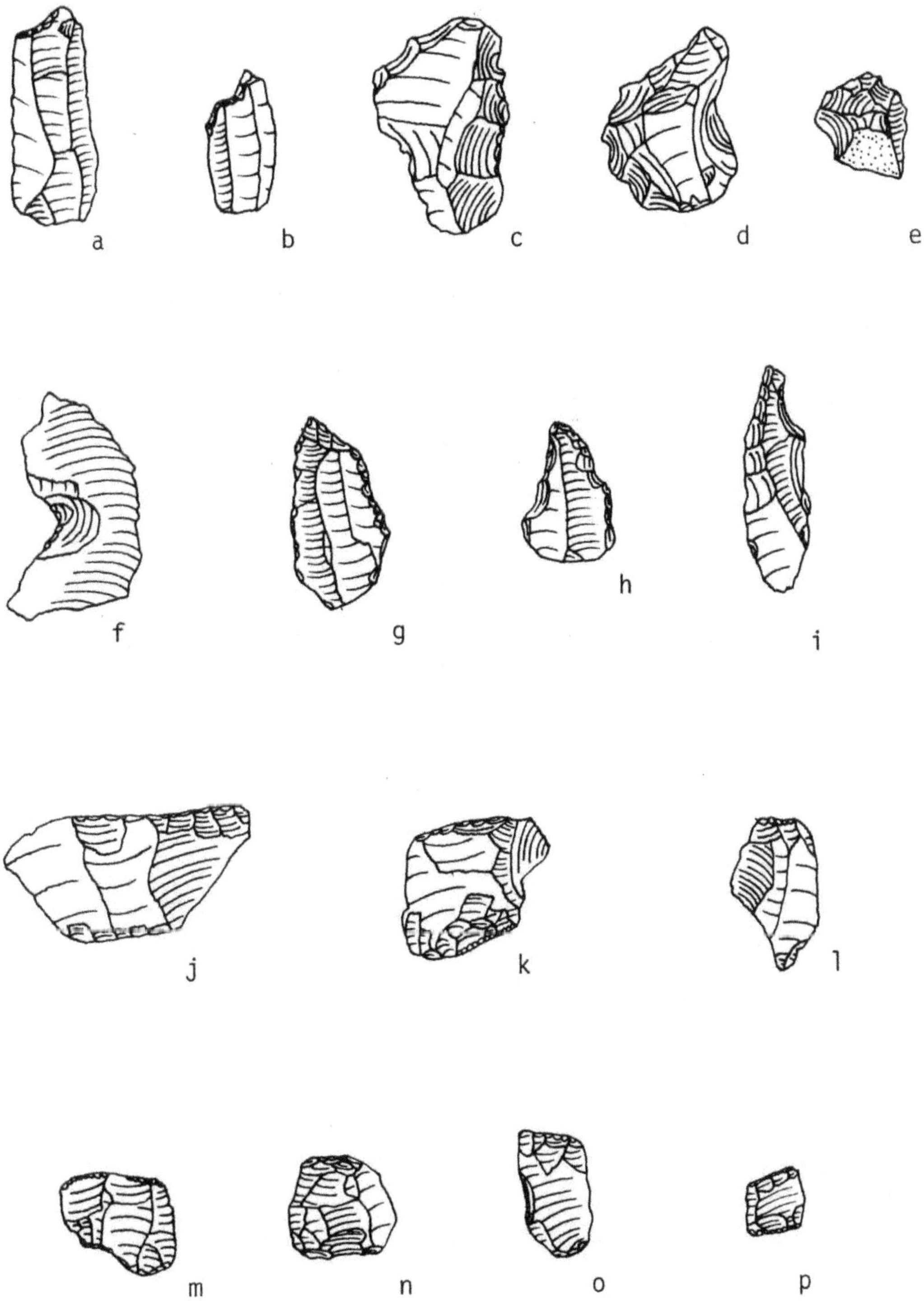

Figure 91. GaJj 15. Stone artifacts.
(a-b) truncations; (c-e) scrapers; (f-g) awls; (h) composite tool - broken; (i-o) outils écaillés. Natural size.

Table 78

GaJj 15

INVENTORY OF CULTURAL REMAINS BY SAMPLING UNIT

Unit	1	2	3	4	5	Total (N)
Stone:						
Microliths	24	10	38	20	35	127
Scrapers	2	2	7	7	3	21
Outils écaillés	4	12	14	6	3	39
Bec	1	1	-	1	-	3
Cas. trimm.	1	1	7	4	6	19
Whole flakes	35	20	122	75	94	346
Broken flakes	69	82	293	193	226	863
Angular frags.	63	89	345	156	275	928
Cores	18	13	98	78	44	251
Bone frags. (NID):						
Mammal	1	7	-	-	-	8
Teeth:						
Mammal	-	-	5	-	-	5
Ostrich e.s. frags.	1	-	-	-	-	1
Raw counts (N)						
Stone	217	230	924	534	686	2597
Bone/teeth	1	7	5	-	-	13

Table 79

GaJj 15: Surface

FREQUENCY OF TRIMMED PIECES

	Chert	Lava	Quartz	Total (N)	(%)
Microliths:				(127)	(60.8)
Triangle	3	-	-	3	1.4
Trapezoid	2	-	-	2	1.0
Crescent	30	-	-	30	14.4
Curved backed flakes	17	-	-	17	8.1
Straight backed flakes	3	-	-	3	1.4
Truncations	4	-	-	4	1.9
Double backed flakes	2	-	-	2	1.0
Micro-percoir (double ended)	5	-	-	5	2.4
Miscellaneous	9	-	-	9	4.3
Fragmentary	52	-	-	52	24.9
Scrapers:				(21)	(10.0)
Side: straight	1	1	-	2	1.0
concave	2	-	-	2	1.0
convex	1	-	-	1	0.5
Double side	1	-	-	1	0.5
End	4	-	-	4	1.9
Notch	9	-	-	9	4.3
Semi-circular	2	-	-	2	1.0
Outils écaillés	38	-	1	39	18.7
Bec	3	-	-	3	1.4
Cas. trimm.	18	1	-	19	9.1
Raw material (N)	206	2	1	209	100.0

Table 80

GaJj 15

FREQUENCY OF CORE TYPES

	Chert	Lava	Quartz	Obsidian	Total (N)	(%)
Single platform	21	-	2	-	23	9.2
Double: alternate	78	-	3	-	81	32.3
other	23	1	-	-	24	9.6
Multi	34	-	1	-	35	13.9
Disc	13	-	-	1	14	5.6
Casual	34	-	1	-	35	13.9
Fragmentary	38	1	-	-	39	15.5
Raw material (N)	241	2	7	1	251	100.0
(%)	96.0	0.8	2.8	0.4		

Table 81

GaJj 15

FREQUENCY OF MAJOR ARTIFACT CLASSES

	Chert	Lava	Quartz	Obsidian	Total (N)	(%)
Trimmed pieces	206	2	1	-	209	8.0
Cores	241	2	7	1	251	9.7
UTFF	1937	48	137	15	2137	82.3
Raw material (N)	2384	52	145	16	2597	100.0
(%)	91.8	2.0	5.6	0.6		

INTRODUCTION

The site of GaJk 2 was situated outside the Plio-Pleistocene sedimentary basin and within the eastern volcanic highlands (Fig. 83). Numerous stone artifacts and fragmentary bone debris were concentrated on the crest of a steep side hill approximately 1/4 km southwest of Nderati Wells (Fig. 92). Pottery sherds were not present. Near the northwestern margin of the hill, patches of fine grained sandy silts and associated mollusc shells were exposed. A radiocarbon shell date of 13,440±640 BP indicated Nderati Wells held a substantial body of water during the late Pleistocene.

STRATIGRAPHY

The area of occupation was located on the crest of a steeply rising 200 m high volcanic hill. The surrounding volcanic mountains are primarily composed of Miocene and Upper Pliocene basalts. Nearby Nderati mountain has been dated to the Lower Miocene and is composed primarily of a rhyolitic plug (Fitch and Vondra 1976). Today, the site's location provides an excellent view of Nderati Wells, the surrounding river drainages and Lake Turkana. Subsurface water, presumably trapped by impermeable volcanic strata, currently provides moisture for the Well's vegetation and also furnishes the principal water supply for base camp at Koobi Fora. Remnants of cemented caliche deposits outcrop discontinuously along the northeast perimeter of the steeply rising hill, indicating the former presence of a fluctuating water table.

The surface of the site was mantled by a thin veneer of light brown silt, large volcanic cobbles and rock fragments. The underlying strata, based on two excavation attempts, consisted almost entirely of volcanic rock. Cultural remains were therefore limited to the uppermost several centimeters of the surface layer.

SAMPLING

A linear transect sampling technique was used at the site. The transect, 180 m by 5 m, was located in an area of concentrated artifacts and was divided into thirty-six 5 m square units. Beginning with the first unit, every other square was collected. A total of 18 units was sampled (Fig. 92).

CULTURAL REMAINS

An inventory of the recovered artifacts by sampling unit is presented in Table 82.

Stone (Tables 83-85)

Microliths accounted for nearly 80% of the shaped tools followed by casually trimmed artifacts and scrapers. A wide variety of

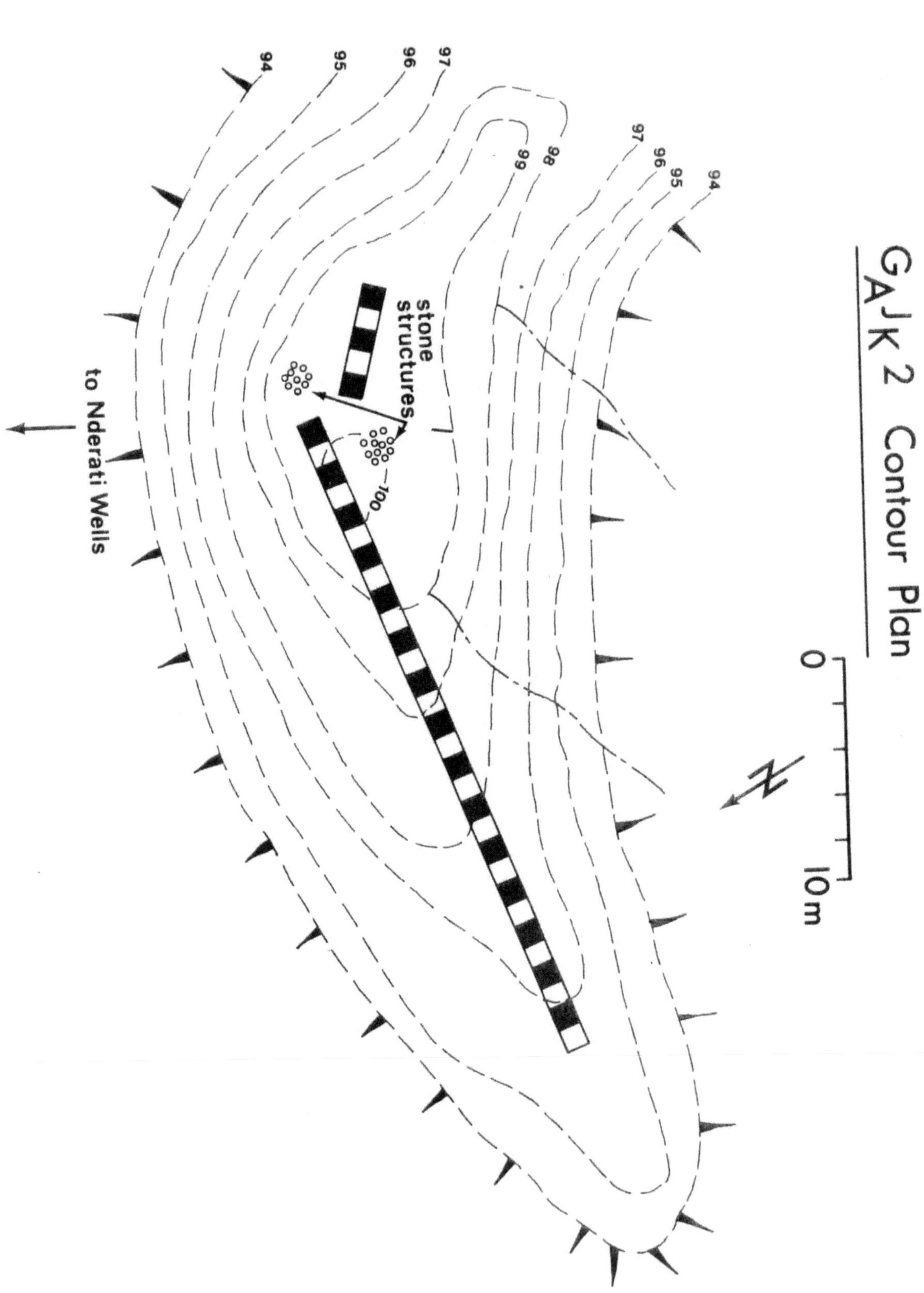

Figure 92. GaJk 2. Contour plan showing location of sampling units.

microlithic types was represented (Fig. 93). Outils écaillés and bipolar cores showed definite intergradation. The stone artifacts were significant because of their small size and almost exclusive selection (96%) of small nodules of chert as raw material. A high proportion of small blades and flake blades had parallel sided flake scars.

A second group of stone artifacts, including core/choppers, polyhedrons, heavy duty scapers and debitage was identified but not collected. All of the artifacts were manufactured from various lavas and were substantially weathered. The pieces were morphologically similar to artifacts from the Karari Industry (J.W.K. Harris 1978) and appear to represent traces of early hominid activity in the area.

Faunal Remains

A small collection of very minute bone fragments was recovered from the sampling units. It was not possible to identify either taxa or body parts from the remains.

DATING

At the base of the steeply rising hill there occurred several horizons of light brown silts with associated unbroken mollusc shells. The molluscs included *Melanoides tuberculata* and *Biomphalaria cf. pfeifferi*. A shell sample yielded a radiocarbon date of:

13,440 ± 640 BP   Geochron Laboratories.

The shell sample was noteworthy for several reasons. First, *Biomphalaria cf. pfeifferi* is the intermediate host of intestinal Bilharsiosis (*Schistoma mansoni*). *Biomphalaria* occurs usually in freshwater pools or slow running small streams, localities often used as drinking places (D. Van Damme pers. comm.). Such water resources, to judge from the sediments at Nderati, were present near GaJk 2. While the occurrence of *Biomphalaria* does not necessarily imply the presence of the parasite, there is a possibility the disease was present during the late Pleistocene and Holocene.

Second, the single radiocarbon date indicates that the Nderati area held a substantial body of water, perhaps several meters deep, during the late Pleistocene. Increased levels of precipitation during the late Pleistocene and early Holocene apparently resulted in raised water tables and standing bodies of water in selected areas within the eastern volcanic highlands.

PRELIMINARY INTERPRETATION

The site of GaJk 2 is thought to represent an aceramic Later Stone Age settlement that was placed on high ground overlooking the oasis of Nderati Wells and located within the Basin's eastern vocanic highlands. The presence of fine grained silts, associated mollusc shells and a radiocarbon date of over 13,000 BP indicates the Nderati

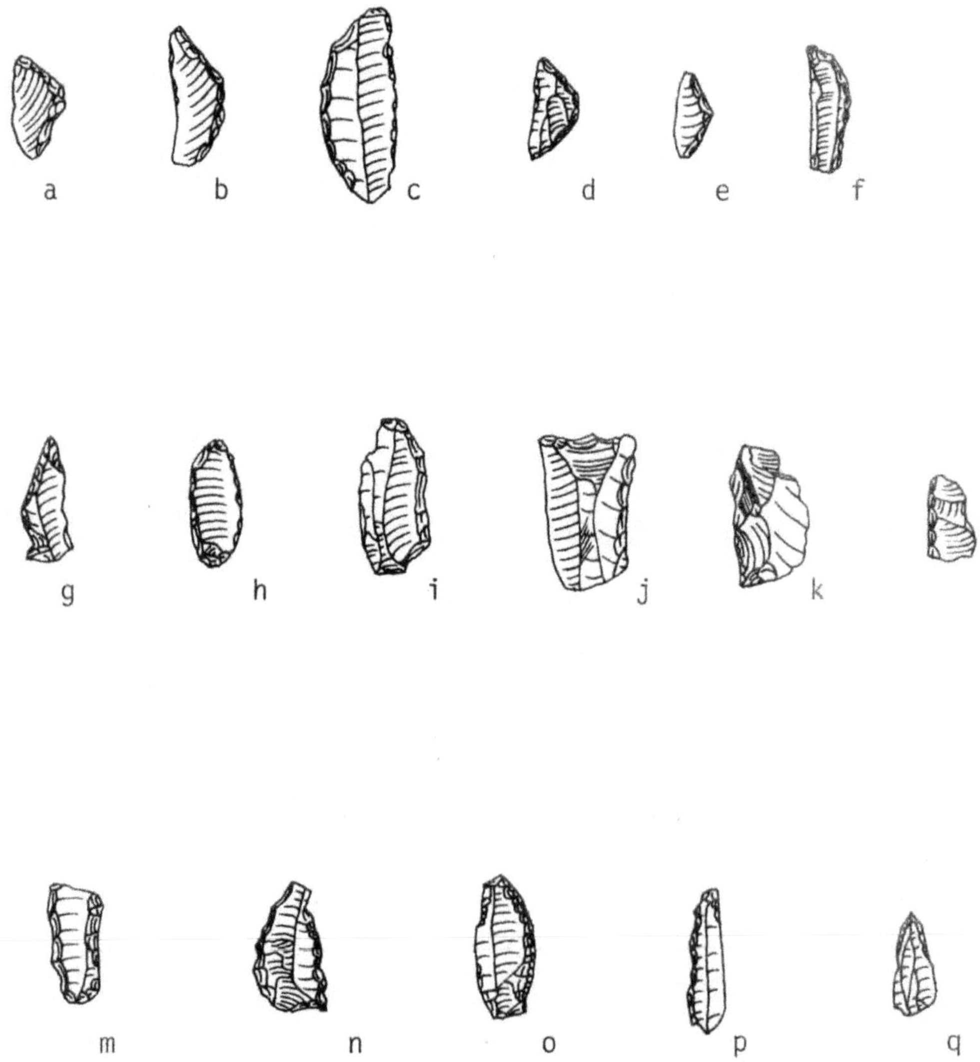

Figure 93. GaJk 2. Stone artifacts.
(a) triangle; (b-f) crescents; (g-i) curved backed flakes; (j-l) straight backed; (m-o) double backed; (p-q) micro-percoir. Natural size.

valley held a standing body of water during the late Pleistocene. Biomphalaria cf. pfeifferi, the intermediate host of intestinal bilharsiosis, was identified among the mollusc remains. Direct evidence for the parasite's existence was lacking however.

Regettably, given the available evidence, it was not possible to reconstruct the subsistance base or duration and number of occupation episodes. Nevertheless, the site and nearby aceramic LSA occupation (GaJk 3) on Nderati Mountain is significant. The consistent low density scatters and concentrated patches of Later Stone Age microlithic chippage among and near the modern oases of the volcanic highlands (Nderati Wells, Karsa Oasis and Buluk Wells) suggest these localities were preferentially situated on elevated ground overlooking inland water sources. It is currently not possible to date the age of the aceramic occupations but they may have been occupied prior to or be contemporaneous with the well documented fishing and pastoralist cultural traditions within the sedimentary basin.

Lastly, the series of core/choppers and other heavily weathered core tools suggest early hominid groups may have established their campsites within the eastern volcanic highlands during the Plio-Pleistocene. As well, these groups, whose home bases and butchery sites were located beside major river channels and smaller deltaic distributaries within the sedimentary basin, may also have periodically visited the Nderati region and other areas along the volcanic perimeter to collect selected lithic raw materials.

Table 82

GaJk 2

INVENTORY OF CULTURAL REMAINS BY SAMPLING UNIT

Unit	1	2	3	4	5	6	7	8	9	10	11	12	13	14	15	16	17	18	Total (N)
Stone:																			
Microliths	5	3	4	12	11	4	-	3	3	1	-	1	7	3	4	1	1	-	63
Scrapers	1	-	-	-	1	-	-	-	-	-	-	1	-	1	-	-	-	-	4
Outils écaillés	-	1	1	-	1	-	-	-	-	-	-	-	-	-	-	-	-	-	3
Awls	-	-	1	-	-	-	-	1	-	-	-	-	-	-	-	-	-	-	2
Becs	-	1	-	2	-	-	-	-	-	-	-	-	-	-	-	-	-	-	3
Cas. trimm.	-	1	-	1	-	1	-	1	-	-	-	-	-	1	-	-	-	-	5
Whole flakes	17	10	10	18	6	4	-	7	2	1	2	3	9	9	15	6	-	-	119
Broken flakes	40	28	34	58	40	25	7	15	10	3	2	10	26	29	23	30	4	5	389
Angular frags.	92	145	163	210	182	94	23	65	36	12	23	71	138	133	103	94	38	5	1627
Cores	13	6	7	9	9	3	1	8	4	1	3	3	9	5	3	4	4	-	92
Raw counts (N)																			
Stone:	168	195	220	310	249	131	31	100	55	18	30	89	184	181	148	135	47	10	2307

Table 83

GaJk 2

FREQUENCY OF TRIMMED PIECES

	Chert	Obsidian	Total (N)	Total (%)
Microliths:			(63)	(78.8)
Triangle	1	-	1	1.3
Trapezoid	1	-	1	1.3
Crescent	5	-	5	6.3
Curved backed flakes	8	1	9	11.3
Straight backed flakes	6	-	6	7.5
Truncations	2	-	2	2.5
Micro-percoir	1	-	1	1.3
Double backed flakes	5	-	5	6.3
Miscellaneous	4	-	4	5.0
Fragmentary	30	-	30	37.5
Scrapers:			(4)	(5.0)
Side	1	-	1	1.3
Nose	1	-	1	1.3
Notch	2	-	2	2.5
Outils écaillés	3	-	3	3.8
Awls	2	-	2	2.5
Bec	3	-	3	3.8
Cas. trimm.	5	-	5	6.3
Raw material (N)	79	1	80	100.0

Table 84

GaJk 2

FREQUENCY OF CORE TYPES

	Chert	Lava	Quartz	Total (N)	Total (%)
Single platform	9	-	2	11	12.0
Double: alternate	40	-	4	44	47.8
other	1	-	-	1	1.1
Multi	2	-	-	2	2.2
Disc	1	-	-	1	1.1
Casual	8	-	1	9	9.8
Fragmentary	24	-	-	24	26.1
Raw material (N)	85	-	7	92	100.0
(%)	92.3	-	7.7		

Table 85

GaJk 2

FREQUENCY OF MAJOR ARTIFACT CLASSES

	Chert	Lava	Quartz	Obsidian	Total (N)	Total (%)
Trimmed pieces	79	-	-	1	80	3.5
Cores	85	-	7	-	90	4.0
UTFF	2050	42	32	11	2135	92.5
Raw material (N)	2214	42	39	12	2307	100.0
(%)	96.0	1.8	1.7	0.5		

INTRODUCTION

The site of GaJj 9 was located in Area 103 along the crest of the Koobi Fora Ridge (Fig. 83). Two adult human burials and fragments of a child were discovered. Portions of each skeleton were exposed by erosion on the surface of a grey diatomaceous silty sand. Mollusc shells, including Melanoides tuberculata and Corbicula africana, were distributed within the sandy unit. Associated grave goods, including two partially reconstructed ceramic vessels and a collection of ostrich eggshell beads, were recovered. Burial 2 has been dated to 3125 ± 210 BP.

BURIAL 1

While site surveying in Transect A, unit 3 (see Chapter VII) several fragments of weathered, partially exposed, human bone were observed. Excavation revealed a poorly preserved incomplete skeleton that lay embedded within unconsolidated diatomaceous silty sands between 5-25 cm below the ground surface. The skeleton was partially mineralized and abundant modern rootlets were present among the fragmentary bones. The skeleton was buried with both legs folded tightly against the chest while the upper limb bones lay beneath the drawn up lower limbs. The skull was broken into numerous small fragments, but nearly one half of the cranium was later reconstructed. The skeleton lay on its right side and faced to the SW.

Emma Mbua and Karen Bell, both associated with the National Museums of Kenya, have undertaken a detailed study of the Galana Boi hominids (Mbua 1984). They report, judging by the size and reduced robusticity of the long bones, that the skeleton belonged to a small female. Analysis of the dental remains indicated the third molars had just come into wear and therefore the individual probably died during her late teens to early twenties. Fragments of a deciduous right lower molar and two germs were discovered with Burial 1. These finds suggest a second individual or portions of a second individual, believed to have been a child between 6-10 years old, may have been interred with the female skeleton.

BURIAL 2

A second burial was identified two meters east of Burial 1 by a small cluster of partially exposed cranial fragments. Subsequent excavation revealed an otherwise complete human skeleton embedded within silty diatomaceous sands (Plate 5). The skeleton lay on its right side and faced NW.

Analysis of both innominates indicated the individual was a male. The teeth were relatively large and well preserved, while the first molars showed extensive wear as did, to a lesser degree, the third molars. The dental evidence tentatively suggests the individual may have been about 30 years of age. Although the skeleton was nearly complete, it was severely weathered. As well, portions of the post cranial skeleton were distorted and crushed.

ASSOCIATED GRAVE GOODS

Grave goods were associated with each adult burial. Two pottery vessels were discovered with Burial 1 while a set of finely crafted ostrich eggshell beads was recovered with the second burial.

Burial 1

Two broken pottery vessels associated with Burial 1 were partially reconstructed. Fragments of Vessel A were scattered about the ground surface as well as adjacent to the burial. The reconstructed vessel possessed a rounded lip and was tempered with coarse gritty sand. Body decoration consisted of an alternating series of deeply incised lines that formed a V or herringbone pattern (Fig. 94). Such decoration was nearly identical to Ileret type pottery from the Ileret Stone Bowl site. The pot had a well defined carinated shoulder and may have originally been shaped as a wide mouth bowl. Carinated bowls, attributed to Nderit Ware, have been collected from the Central Rift (L.S.B. Leakey 1931). Vessel A, however, lacked internal decoration.

Most of the second fragmentary pot was discovered in situ directly in front of the facial region. Vessel B was globular in shape with two projecting shoulders, a constricted neck and a small mouth (Plate 4). The rim area consisted of four small incised triangles that were equally spaced between oblique parallel incised lines. Rows of cord-like impressions occurred within the triangles. A single row of circular impressions lay below the rim followed by five rows of broad shallow incised lines. Decoration continued on the vessel's shoulders and midsection. A series of 4-6 broad shallow parallel incised lines surrounded four triangular or rectangular geometric figures. Rows of cord-like impressions, similar to those on the vessel's rim occurred within the geometric figures. Further decoration was not present below the pot's midsection. Significantly, the base of the vessel was broken and a thorough search of the surface and excavated deposits did not yield the missing pieces. Apparently, the vessel's base was deliberately broken prior to internment with the individual. A large slab of indurated bioclastic sandstone was recovered just 35 cm NE of the facial region. Its function was not ascertained.

Burial 2

A collection of 22 well crafted ostrich eggshell beads was associated with the second burial. The beads were recovered from the shoulder and facial region and may have originally been worn around the neck.

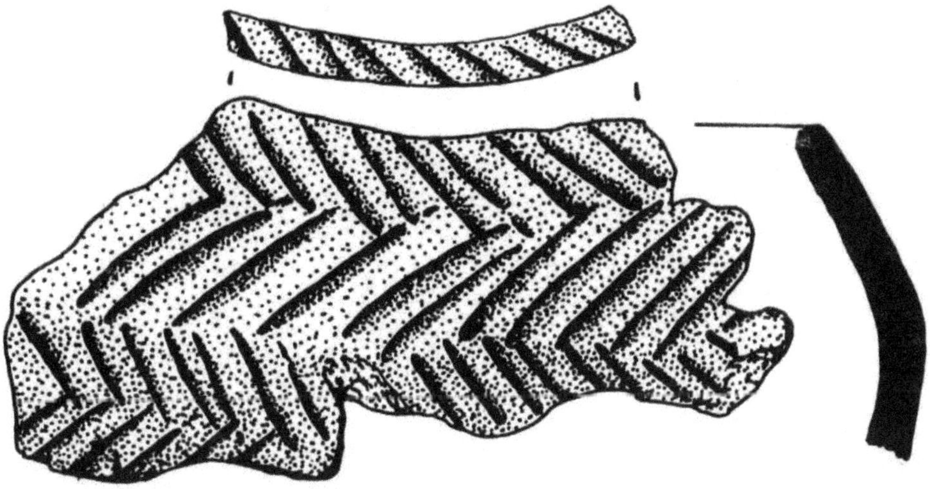

Figure 94. GaJj 9. Reconstructed pot from Burial 1. Natural size.

DATING

A radiocarbon determination on the bone apatite fraction from Burial 1 yielded a modern age reading. However, the near absence of bone collagen and an anomalously high C^{13} value (-10.3) both strongly indicate the carbon content of the skeleton has been altered, probably by subaerial exposure. A second human bone sample from Burial 2 yielded a radiocarbon date of:

3125 ± 210 BP Geochron Labs.

PRELIMINARY INTERPRETATION

Two adult burials with associated grave goods as well as the fragmentary remains of a young child were recovered from the Galana Boi beds in Area 103. It is possible, given their close proximity and similar bone preservation, that the individuals were buried together or interred within a short time of each other. No cause of death was evident for any of the three individuals.

With the exception of the herringbone design on Vessel A and the body shape of Vessel B, the associated burial pots showed few structural or decorative features similar to the pottery collection from any of the site groups. The reconstructed pots may well represent funerary, non-utilitarian vessels that lack shape and design affinities with ceramic collections from the area's occupation sites. The vessels may also reflect the existence of ceramic traditions not yet identified within the research area. The latter interpretation is supported by the comparatively recent radiocarbon age for Burial 2. It suggests the burials, or at least the second burial, date to just over 3000 BP. If correct, and single bone apatite dates should be viewed as tentative, the human remains belong to a phase of occupation at East Turkana for which little archaeological evidence has so far been collected.

Plate 4. Reconstructed pot from Burial 1.

Plate 5. Burial 2.

LOCALITIES

Archaeological localities consisted of one or more artifacts that were collected from an uncertain provenience or other questionable contexts (Nelson 1973). In practice, the majority of the artifacts recovered from the 74 localities at East Turkana came from single isolated occurrences. Artifact descriptions for 16 localities are given below.

Bone Artifacts (Figs. 95-96)

During exploratory and transect survey a number of barbed harpoons and other bone artifacts were collected from Holocene beach sands and erosion surfaces. Fig. 95 illustrates several finds including a broken harpoon with remnants of marrow cavity still visible, one harpoon with a single barb and a bone artifact with double sided notching. As well, a large biserial harpoon or spearhead was collected from the 75-80 m beach sands in Area 100 (Fig. 96). The complete specimen originally measured between 150-200 mm. It represents the largest bone artifact discovered at East Turkana.

Stone and Shell Artifacts (Fig. 97)

A unifacially flaked lava limace was collected from beach sands near GaJj 12 in Area 127. Limaces formed part of the core tool inventory at several fishing settlements. An isolated bifacial chert point was collected in Area 129 (Fig. 97, b). The point had a relatively thin cross section and may have been associated with either MSA or LSA stone industries.

The circular mollusc shell disc illustrated in Fig. 97 c was collected in Area 10 and closely resembled the stone disc excavated from the fishing station at GaJi 3. Both artifacts probably served as body jewelry.

Pottery (Figs. 97-100)

A large collection of isolated decorated pottery sherds was recovered from the research area. Representative samples of some of the sherds are illustrated including Nderit Ware (Figs. 97, a; 98, a-c) and pieces with closely spaced impressions and paneling (Fig. 98, e-f). Turkwell tradition pottery (Robbins 1980) was also present (Fig. 99, c-d).

Ileret pottery was collected approximately 1/2 km west of GaJj 11 in Area 127. Galana Boi deposits were traced for several kilometers and artifacts, especially pottery, were discontinuously distributed over the entire length of the outcrop. Exploratory survey did not reveal any recognizable clusters of artifacts and the finds have therefore been assigned a locality number.

The pottery assemblage included Ileret type pottery with characteristic zig-zag or herringbone decorative patterns (Fig. 100, c; 101, d, f). Other Ileret sherds displayed crescentic or lozenge shaped comb stamped impressions (Fig. 100, a-b) that sometimes overlapped (Fig. 101, b, e). Several vessel shapes were represented including shallow and deep open mouthed bowls and possible beakers (Fig. 101, a). One sherd from a carinated bowl (Fig. 99, b) had been burnished and slipped. Comparable examples of burnished and slipped sherds were identified from the Ileret Stone Bowl site.

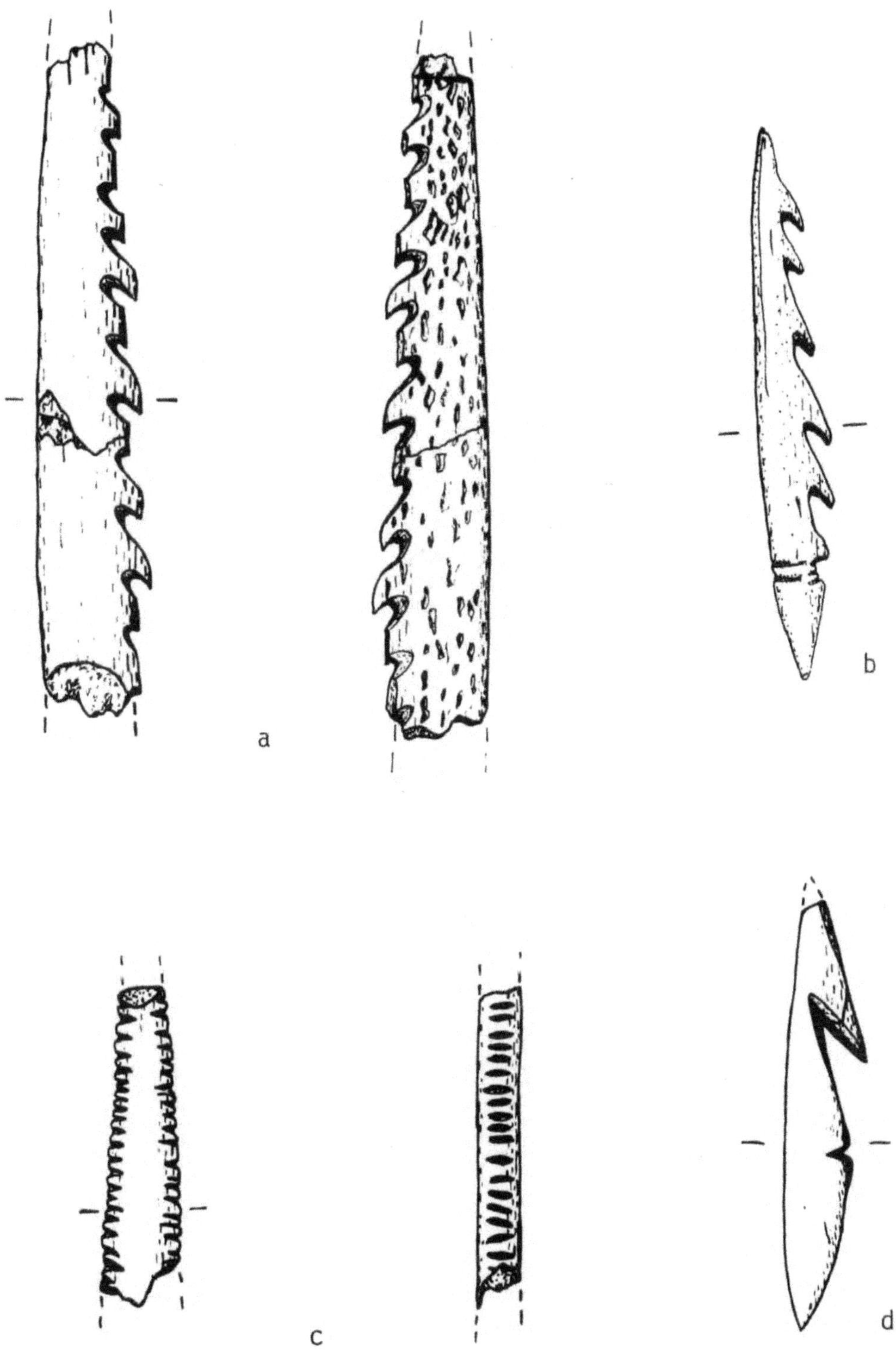

Figure 95. Bone artifacts.
(a-c) uniserial harpoons; (d) bilateraly notched. Natural size.

Figure 96. Biserial bone harpoon or spearhead. 3:4 size.

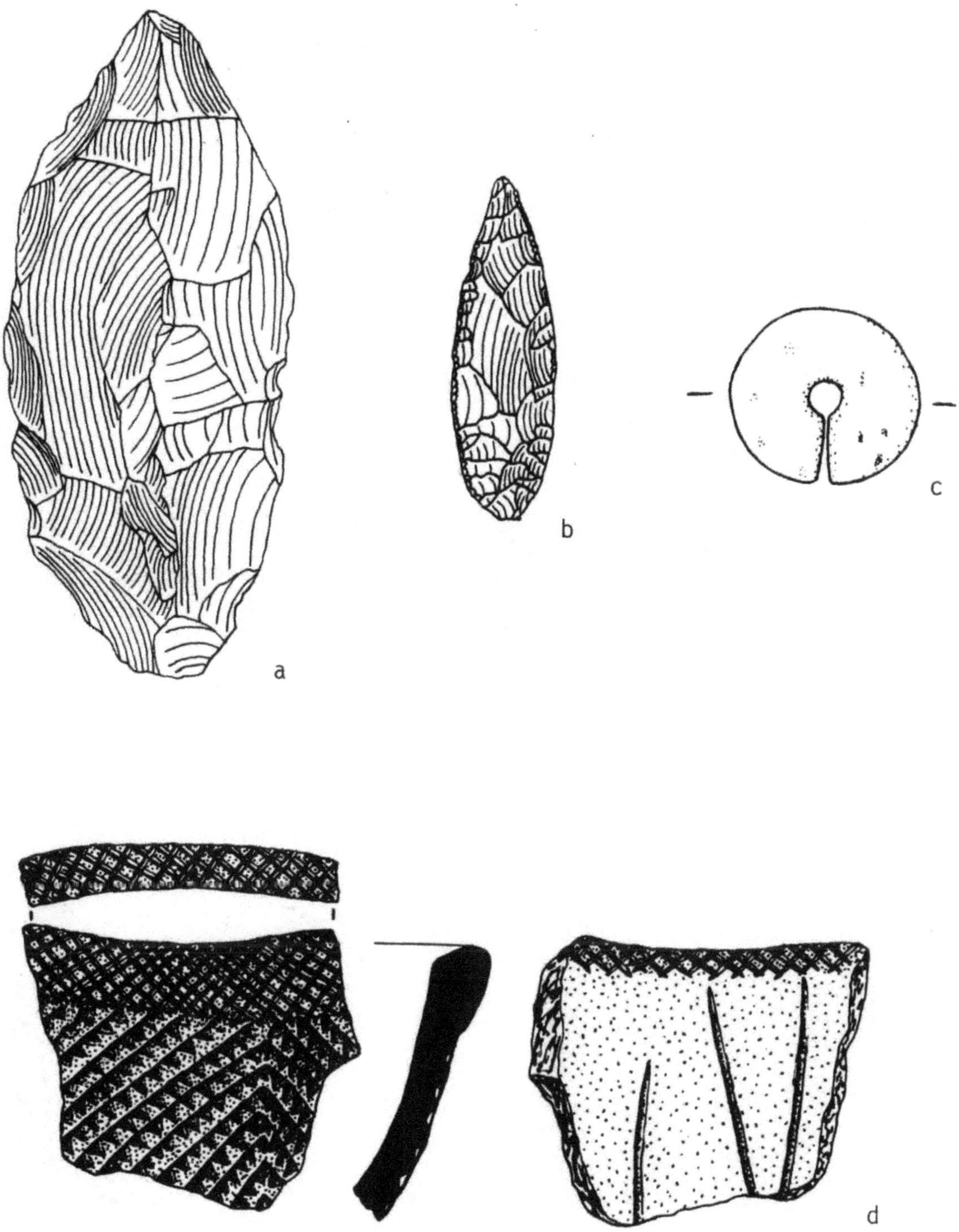

Figure 97. Stone, bone and shell artifacts.
(a) limace; (b) MSA bifacial point; (c) shell disc; (d) decorated rim sherd. Natural size.

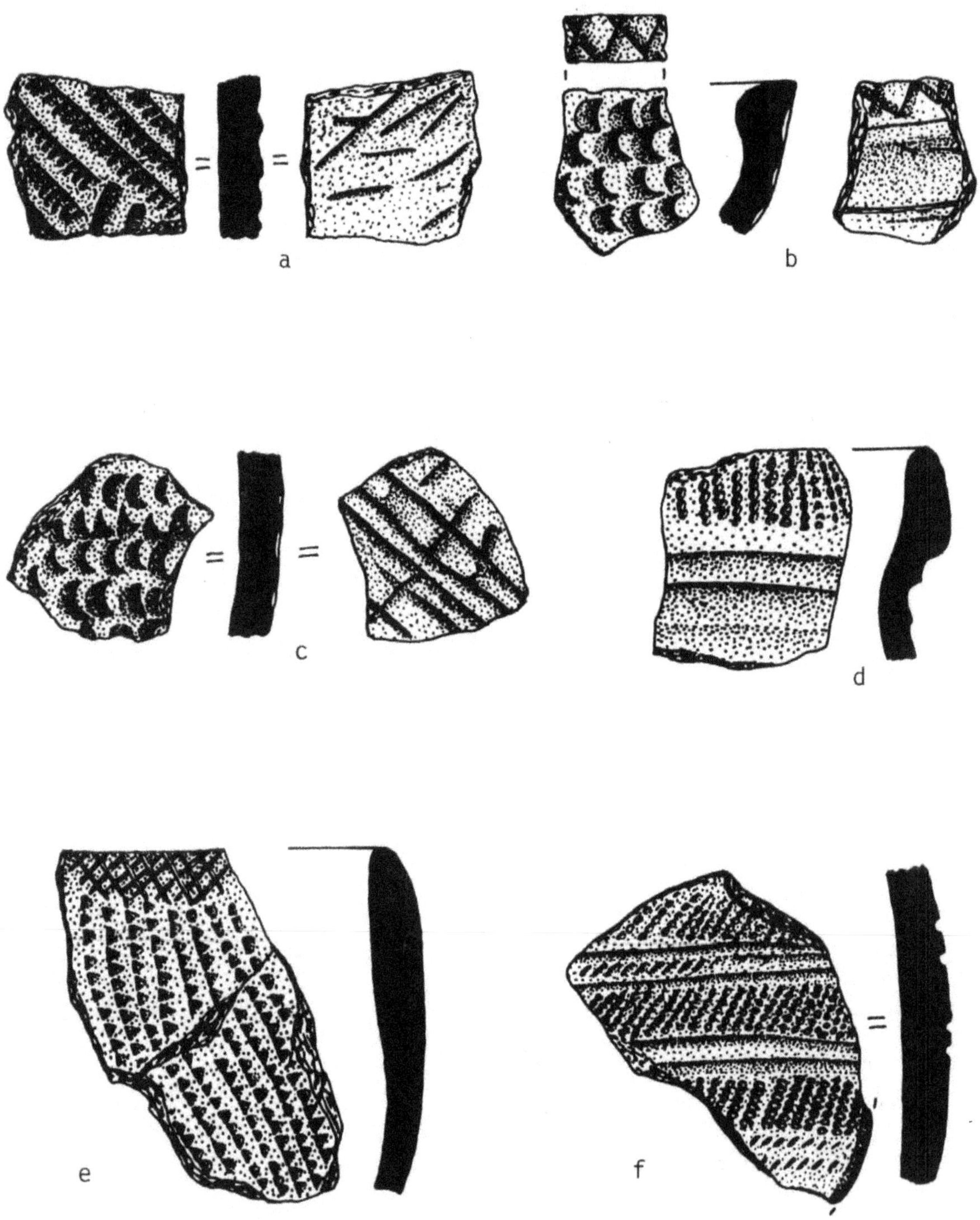

Figure 98. Pottery.
(a-c) Nderit ware; (d) narrow mouth rim; (e) decorated sherds.
Natural size.

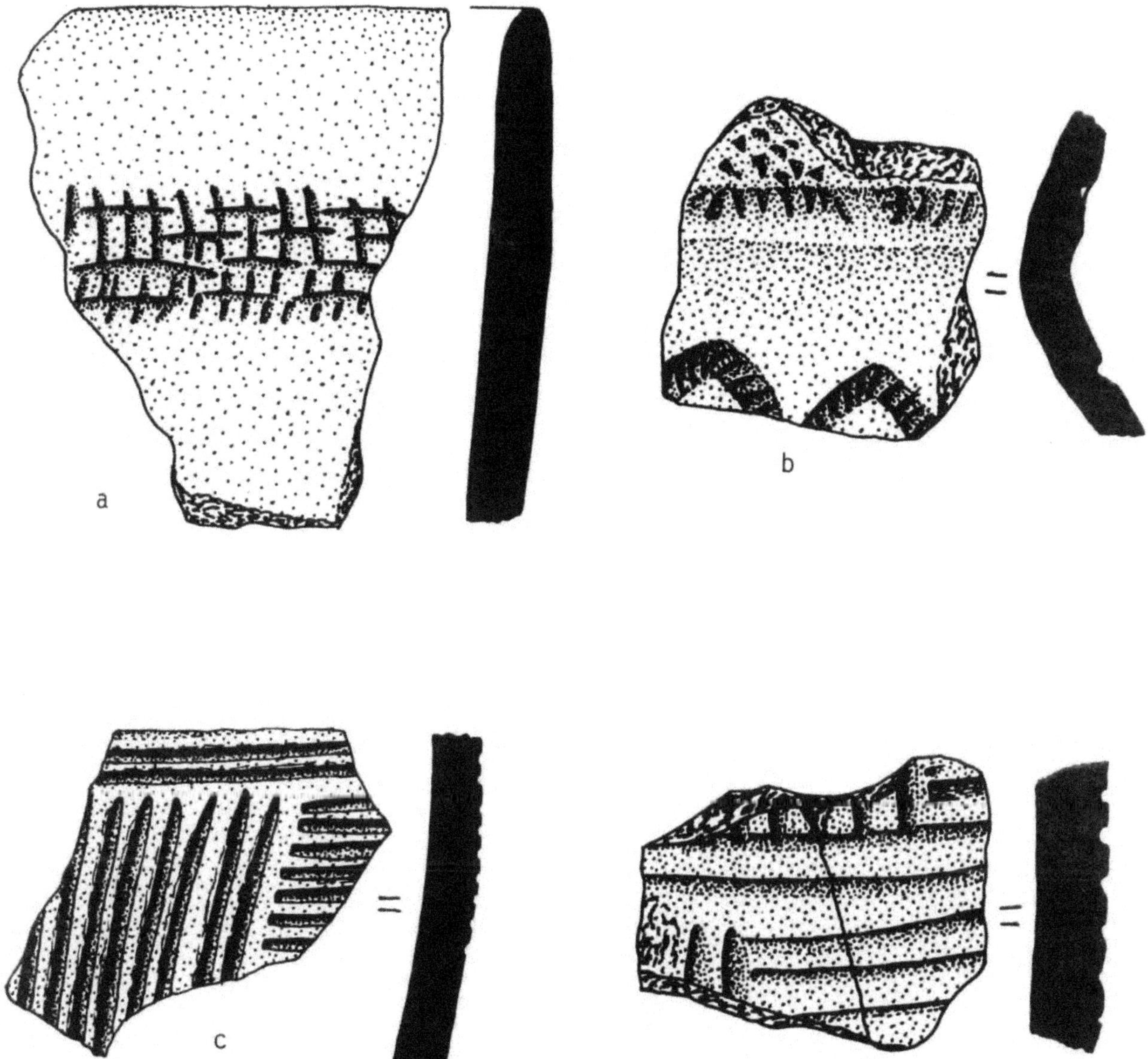

Figure 99. Pottery.
(a) rim sherd; (b) carinated vessel with red slip; (c-d) Turkwell sherds. Natural size.

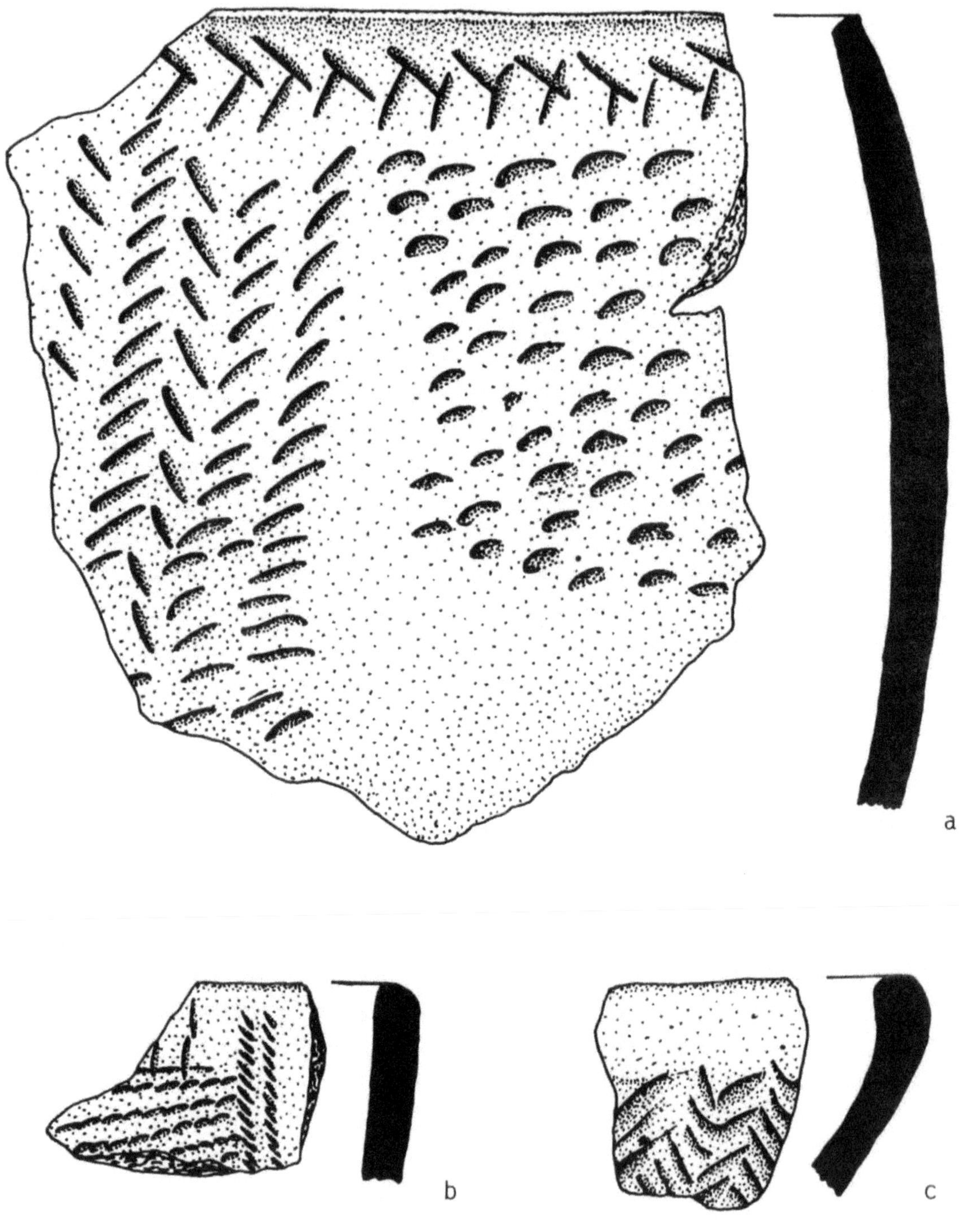

Figure 100. Ileret type pottery.
(a-c) rim sherds. Natural size.

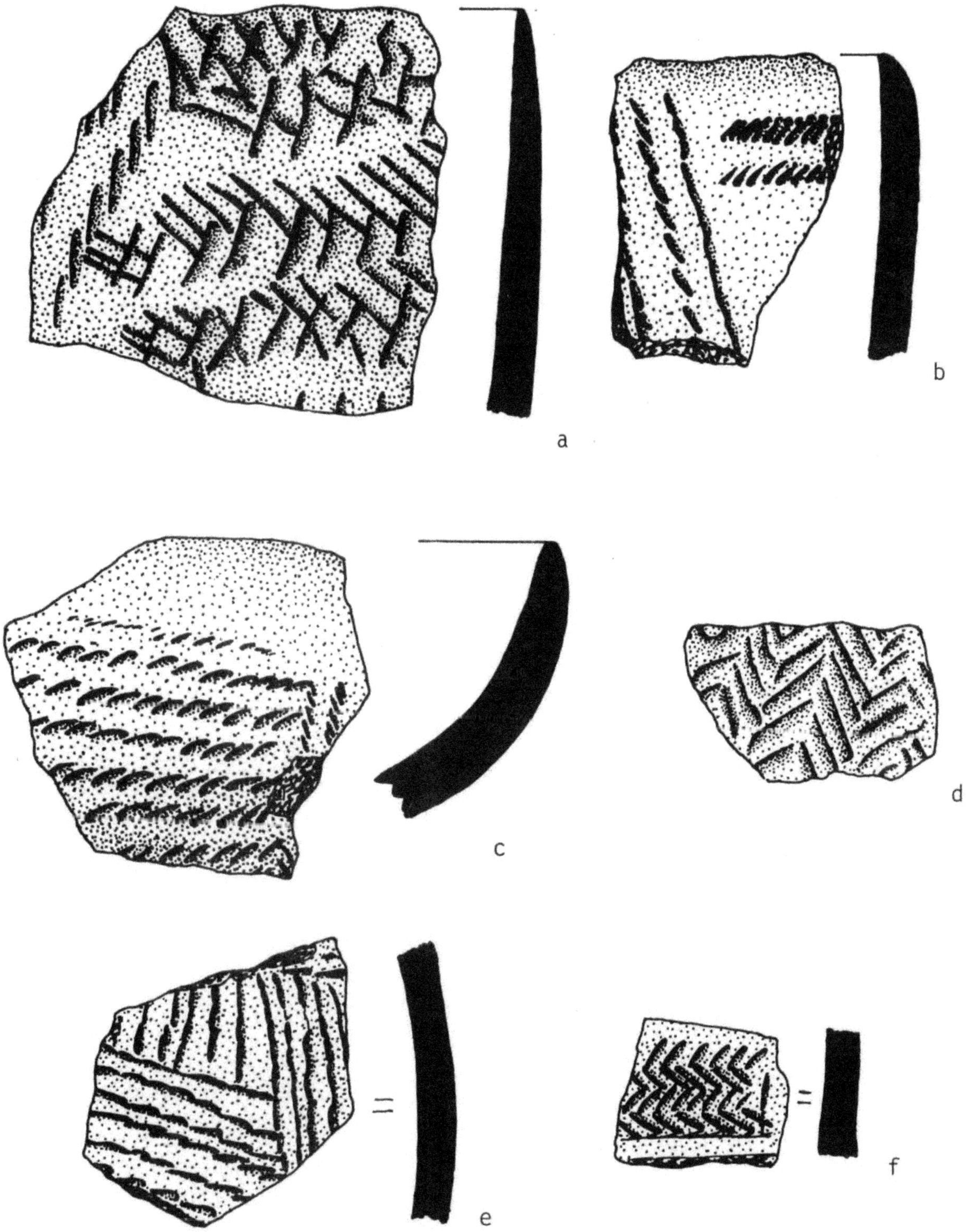

Figure 101. Ileret type pottery.
(a-c) rim sherds; (d-f) body sherds. Natural size.

CHAPTER VI

ARTIFACT ANALYSIS

Detailed artifact analysis of the major stone, bone and pottery shaped tool categories was conducted for each of the excavated and sampled sites. Three principal research objectives were followed: 1) documentation of artifact similarities and differences within each site group. For example, among the fishing settlements significant differences occurred in the percentage composition of core tools and lithic raw materials; 2) comparison of artifact assemblages between site groups. Pastoral Neolithic sites for instance, contained higher frequencies of utilized micro-blades and obsidian than fishing settlements; and 3) key artifact patterns of the East Turkana sites were compared with similar artifact assemblages in Kenya and elsewhere in Eastern Africa.

STONE ARTIFACTS

GROUP I FISHING SETTLEMENTS (Table 86)

Shaped Tools (Tables 88-92; Figs. 102-104)

For Group I sites, backed microliths were the predominate shaped stone artifacts. Microliths accounted for 58.4% of the shaped tools followed by *outils écaillés* (10.8%), core tools (10.3%) and scrapers (8.9%). Other shaped tools, including becs and awls, represented 11.4% of the sample. Crescents and curved back flakes were the most common microlithic types. A low but persistent quantity of triangles, trapezoids and straight backed pieces also occurred. Fragmentary pieces were well represented in both the surface and excavated collections. The length of crescentic forms and curved backed flakes ranged from 11 mm to over 60 mm. Mean length was between 25-30 mm. Different varieties and nodule sizes of chert accounted for the wide range of measurements. Significantly, the entire collection of microlithic tools from Group I sites was manufactured from chert. Obsidian, with the exception of two crescents recovered from the surface of FxJj 12 North and probably not associated with the occupation, was not used in the production of microlithic forms.

Outils écaillés constituted an important component of the shaped tool inventory. They showed considerable variation in percentage composition between sites but usually comprised between 15-20% of all shaped artifacts. Mean length varied between 18-22 mm with a range of between 10 and nearly 40 mm. The great majority of *outils écaillés* were manufactured from chert although infrequent quartz examples also occurred. No specimens made from lava or obsidian were collected.

Core tools were widespread and abundant. These artifacts, chiefly polyhedrons, core/choppers and discoids, were morphologically and typologically similar to core tools recovered from the Lower Member sites and most sites (especially FxJj 18 NS and FxJj 20 M and E) of the Upper Member Karari Industry at Lake Turkana (Isaac 1973; J.W.K. Harris 1978). They were recovered from all Group I sites and accounted for over 20% of the shaped tools from two sites. All examples were made from volcanic materials including welded tuff. Three examples of a limace, also morphologically akin to artifacts from the Karari Industry, were recovered from fishing settlements.

A diverse series of scrapers and other shaped tools were collected from the Group I sites. Side scrapers were the most common type followed by double side, convergent and notched forms. Notably, no scrapers were recovered from the three excavations; the entire sample of scraper types was derived from surface collections. They represented a low percentage but persistent tool category. Among the group of additional shaped tools, casually trimmed and retouched pieces were the most common followed by becs, technical burins, and awls. Purposefully struck angle and dihedral burins were absent.

Cores (Fig. 105)

A wide range of core forms was identified from the fishing settlements. Double platform cores were the most common variety followed by casual, multi, and single platform cores. Considerable variation in core types existed between individual sites. Double platform cores included examples that were alternate ended and others with two striking platforms at right angles.

The series of core forms was readily distinguished by raw material and size. Lava cores were invariably larger than chert examples. Many of these large cores were designated as core tools but probably functioned as double and multi platform cores. The high percentage of casual cores at GaJj 1 and GaJj 12 reflects the selection of hydrothermally produced chert as the primary raw material. The chert nodules were small (usually > 40 mm), irregular in shape, and often possessed internal fracture weaknesses. Apparently, the nodule's flaking properties were first determined by removing 1-2 flakes. If the chert nodule was internally flawed or of a highly irregular shape, the piece was discarded.

The great majority of the core varieties were produced by direct percussion. However, parallel scar ridge patterns on single and double platform cores may have reflected the occasional use of the bipolar technique.

Debitage: Whole Flakes (Figs. 106-108)

The large sample of whole flakes was manufactured primarily from chert or lava. Smaller quantities of quartz and obsidian were also present. Figure 106 illustrates the variation in maximum flake length by raw material for the Group I sites.

The size of whole chert flakes was relatively small with mean length of less than 20 mm and a breadth/length ratio of .790. The great majority of chert flakes had single platform facets and were probably struck from unprepared cores. Several flakes exhibited prepared or crushed platforms. In contrast, lava flakes showed a mean length of < 40 mm and a B/L ratio of .930. These flakes were nearly as wide as they were long. Virtually all lava flakes had single platform facets indicating that they were struck from large unprepared cores.

Figures 107 and 108 illustrate frequency distributions by raw material for maximum length and B/L ratios from two Group I flake samples. The disparity in flake lengths was significant while each raw material showed similar B/L ratios. The marked contrast between raw material and flake size suggests different raw materials were purposefully chosen to produce different sized flakes. These flakes in turn may well have been fashioned into different tools designed for different purposes.

Raw Material (Fig. 109)

Various types of chert and lava provided the chief raw materials for the Group I sites. Chert was used primarily in the manufacture of microliths, scrapers and *outils écaillés*. Lava, on the other hand, was chosen for the production of scrapers, large casually trimmed pieces and core tools. Quartz and obsidian (> 1%) were also employed. Considerable variation in raw material selection occurred between the fishing sites. For example, at some sites chert accounted for over 95% of all artifacts and less than 50% at other sites. Lava constituted greater than 50% and less than 5% of the raw material at additional sites. Such disparity probably reflects a significant difference in the preference for a particular lithic raw material.

GROUP II PASTORAL NEOLITHIC SITES (Table 87)

Shaped Tools (Tables 88-92; Figs. 102-104)

Backed microliths accounted for 47.5% of the total Group II shaped tools followed by *outils écaillés* (26.6%), scrapers (14.7%) and other tools (11.1%). Figure 113 illustrates the variable frequencies by site for each category of shaped stone artifact.

Fragmentary, truncated, and double backed examples were the most common microlithic types. Crescents and curved backed flakes were present in relatively low numbers while geometrics were completely absent from the excavated sample at Dongodien. The microliths had a mean maximum length of between 20-25 mm. Chert and obsidian were used in the production of microliths. No examples of either quartz or lava microliths were identified.

Outils écaillés constituted an important component of the shaped tools from the Pastoral Neolithic sites. They accounted for over 30% of the tool inventory from the excavated samples at Dongodien and the Ileret Stone Bowl site. The majority of *outils* were manufactured from obsidian and chert but quartz specimens were also recovered from the Ileret Stone Bowl site. The obsidian examples were relatively small with a mean maximum length of 15-20 mm. Chert outils were larger and had a mean maximum length of nearly 25 mm. A number of small obsidian outils intergraded with alternate ended cores and may in fact have represented exhausted or spent core forms. Some of the diminuative examples (> 10 mm) were probably too small to effectively yield flakes and are best considered as shaped tools.

Scrapers comprised a significant component of the shaped tools at the Ileret Stone Bowl site but were absent or nearly absent at GaJi 2 and Dongodien. At Ileret, a wide range of scrapers was represented including side, end, convergent, notch and circular forms. Chert was the preferred raw material followed by obsidian and quartz.

A small series of other shaped tools including casually trimmed debitage, becs and technical burins were also collected. A single obsidian dihedral burin, the only true burin discovered from the entire set of sampled and excavated Holocene sites, was recovered from the surface at Dongodien.

Cores (Fig. 110)

Relatively few cores were identified from the Group II sites. Of those recovered, double, fragmentary and multi platform types were the most numerous. The single and double platform cores at Dongodien exhibited prominent parallel sided flake scars and crushed striking platforms. Obsidian and chert were the most frequently used raw materials.

Debitage: Whole Flakes (Figs. 106, 111-112)

Whole flakes from the Pastoral Neolithic sites were smaller in length when compared to the other site groups. Chert flakes had a mean length of 19.6 mm and B/L ratio of .923. Obsidian flakes were still smaller with a mean length of 14.6 mm and a B/L ratio of .727. Figures 111 and 112 illustrate frequency distributions by raw material for selected whole flake attributes from Dongodien and the Ileret Stone Bowl site. Raw materials from each site had similar mean length distributions but showed differences in their B/L ratios. Obsidian flakes from both sites had well defined parallel dorsal scar ridges and crushed or faceted platforms while more than 40% of the flakes had a B/L ratio of less than 0.5. These data suggest that the production of blades and flake-blades, in contrast with Group I sites, was an important component of the Pastoral Neolithic stone industries.

Raw Materials (Fig. 109)

Obsidian was the predominate raw material utilized at the Group II sites and accounted for over 95% of the artifacts from Dongodien. Outcroppings of obsidian occur along the northeastern perimeter of basin especially within the Miocene Suregei-Asilla volcanics (Watkins 1981). The obsidian usually occurs as thin discontinuous lenses from which small cobbles and pebbles may be collected. The presence of significant amounts of obsidian from the Pastoral Neolithic settlements indicates the raw material was selected and transported 45-55 kms and perhaps further into the basin.

Merrick and Brown (1984) have recently analyzed obsidian source areas in northern Kenya using x-ray fluorescence. Their initial results indicate local sources accounted for most of the obsidian recovered from the Pastoral Neolithic sites. Several flake fragments from GaJi 2 and the Ileret Stone Bowl site however, appear to represent unknown source areas.

GROUP III SITES (Other Later Stone Age sites with ceramics) (Tables 87-92; Figs. 102-104)

The great majority of the shaped tools from the Group II sites consisted of a wide assortment of scraper types, crescents and fragmentary microliths that were primarily manufactured from chert. *Outils écaillés* were present in relatively low frequencies but included several large quartz specimens over 40 mm in length. Multi platform and casual cores were the most common types identified.

Whole flakes and other debitage were manufactured primarily from chert, quartz and obsidian. The mean length and B/L ratios for Group III whole flakes were 25.7 mm and .856 respectively. The whole flake sample from FxJj 102 yielded a mean length of 26.7 mm. This figure represents the highest mean length value for whole flakes among the 16 sites investigated and reflects the use of relatively large angular blocks, not pebbles , of chert.

GROUP IV SITES (Aceramic late Pleistocene-Holocene sites without bone harpoons) (Tables 87-92; Figs. 102-104)

Backed microliths, especially crescents, were the most abundant shaped tools among the Group IV sites. The mean length of all microlithic types was small, between 16-22 mm. *Outils écaillés* were manufactured principally from chert and had a mean length of only 16.6 mm. Few scraper types were present and core tools were completely absent.

Double platform cores, primarily alternate ended with parallel sided scar ridges, and casually flaked specimens were the common core varieties. Debitage from the Group IV sites was dominated by small flakes and flake blades struck from diminuative chert cores. Whole flakes, for example, had a mean length of 16.9 mm and a B/L ratio of .828.

Chert accounted for over 90% of the raw material utilized at the aceramic LSA sites. The location of the sites, near the eastern margin of the sedimentary basin and surrounding volcanic highlands, coincided with abundant quantities of small nearly transluscent chert nodules. These nodules are formed as hydrothermal precipitates within the vesicular crevices of certain volcanic rocks. Their local abundance and excellent flaking properties indicate, within the highlands, chert nodules were selected over obsidian as the preferred raw material.

SUMMARY - GROUPS I-IV (Figs. 113-115)

Backed microliths were the most common shaped stone tool manufactured during the late Pleistocene to middle Holocene in the northeastern sector of the Lake Turkana Basin. Their frequencies were highest among Group I and IV sites. Crescents and curved backed flakes were the most frequent of the microlithic types. *Outils écaillés*, the next largest shaped stone tool category, were recovered from each site group but were most numerous within the Group II sites. Scrapers comprised an unstandarized and relatively infrequent set of shaped tools except for Group III. Core tools, many of which may have functioned primarily as cores, occurred almost entirely within the Group I sites. A small but persistent assortment of other shaped and modified stone tools was present within each site group.

Whole flakes and other debitage were primarily struck by direct percussion from unprepared single, double, multi and casual platform cores. Blades and flake blades constituted a high proportion of the debitage from the Group II sites and probably indicate the use, however limited, of the punch technique.

Significant differences in raw material utilization occurred within as well as between each site group. Chert constituted the primary raw material for all site groups. Lava was used extensively within most of the Group I sites while obsidian was employed primarily within the Group II sites and used intensively at Dongodien. Quartz was utilized in low but stable frequencies throughout the research area.

SUMMARY: COMPARISON WITH OTHER ASSEMBLAGES

Group I (Table 93)

The shaped stone tool assemblages from the fishing settlements at East Turkana consisted of microliths, *outils écaillés*, core tools, scrapers and a number of sundry forms. When these assemblages were combined and compared with other fishing sites located along the basin's southeastern and southwestern margins, several important differences were noted. First, even though all the fishing sites shared similar technological, economic and settlement features, their stone tool inventories showed marked differences. For example, at Lothagam nearly 50% of the shaped tools were identified as scrapers (Robbins 1974:177) whereas scrapers were entirely absent from Lowasera (Phillipson 1977a) and occurred in low frequencies among the Group I sites. As well, *outils écaillés* and core tools were relatively common within Group I sites while at Lothagam and Lowasera they were absent or occurred in limited numbers. Such distinctions can, for the most part, be explained by the fact that different researchers have used different typological frameworks. For example, many of the anvil, radial and polyhedral cores at Lowasera (Phillipson 1977a:12-13) have been classified as core tools (choppers, discoids and polyhedrons) by the present author. At Lothagam, the complete absence of outils écaillés apparently reflects their absorption into the category of cores, whereas the numerous core burins identified from Lothagam appear to have been small flake-blade and bladelet cores with sharp scar ridges. Microliths, on the other hand, consistently occurred in high frequencies at all fishing sites. At Lowasera and the Group I settlements, microliths represented the predominate shaped tool category.

An important feature of the Lake Turkana fishing sites was the marked difference in the availability of and selection for certain raw materials. At Lothagam, 87% of the artifact assemblage was manufactured from fine grained lavas in contrast to 55% from Lowasera. Among the Group I sites, lava accounted for a high of 54.2% and a low of 2.1%. The high percentage of lava at Lothagam may be attributed to the widespread availability of lava cobbles as well as the poor quality of local chert supplies (Robbins 1974:169). As a consequence, 88% of all Lothagam's backed pieces were manufactured from lava. While fine grained lavas were readily available at Lowasera and the Group I fishing sites, their fluctuating percentages appear to reflect purposeful selection for particular tool forms. For instance, core tools, cores and various scrapers were consistently manufactured from lava. Finer grained materials, including chert, quartz and, infrequently, obsidian, were primarily used to manufacture microliths, scrapers and outils écaillés.

Comparisons of stone tool assemblages with fishing sites that lay outside the Basin were also undertaken. These assemblages, chiefly from Khartoum and Ishango, revealed major differences in the proportion of microliths, scrapers and core tools. The large set of other tools from these sites was made up primarily of utilized flakes, borers and burins. The principal shaped stone tools associated with the early Holocene fishing sites in Eastern and Central Africa consisted of a diverse series of geometric and non-geometric microliths, various scrapers, core tools and casually trimmed and retouched pieces.

GROUP II (Table 94)

Comparisons of the Group II tool assemblages from East Turkana with those from selected Pastoral Neolithic sites in central and western Kenya revealed an interesting set of patterns. The comparisons indicated that, while microliths and scrapers constituted the predominant shaped stone implements, their percentage composition fluctuated sharply from region to region and within individual regions. Outils écaillés also demonstrated significant variability and were most frequent in western Kenya. Such variability suggests increased reliance on the bipolar core technique in this area. In contradiction to the East Turkana sites, burins formed a significant component of the Pastoral Neolithic sites in central Kenya. Core tools were uncommon from all the sites irrespective of region. The increased number of scraping implements may well indicate greater concern with domestic cattle and sheep/goat hide preparation as compared to the hunting of wild animals with microliths. For example, at Narosura in central Kenya, the faunal remains consisted predominantly of domestic stock while at Prolong Drift, also in central Kenya, a mostly wild fauna was represented (Onyango-Abuje 1978).

BONE ARTIFACTS

Barbed Bone Harpoons (Tables 95-98)

A total of 265 barbed bone harpoons were collected during exploratory survey, sampling, and excavation of the Group I sites. The great majority of the harpoons were broken; only 7% were complete. Nearly 90% of the harpoons were uniserially barbed but biserial and triserial forms also occurred (Barthelme 1977).

The harpoons were primarily manufactured from mammal long bone shafts except for a single permineralized wooden example. The shafts were apparently split and suitable splinters were selected and carefully ground to a sub-rounded cross section. All traces of the marrow cavity were removed except in a few specimens (Figs. 95, b). Grinding striations were conspicuous on both the interstices betwen the barbs and often on the main shaft. No evidence of the groove and splinter technique, commonly utilized with antler in Northern Europe (J.G.D. Clark and M.W. Thompson 1953), was identified.

Complete harpoons exhibited a wide range of length, number of barbs and distinctive basal markings. The specimens ranged in size from 43-128 mm and had a mean length of 78.5 mm. Eight harpoons measured over 90 mm in length while one broken biserially barbed harpoon or spearhead originally measured over 200 mm (Fig. 96). The bone artifacts displayed considerable variation in their number of barbs. Two complete specimens possessed but a single barb although additional broken single barbed harpoons were recovered from GaJj 11. The number of barbs ranged from 1-17 with a mean number of 6.2.

Analysis of the barbed harpoons revealed two distinctive basal markings both of which presumably facilitated the attachment of line. Of the two, notched bases were almost three times as common as circular grooved bases. Several harpoons had a combination of both features. Significantly, only harpoons with notched bases were collected from two sites. At GaJj 11, notched bases were associated with an aceramic occupation while at FxJj 12 North notched bases were found together with wavy-line pottery. The majority of the fishing settlements, including a second aceramic occupation at FxJj 12, contained harpoons with both notched and grooved bases. Harpoons with grooved bases tended to be larger and had more barbs than their notched counterparts. This apparent correlation cannot be fully explained and may be due to either unknown functional or stylistic factors. No examples of basally grooved catfish spines, comparable to the specimen from Lowasera (Phillipson 1977a, Fig. 15, no. 4), were identified.

Barbless Bone Artifacts

Seven unbarbed artifacts were collected from the Group I sites. Four of the bone artifacts probably functioned as barbless points or as blanks for barbed harpoons. The two remaining pieces had a series of deeply incised parallel circular rings along their entire length (Fig. 95, f). These artifacts may have functioned as holders of reserve fishing line.

COMPARISON WITH OTHER ASSEMBLAGES (Table 99)

Within the Turkana Basin, the barbed bone harpoons collected from Lothagam, Lowasera and the Group I sites consisted overwhelmingly of uniserially barbed specimens. Biserial and barbless forms occurred in small numbers. Triserially barbed harpoons were identified from only a single site (GaJj 11). These artifacts are believed to be the only triserially barbed specimens yet identified from the early Holocene fishing settlements in north, east, or central Africa. Other triserial bone harpoons are known from Nigeria but apparently date no earlier than 2700 BP. (Connah 1976). Uniserial harpoons were the most common barbed form collected from Khartoum (Arkell 1949) and from the numerous Saharan fishing sites (G. Camps 1969, 1974; Smith 1974; Petit-Maire 1979). At the site of Ishango, located along the western edge of Lake Edward in Zaire, biserially barbed harpoons both outnumbered, and occurred stratigraphically below, uniserial forms (de Heinzelin 1957). It has been proposed that Ishango was the fount from which the technique and use of barbed harpoons spread into Eastern and Northern Africa (de Heinzelin 1962). The site is inconclusively dated however, and this hypothesis has been questioned (Phillipson 1977b:49).

The barbed harpoons collected from the Turkana Basin exhibited considerable variation in overall length and number of barbs. The harpoons varied in length from 20-218 mm at Lothagam (Robbins 1974:191), 40-124 mm at Lowasera (Phillipson 1977b), and 43-128 mm from the fishing sites at East Turkana. Similarly, the number of barbs per harpoon ranged from 2-20 at Lothagam, 2-10 at Lowasera, and 1-17 from the Group I sites. Notably, several complete harpoons collected at GaJi 11 had but a single barb. Other single barbed harpoons have not been conclusively identified from the Turkana sites, but possible broken examples may be present at Lothagam (Robbins 1974:Fig. 30, no. e and 4) and Lowasera (Phillipson 1977b:Fig. 22, no. 2). Both notched and circular grooved bases occurred at Lowasera and Lothagam and the majority of the Group I sites. Several East Turkana sites revealed only notched bases and were associated with aceramic as well as ceramic fishing settlements. Importantly, neither barbed harpoons with basal circular openings nor fishhooks manufactured from shell or bone were recovered from any of the East Africa fishing sites nor from Ishango. Both kinds of artifacts have been reported from Shaheinab in the Sudan (Arkell 1954) and from certain Saharan sites (G. Camps 1974; Smith 1974) and are probably associated with middle to late Holocene fishing occupations.

STONE BOWLS (Table 100, Fig. 116)

The area to the northeast of Lake Turkana contained a large number and diversity of shaped stone bowls. A total of 17 broken and 1 unbroken stone bowls were collected from artifact localities and the Pastoral Neolithic occupation at the Ileret Stone Bowl site. The sample was recovered from the surface and no bowls were associated with caves, rockshelters or burial cairns.

A number of body shapes and raw materials were present. The majority of the specimens consisted of deep bowls with rounded or flattened bases. Shallow bodied vessels were rare and did not resemble the wide circular platters frequently recovered in the Central Rift (Merrick 1973). Several specimens had milled or beaded rims and one bowl had a basal projection which may have served as a resting stand. No evidence for either firing or charcoal staining was present. Various raw materials were used in the manufacture of stone bowls including lavas, welded tuff, quartzite and indurated sandstone. Basalt was the predominant raw material utilized.

POTTERY

The pottery remains from northeast Turkana revealed an unanticipated wealth of different wares and distinctive design features. Over 4600 sherds were collected from surface sampling and excavation.

Wavy-line pottery was identified from two early Holocene fishing settlements (FxJj 12 North and FxJj 108) in the Kokoi area. Nderit Ware and a distinctive slipped and burnished pottery collection were recovered from the Pastoral Neolithic occupation at Dongodien. Ileret pottery, a new category from the Ileret Stone Bowl site, was described. Ileret pottery, as well as other sherds from this site, showed decorative affinities with the widespread and well known Narosura Ware from central Kenya. Several partially reconstructed pots from additional sites displayed strong similarities with late historic and modern Sambury pottery. The vast majority of the pottery sherds from East Turkana were fragmentary and associated with open air settlements. Partially reconstructed vessels were recovered at the Ileret Stone Bowl site and from open air burials at GaJj 9. A more detailed discussion of each pottery ware, type and assorted sherds may be found in the individual site reports within Chapters III and IV.

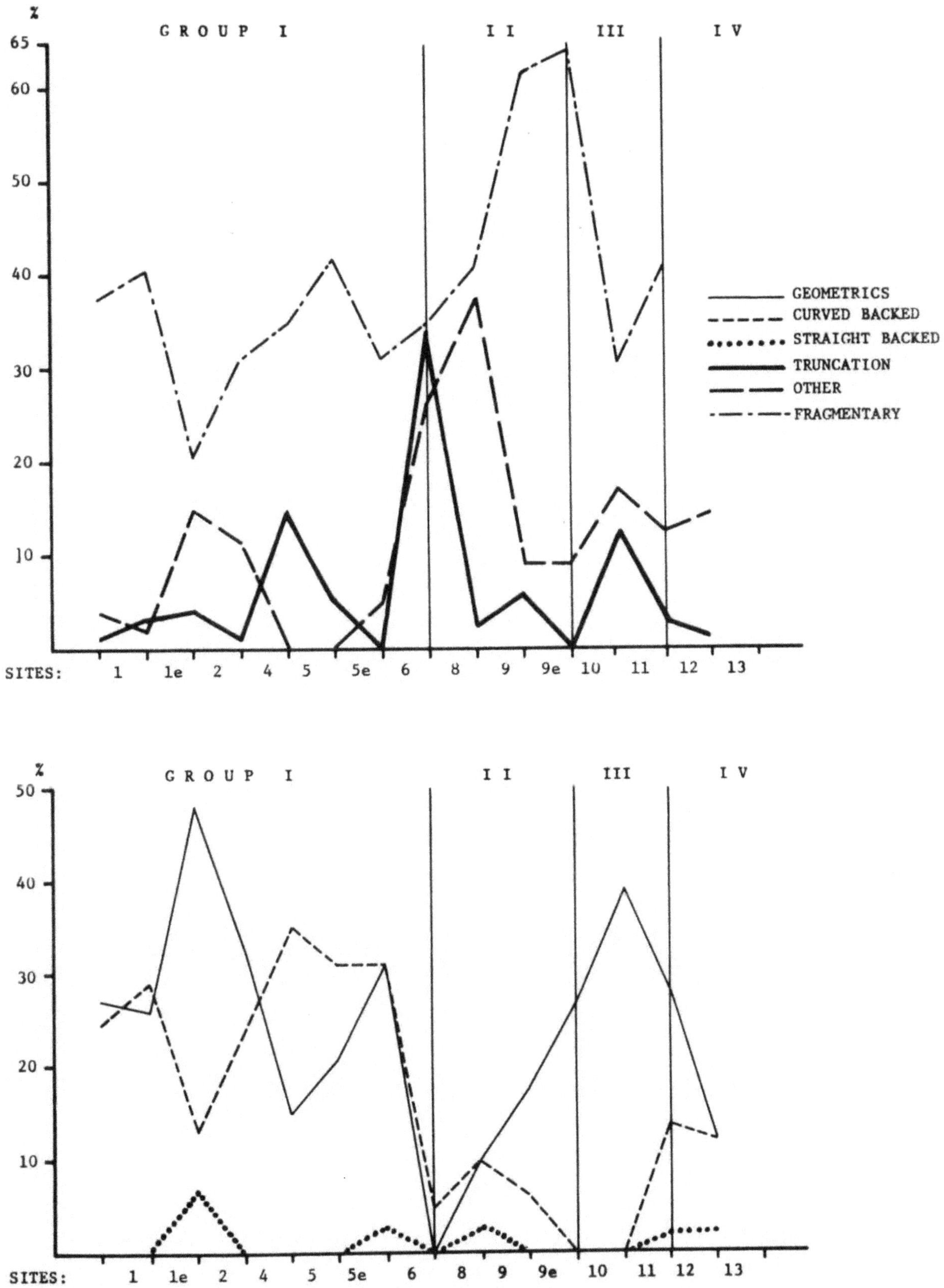

Figure 102. Percentage frequency for backed microliths by Site Group.

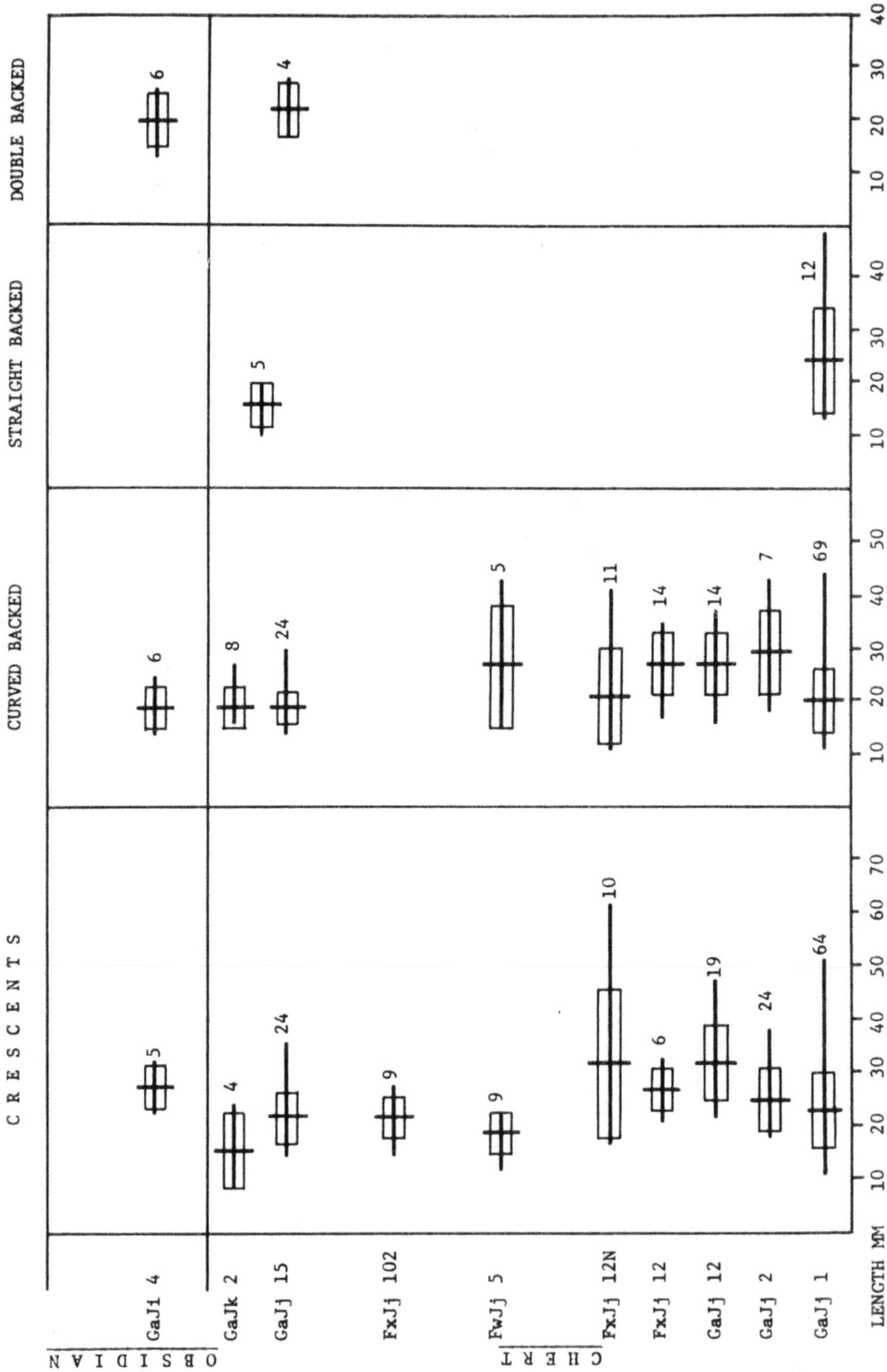

Figure 103. Modified Dice-Leraas diagrams for backed microliths parameters.

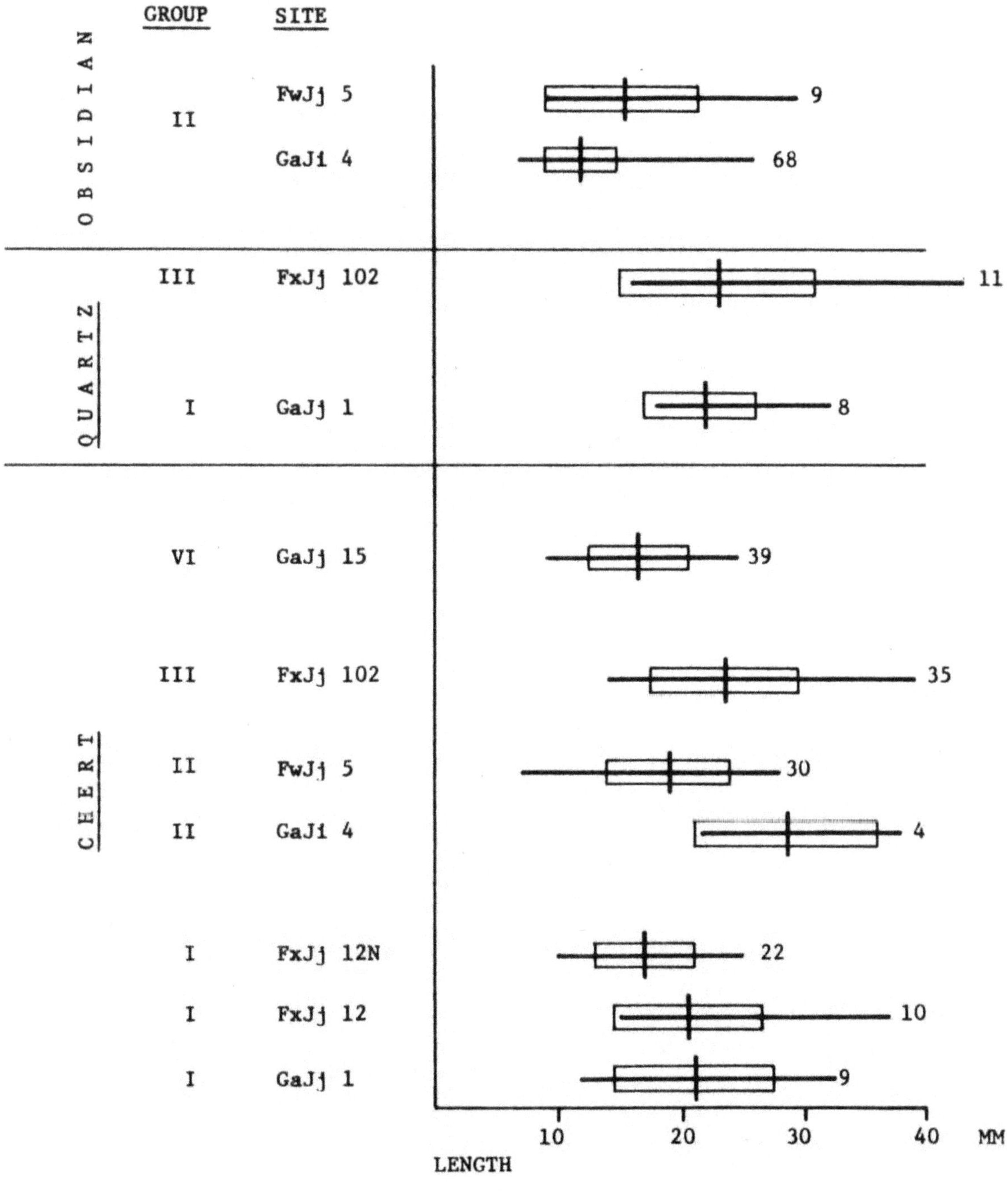

Figure 104. Modified Dice-Leraas diagrams for <u>outils écaillés</u> parameters.

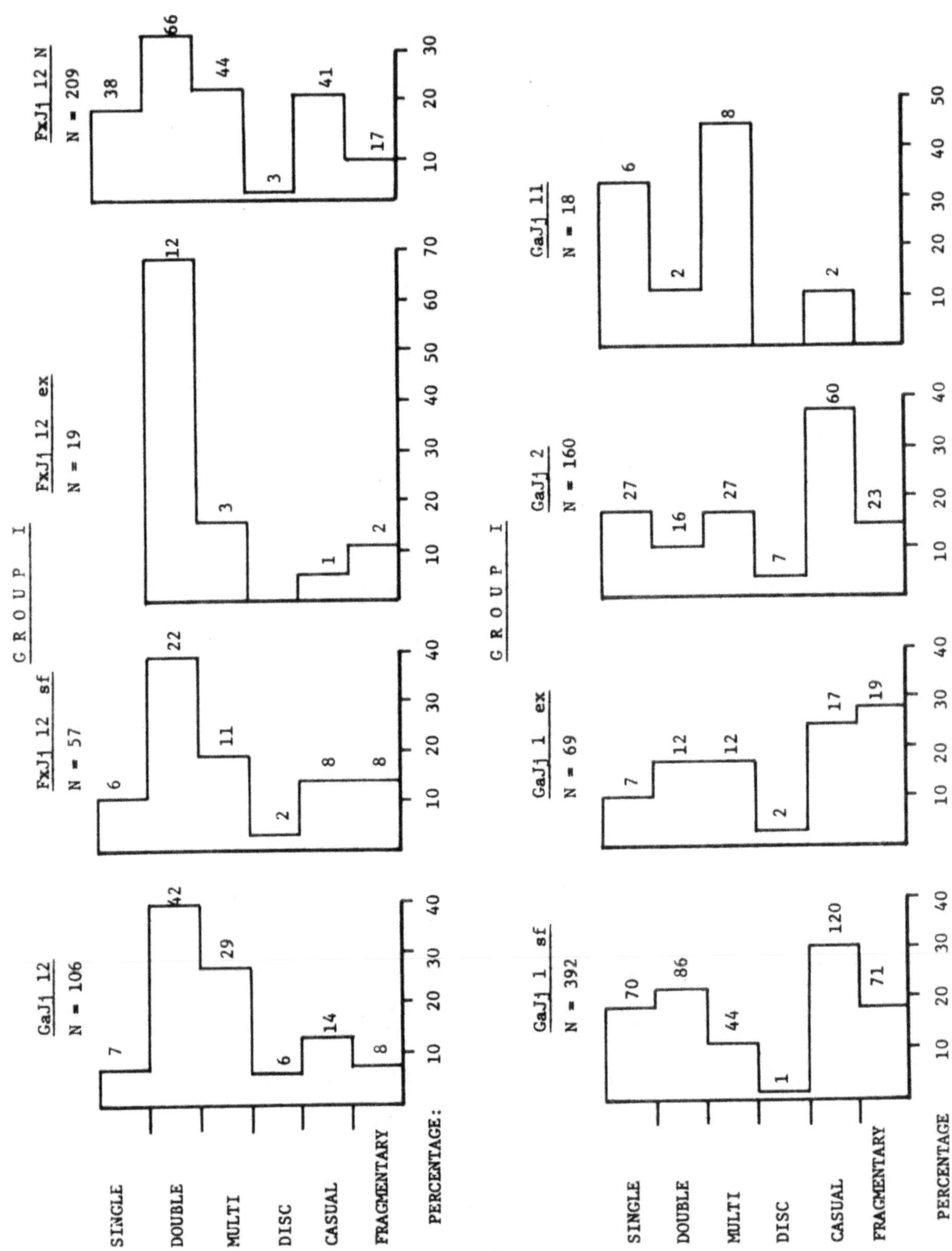

Figure 105. Histograms illustrating frequency of core forms from Group I sites.

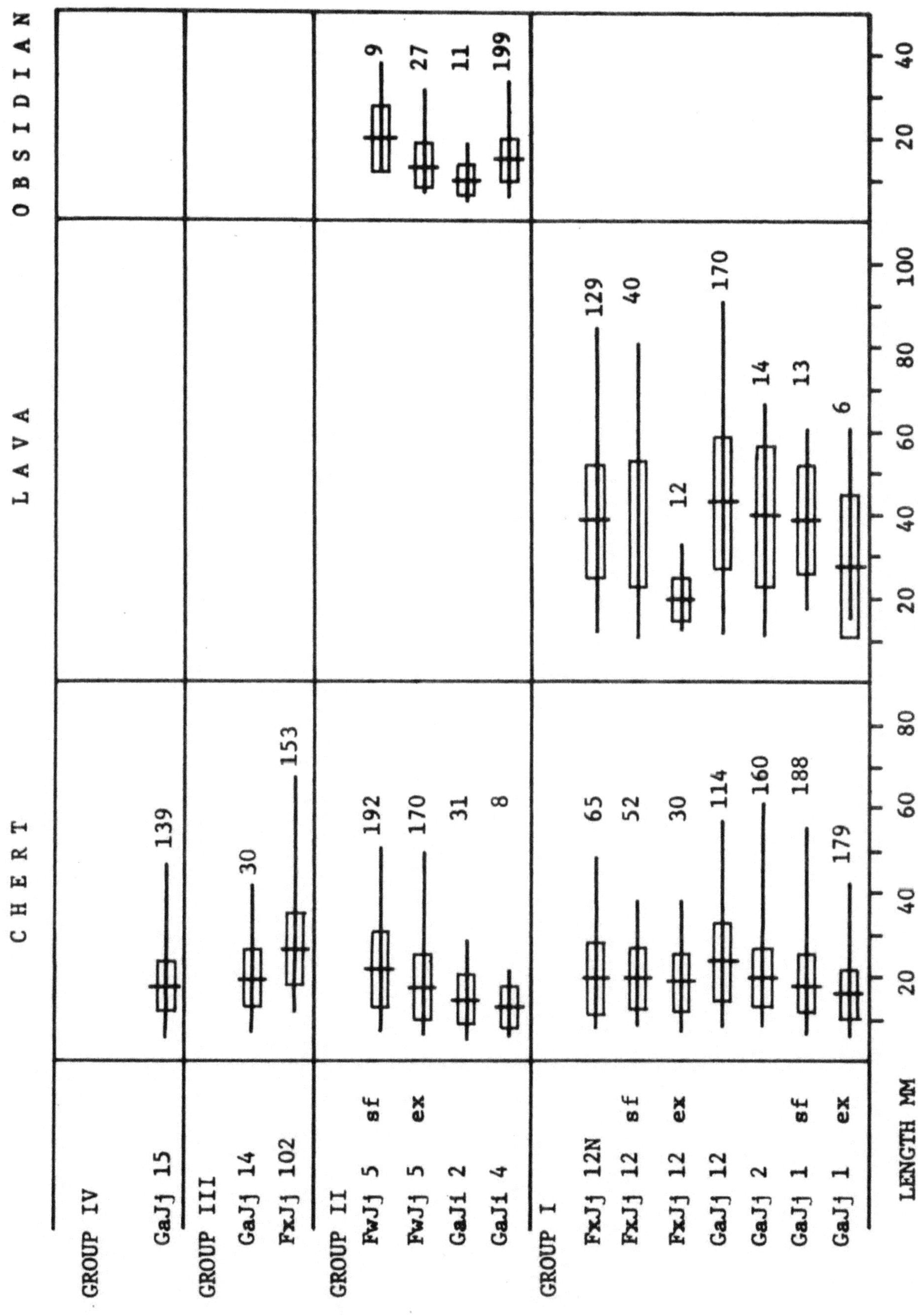

Figure 106. Modified Dice-Leraas diagrams for whole flake parameters.

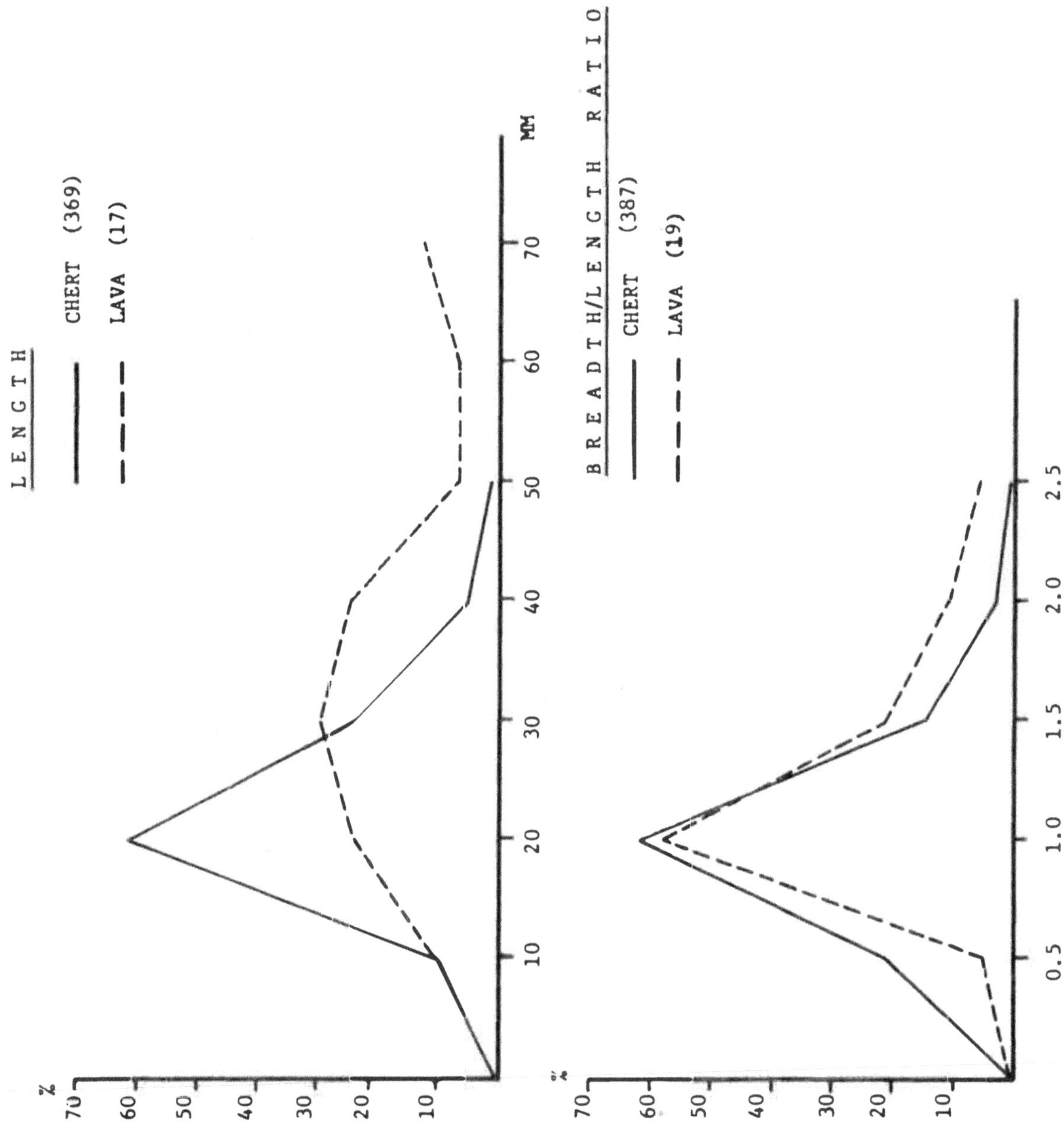

Figure 107. Selected whole flake attribute frequencies from GaJj 1 (excavated).

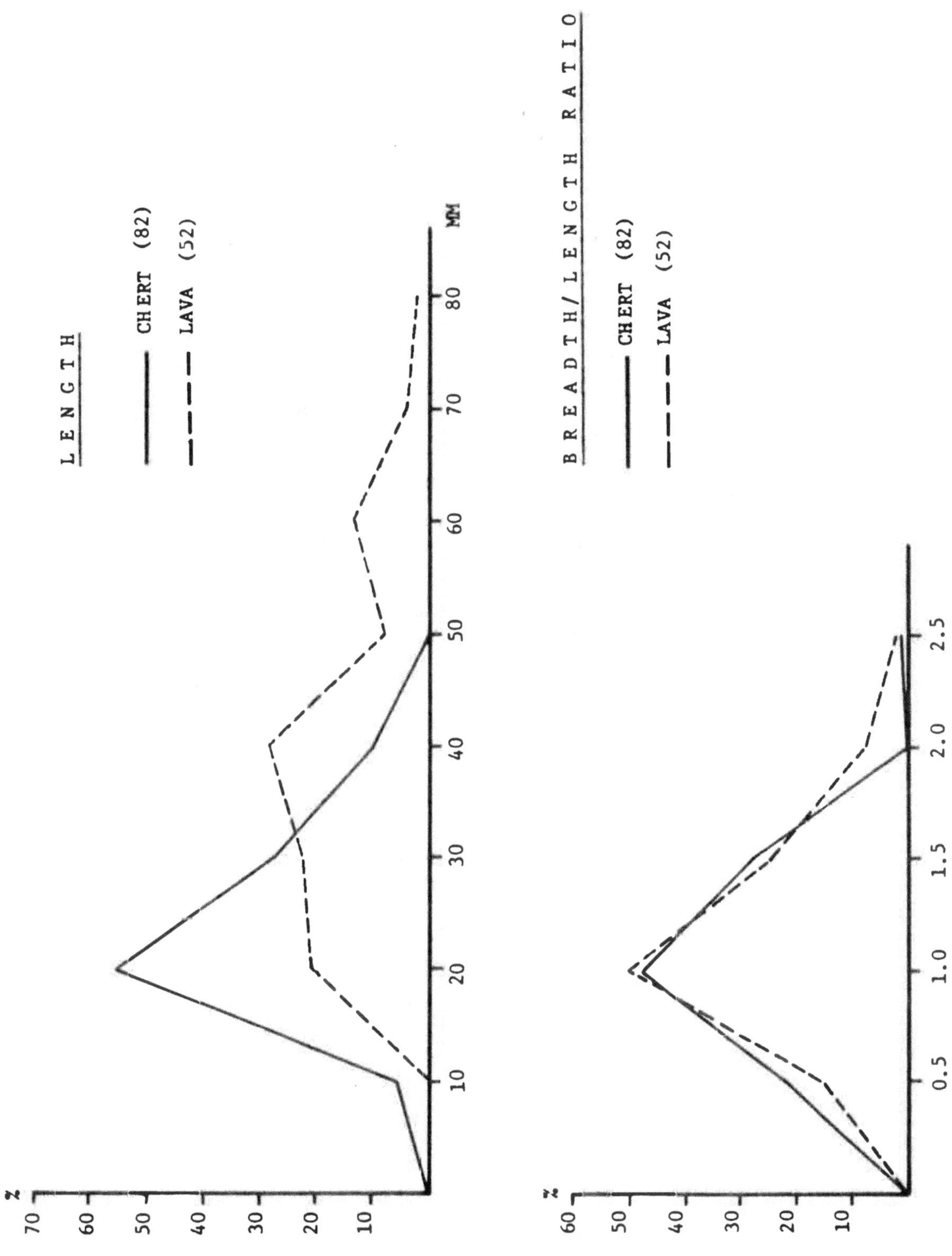

Figure 108. Selected whole flake attribute frequencies from FxJj 12 (surface).

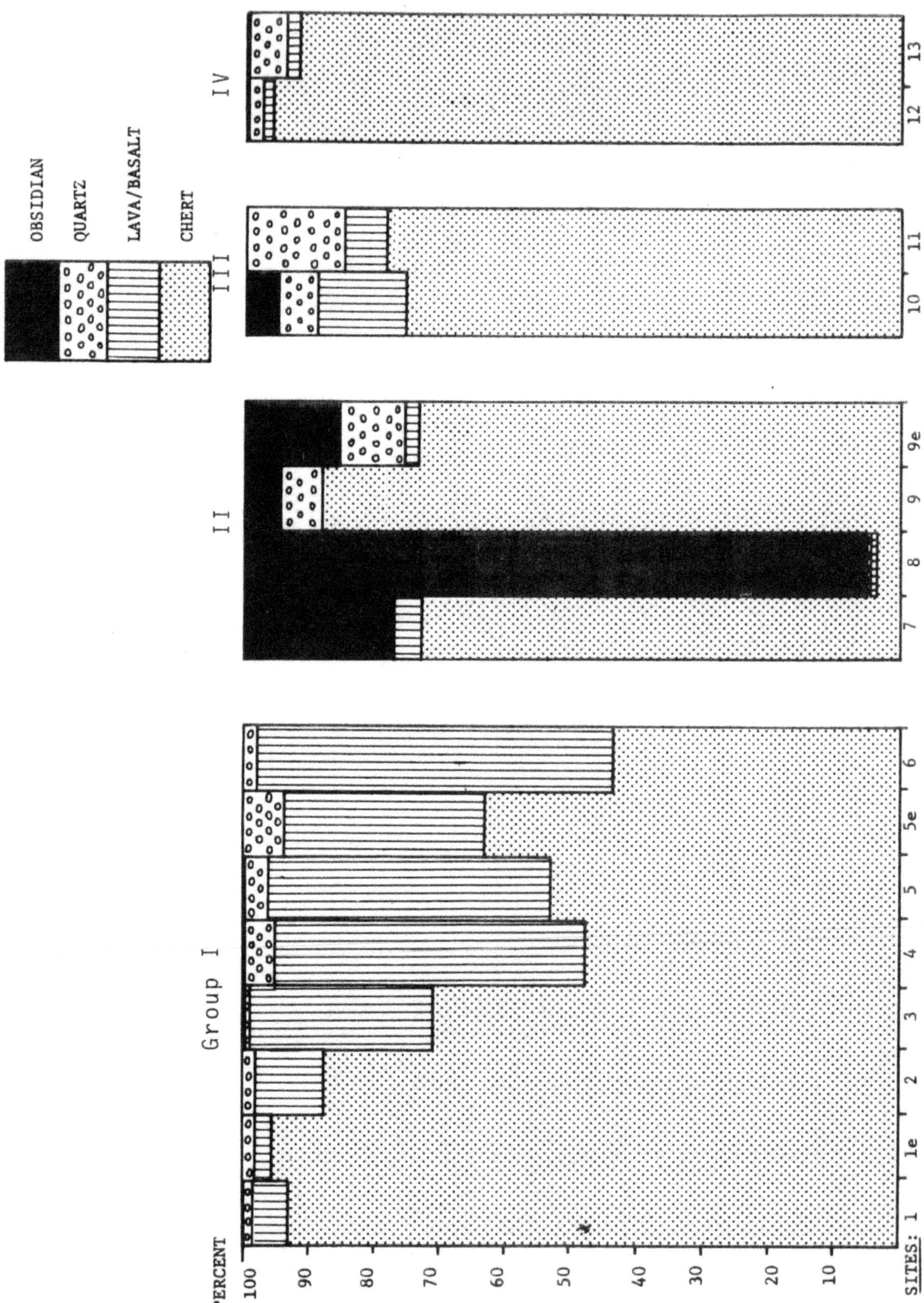

Figure 109. Cumulative percentage diagram of principal lithic raw materials.

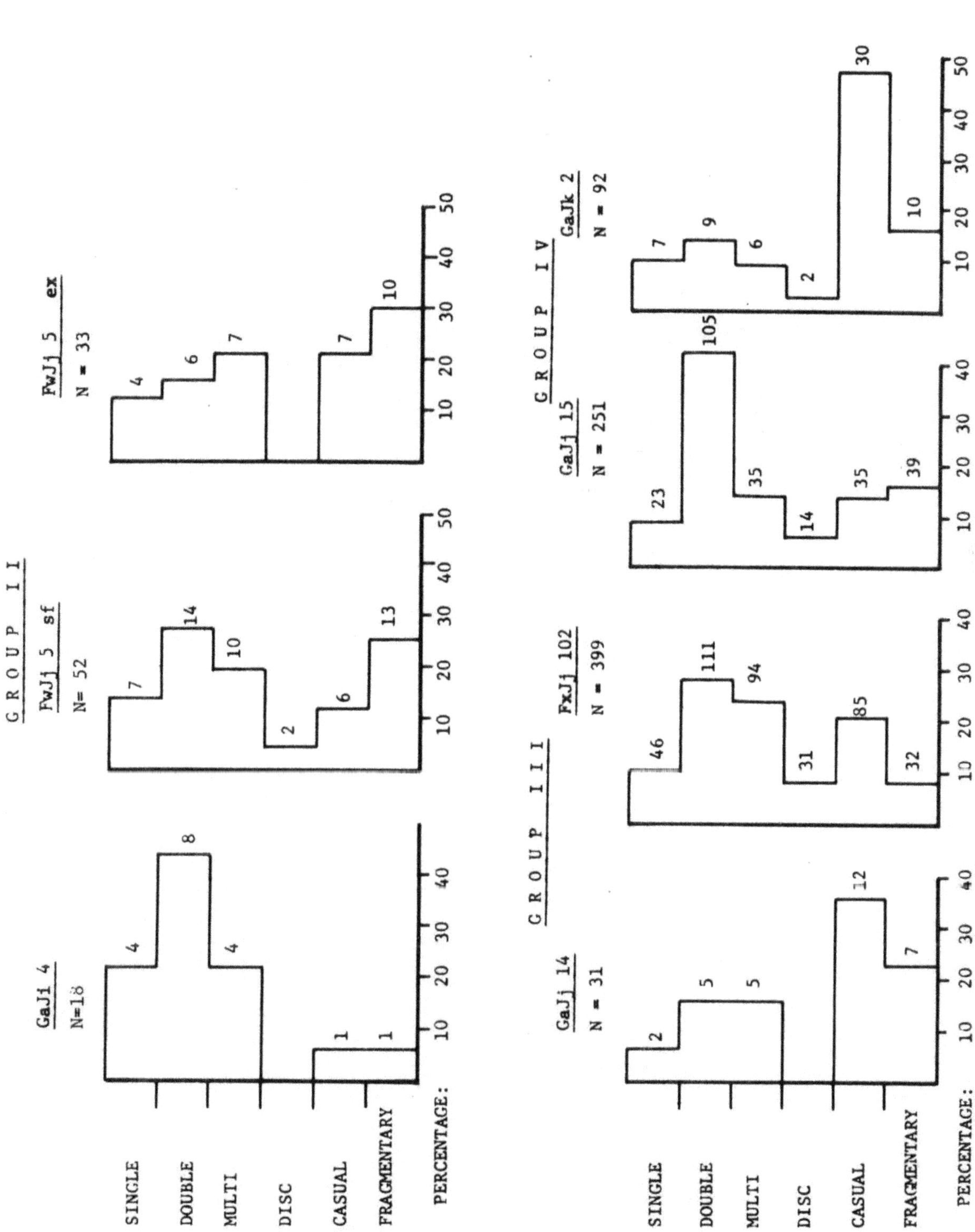

Figure 110. Histograms illustrating frequence of core forms from Site Groups II-IV.

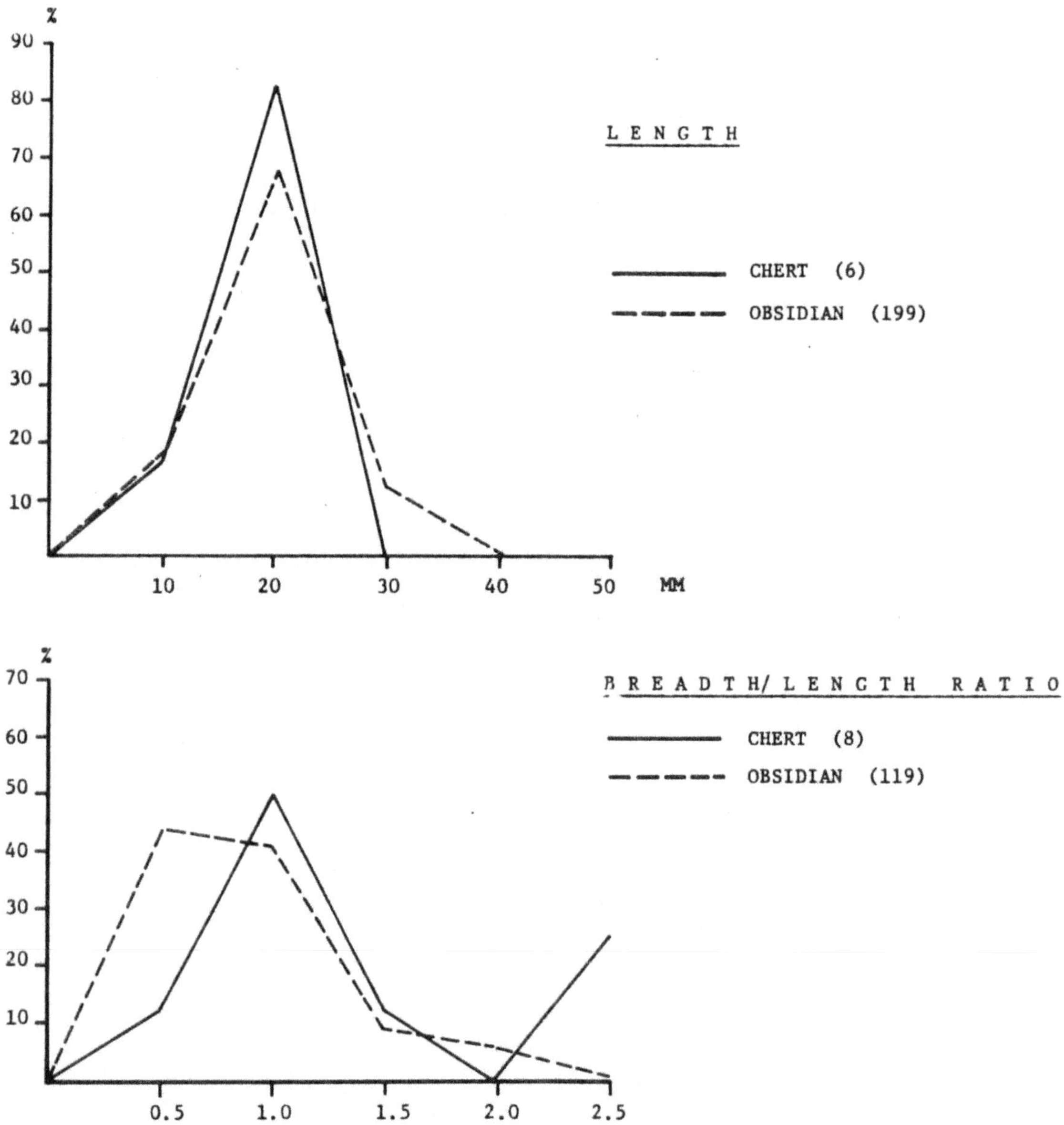

Figure 111. Selected whole flake attribute frequencies from Dongodien (GaJi 4).

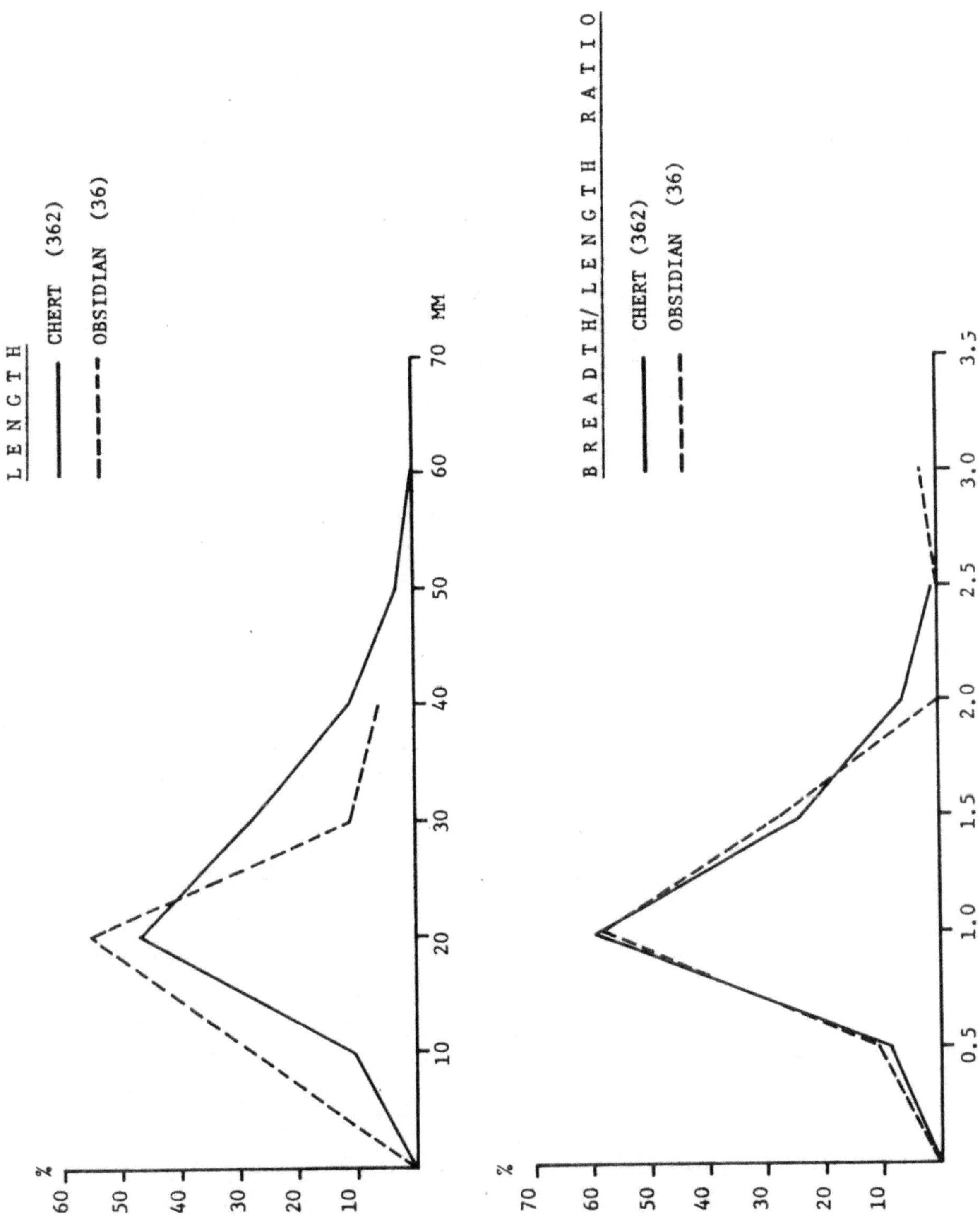

Figure 112. Selected whole flake attribute frequencies from the Ileret Stone Bowl Site (FwJj 5).

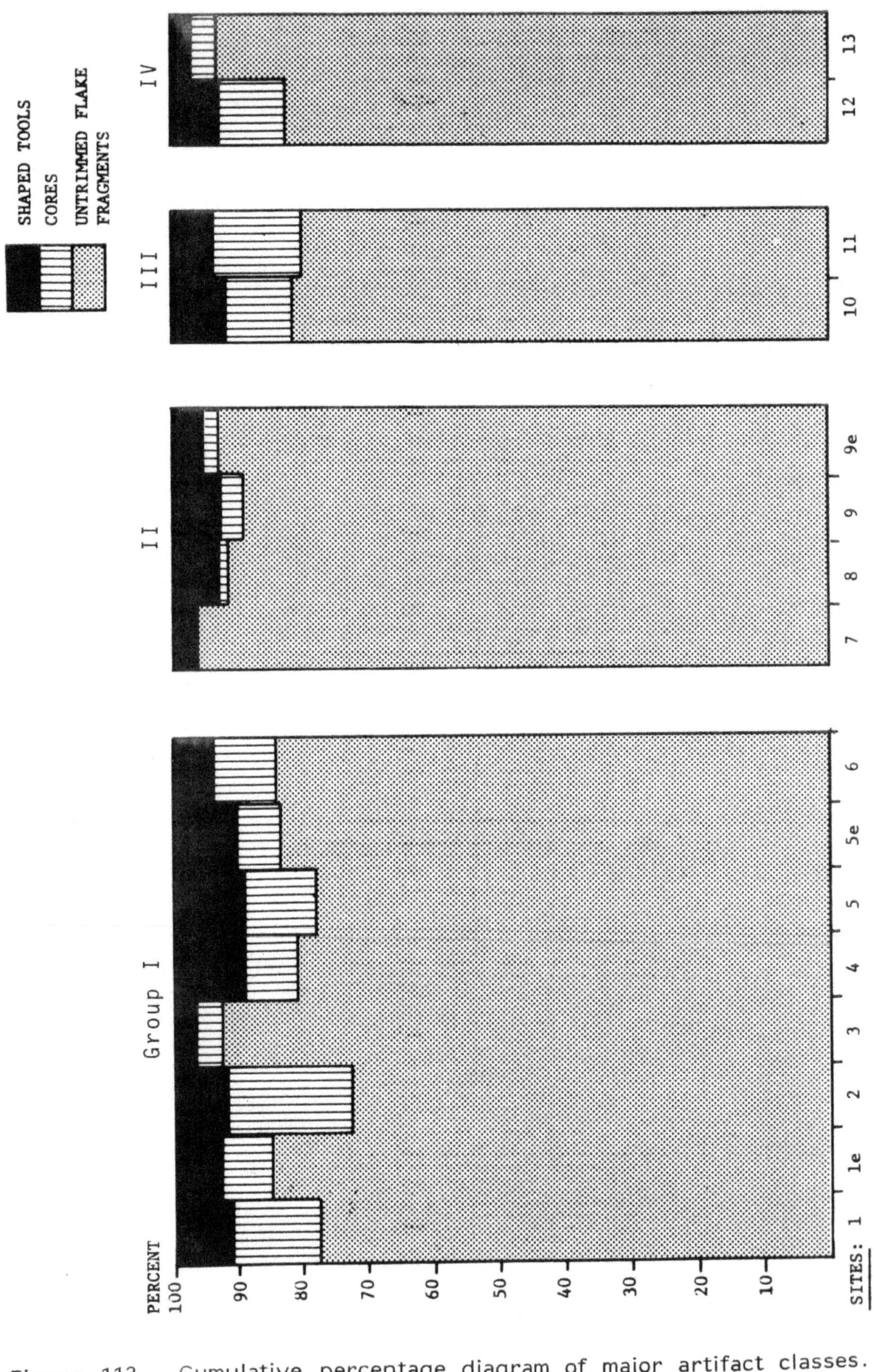

Figure 113. Cumulative percentage diagram of major artifact classes.

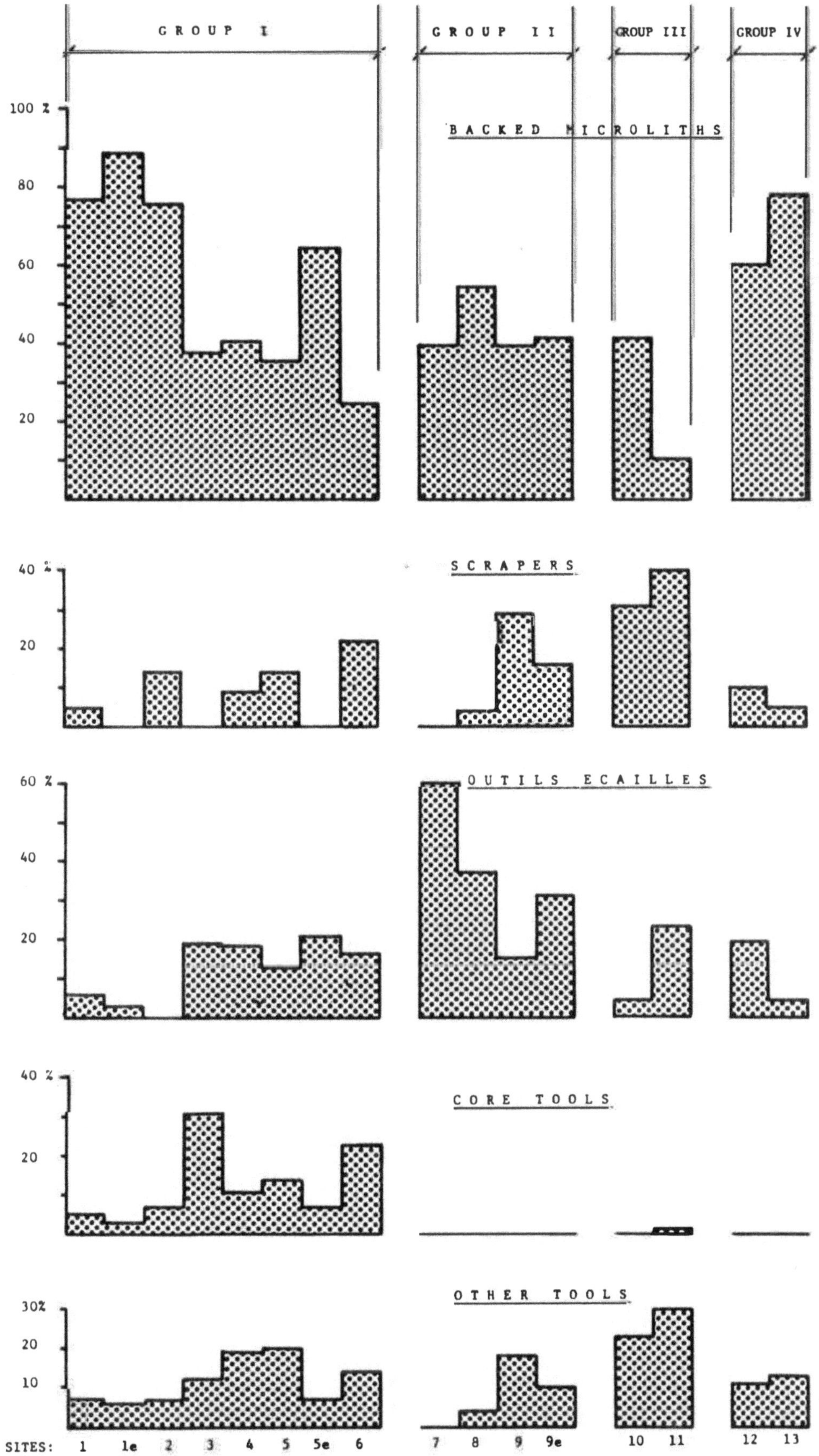

Figure 114. Frequency distributions of shaped stone artifacts by Site Group.

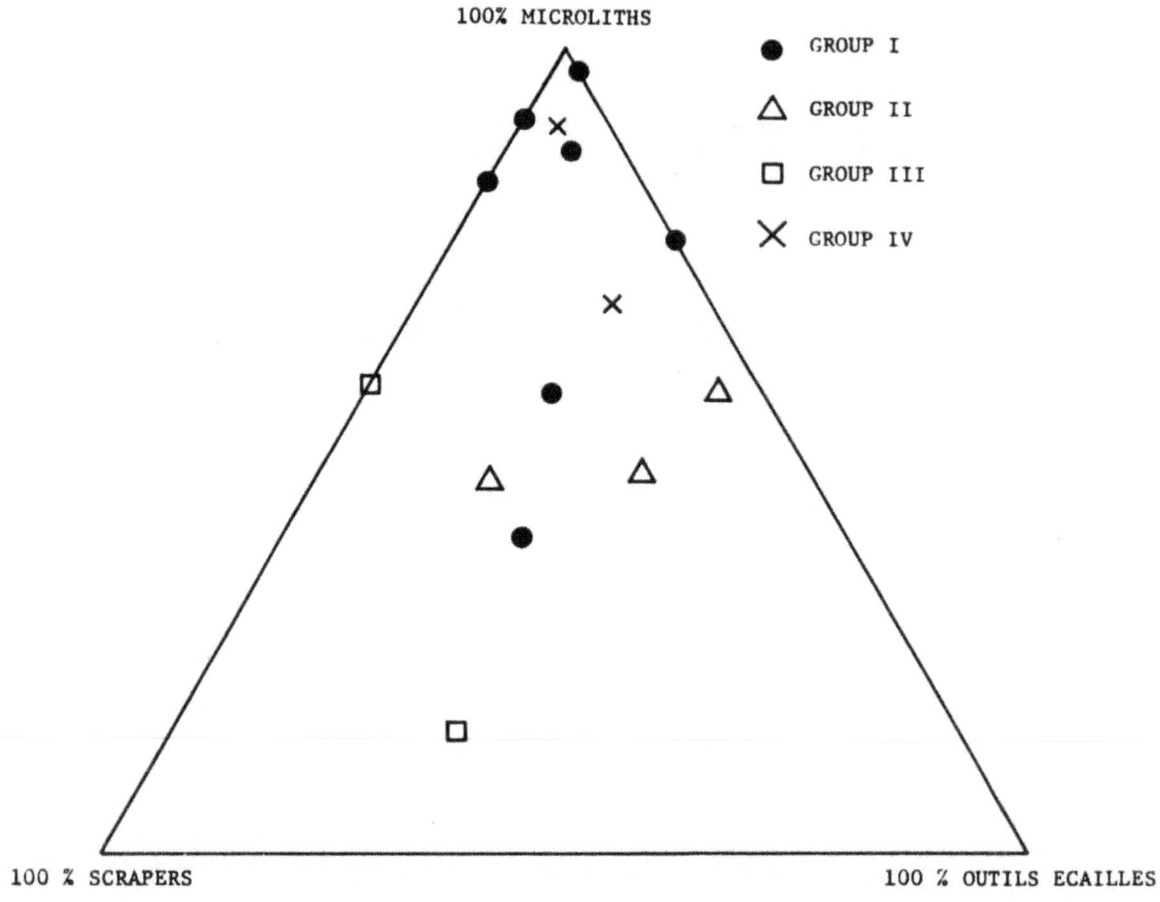

Figure 115. Relative frequencies of microliths, outils écaillés and scrapers by Site Group.

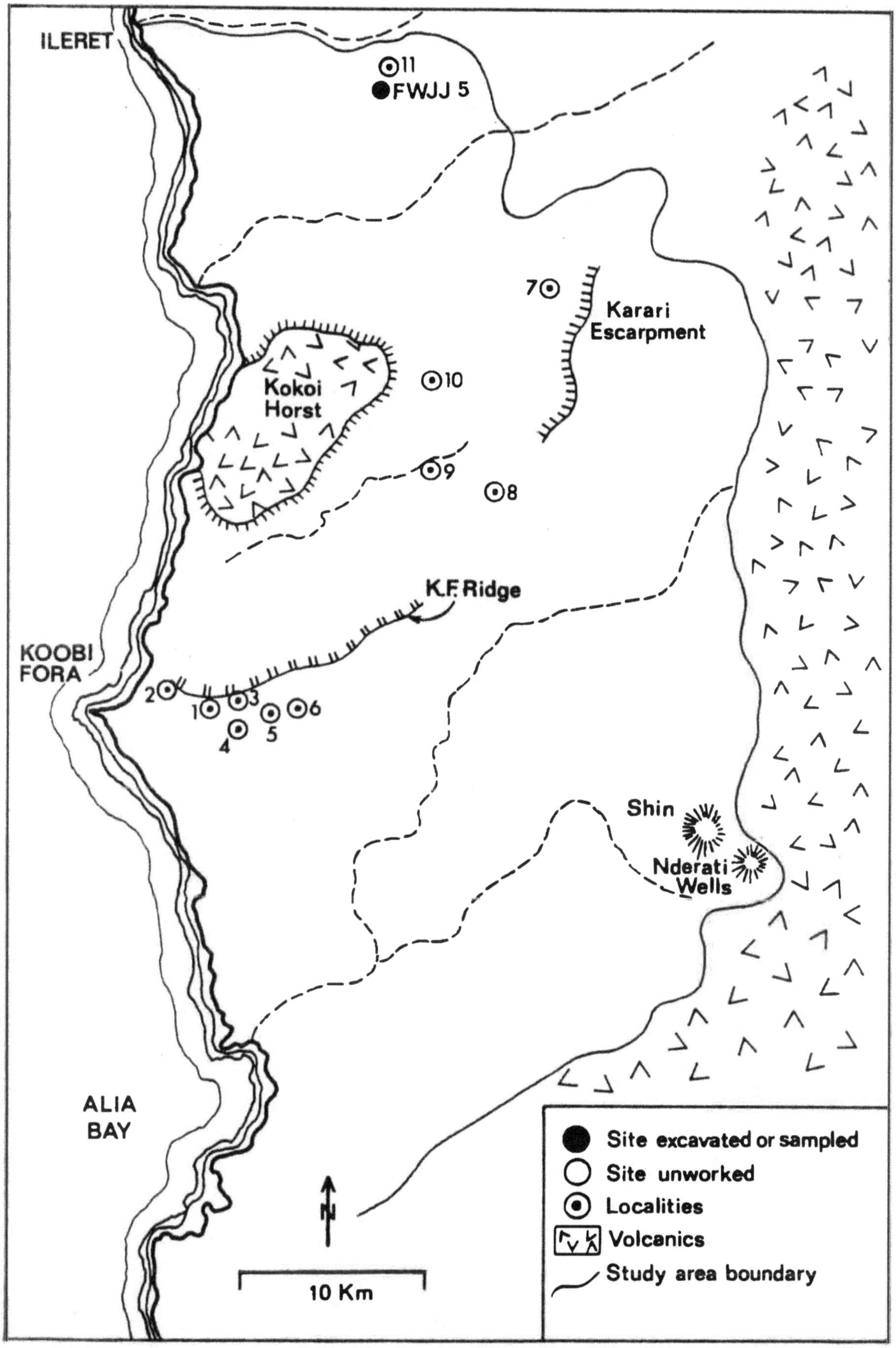

Figure 116. Location of sites and localities with stone bowls.

Table 86

INVENTORY OF CULTURAL REMAINS FROM BONE HARPOON SITES (Group 1)

	Excavated					Sampled							
	GaJj 1	GaJj 11	GaJi 3	FxJj 12	Total (N)	GaJj 1	GaJj 2	GaJj 11	GaJj 12	FxJj 12	FxJj 12N	GaJi 3	Total (N)
Stone:													
Microliths	62	-	-	19	81	209	54	6	58	20	36	-	383
Scrapers	-	-	-	-	-	11	7	-	13	8	32	-	71
Outils écaillés	2	3	-	6	8	17	-	3	26	7	24	1	81
Burin	1	-	-	-	1	3	-	-	1	2	-	-	6
Bec	-	-	-	-	-	-	1	1	1	2	2	-	7
Awl	-	-	-	-	-	2	-	-	1	-	-	-	3
Core/chopper	2	-	-	2	4	14	4	4	14	6	33	-	75
Pick	-	-	-	-	-	-	1	-	1	-	-	-	2
Limace	-	-	-	-	-	-	-	1	1	-	-	-	2
Cas. trimm.	3	-	-	2	5	8	3	1	12	7	19	-	50
Cas. retouch	-	-	-	-	-	3	1	-	12	3	-	-	19
Whole flakes	193	15	-	46	239	540	179	180	288	100	443	-	1730
Broken flakes	301	6	-	106	407	870	253	113	498	207	890	1	2831
Angular frags.	284	8	2	92	378	865	181	131	261	119	562	9	2128
Cores	80	-	-	19	99	392	160	18	106	57	209	4	939
Grindstone	-	-	-	-	-	-	-	-	-	-	-	-	-
Anvil	1	-	-	-	1	2	-	-	-	-	-	-	2
Hammerstone	2	-	-	-	2	3	2	-	-	-	-	-	5
Bone harpoons:													
Uniserial	8	1	1	1	11	47	22	121	8	6	9	3	216
Biserial	-	-	1	-	1	1	1	4	-	-	-	-	6
Triserial	-	-	-	-	-	-	-	5	-	-	-	-	5
Barbless	-	-	-	-	-	1	2	1	-	-	-	-	4
NID	-	-	-	-	-	-	-	8	-	-	-	-	8
Pottery:													
Wavy-line	-	-	-	-	-	-	-	-	-	-	9	-	9
Other: decorated	-	-	-	-	-	-	6	-	17	-	35	1	59
undecorated	16	-	-	-	16	52	46	1	24	-	134	8	265
Bone frags. (NID):													
Mammal	223	-	-	333	556	1	275	1	248	396	85	2	1008
Fish	790	-	-	1769	2559	8	144	20	46	1029	128	-	1375
Crocodile	-	-	-	1	1	-	-	2	9	-	1	-	12
Turtle	6	-	-	69	75	-	8	5	27	106	7	6	159
Bones w/art. sur.	2	-	-	51	53	4	23	9	13	119	9	1	178
Horn core	-	-	-	-	-	-	-	-	-	4	-	-	4
Teeth:													
Mammal	11	-	-	30	41	8	47	5	24	26	18	1	129
Fish	12	-	36	-	48	-	1	-	-	-	3	-	17

Table 87

INVENTORY OF PASTORAL NEOLITHIC (Group 2) AND OTHER SITES (Groups 3 and 4)

	Group 2: Excavated				Group 2: Sampled				Group 3		Group 4			
	GaJi 2	GaJi 4	FwJj 5	Total (N)	GaJi 2	GaJi 4	FwJj 5	Total (N)	GaJj 14	FxJj 102	GaJj 15	GaJk 1	GaJk 2	Total (N)
Stone:														
Microliths	2	85	34	121	12	40	54	106	11	23	127	3	63	227
Scrapers	-	7	13	20	-	9	41	50	8	85	21	-	4	118
Outils écaillés	3	57	25	85	3	16	22	41	1	49	39	1	3	93
Bec	-	-	2	2	-	-	8	8	-	13	3	-	3	19
Burin	-	-	-	-	-	2	2	4	1	-	-	-	-	1
Awl	-	-	-	-	-	-	-	-	-	6	-	-	2	8
Core/chopper	-	-	-	-	-	1	-	1	-	1	-	-	-	1
Cas. trimm.	-	3	4	7	-	7	14	21	4	19	19	-	5	47
Cas. retouch	-	2	2	4	-	1	4	5	2	15	-	-	-	17
Whole flakes	23	547	217	787	22	151	226	505	36	517	346	34	119	1052
Broken flakes	46	650	532	1228	30	109	326	465	90	919	863	65	389	2326
Angular frags.	49	511	615	1175	58	123	847	1028	130	924	928	396	1627	4005
Cores	-	18	33	51	2	19	52	73	31	399	251	13	92	786
Stone bowls	-	-	-	-	-	1	8	9	-	-	-	-	-	-
Groundstone	-	-	-	-	-	1	-	1	-	1	-	-	-	1
Anvil	-	-	-	-	-	-	-	-	-	1	-	-	-	1
Grindstone	-	-	-	-	-	1	-	1	-	1	-	-	-	1
Pottery:														
Nderit Ware	-	28	-	28	-	22	-	22	-	-	-	-	-	-
Burnished/slipped	-	216	115	331	-	76	52	128	-	-	-	-	-	-
Other:														
Decorated	16	34	419	469	8	103	139	250	13	18	-	13	-	44
Undecorated	46	78	1591	1715	18	393	386	797	57	2	-	8	-	67
Disc	-	16	5	21	-	10	-	10	-	-	-	-	-	-
Handles	-	-	-	-	-	-	2	2	-	-	-	2	-	2

Table 87 Continued

INVENTORY OF PASTORAL NEOLITHIC (Group 2) AND OTHER SITES (Groups 3 and 4)

	Group 2: Excavated				Group 2: Sampled				Group 3		Group 4			
	GaJi 2	GaJi 4	FxJj 5	Total (N)	GaJi 2	GaJi 4	FwJj 5	Total (N)	GaJj 14	FxJj 102	GaJj 15	GaJk 1	GaJk 2	Total (N)
Bone frags. (NID):														
Mammal	312	1792	420	2524	127	121	201	449	31	4	8	-	-	43
Fish	249	3467	319	4035	252	191	4	447	-	-	-	-	-	-
Turtle	-	-	-	-	-	4	-	4	-	-	-	-	-	-
Bones w/art. sur.	17	332	12	361	9	27	8	43	2	1	-	-	-	3
Horn core	-	-	-	-	-	1	-	1	-	-	-	-	-	-
Teeth:														
Mammal	29	189	58	276	-	27	43	70	-	1	5	-	-	6
Fish	1	10	5	16	-	2	-	2	-	-	-	-	-	-
Crocodile	1	-	-	1	-	-	-	-	-	-	-	-	-	-
Beads:														
Ostrich eggshell	1	14	1	16	1	-	-	1	-	-	-	-	-	-
Stone	1	-	-	1	-	-	-	-	-	-	-	1	-	1
Glass	-	-	-	-	-	-	-	-	-	-	-	3	-	3
Bone tool	-	1	-	1	-	-	-	-	-	-	-	-	-	-
Ostrich e.g. frags.	8	22	35	65	15	-	-	15	10	1	1	-	-	12
Fire cracked rock	9	17	110	136	-	-	-	-	-	-	-	-	-	-
Angular silt blocks	-	95	-	95	2	-	-	2	-	-	-	-	-	-

Table 88

SUMMARY TOTALS OF CULTURAL REMAINS FROM ALL SITES AND LOCALITIES

	Bone Harpoon Sites (Group 1)		Pastoral Neolithic Sites (Group 2)		Other Sites		Site Survey		Total (N)
	Excavated	Surface	Excavated	Surface	Group 3 Surface	Group 4 Surface	Localities	Sites	
Stone:									
Microliths	383	81	106	121	34	190	12	9	939
Scrapers	71	-	50	20	93	25	5	4	268
Outils écaillés	81	8	41	85	50	42	-	-	308
Burins	6	1	4	-	-	-	-	1	13
Becs	7	-	8	2	13	6	-	-	36
Awls	3	-	-	-	6	2	-	1	12
Bifacial point	-	-	-	-	-	-	1	-	1
Pick	1	-	-	-	-	-	-	-	1
Limace	2	-	-	-	-	-	1	-	3
Core/chopper	75	4	1	-	1	-	-	-	81
Cas. trimm.	50	5	21	7	23	24	2	3	135
Cas. retouch	19	-	5	4	17	-	-	-	45
Whole flakes	1730	239	505	787	553	465	9	18	4340
Broken flakes	2831	407	465	1228	1009	1252	11	25	7293
Angular frags.	2128	378	1028	1175	1055	2555	8	30	8752
Cores	939	99	73	51	430	343	2	7	1957
Stone bowls	-	-	9	-	-	-	9	-	18
Groundstone	2	-	1	-	1	-	3	-	7
Anvil	1	1	-	-	1	-	-	-	3
Hammerstones	5	2	1	-	9	-	-	-	17
Bone harpoons:									
Uniserial	216	11	-	-	-	-	12	6	245
Biserial	6	1	-	-	-	-	1	1	7
Triserial	5	-	-	-	-	-	-	-	5
Barbless	4	-	-	-	-	-	-	-	4
NID	8	-	-	-	-	-	-	-	8
Pottery:									
Nderit Ware	-	-	22	28	-	-	14	-	64
Wavy-line	-	9	-	-	-	-	-	1	10
Burnished/slipped	-	-	128	331	-	-	-	17	476

Table 88 Continued

SUMMARY TOTALS OF CULTURAL REMAINS FROM ALL SITES AND LOCALITIES

	Bone Harpoon Sites (Group 1)		Pastoral Neolithic Sites (Group 2)		Other Sites		Site Survey		Total (N)
	Excavated	Surface	Excavated	Surface	Group 3 Surface	Group 4 Surface	Localities	Sites	
Pottery: (continued)									
Other:									
Decorated	60	-	250	469	21	-	186	19	1028
Undecorated	264	16	797	1715	69	-	78	32	2969
Disc	-	-	10	21	-	-	-	-	31
Handle	-	-	2	2	-	-	11	-	17
Reconstructed pots	-	-	-	-	-	-	3	5	8
Bone frags. (NID):									
Mammal	1008	556	449	2524	35	8	-	-	4580
Fish	3243	2559	447	4035	-	-	-	-	10284
Crocodile	12	1	-	-	-	-	-	-	13
Turtle	164	75	4	-	-	-	-	-	243
Bird	-	-	-	-	-	-	-	-	-
Bones w/art. sur.	178	53	43	361	3	-	-	-	638
Horn core	-	4	1	-	-	-	-	-	5
Teeth:									
Mammal	129	41	70	276	1	5	-	-	522
Fish	21	48	2	16	-	-	-	-	87
Crocodile	-	-	-	1	-	-	-	-	1
Beads:									
Ostrich eggshell	7	4	1	16	-	-	-	-	28
Stone	1	-	-	1	-	-	-	-	3
Bone	-	2	-	-	-	-	-	-	2
Glass	-	-	-	-	-	-	5	-	5
Ostrich e.s. frags.	13	2	15	65	11	1	-	-	107
Homo sapiens									
Fragmentary	54	2	9	-	-	-	-	-	63
Burials	-	-	-	-	-	-	-	2	2

Table 89

PERCENTAGE COMPOSITION OF SHAPED STONE TOOLS BY SITE

Site	(N)	Microliths	Scrapers	Outils Écaillés	Burins	Becs/Awls	Core Tools	Cas. Trimm.	Cas. Retouch
Group 1									
GaJj 1	271	77.0	4.8	6.3	1.2	1.9	5.2	3.0	1.1
GaJj 1: Ex.	70	89.0	-	2.8	1.4	-	2.9	4.3	-
GaJj 2	71	76.1	13.8	-	-	1.4	7.0	4.3	1.4
GaJj 11	16	37.6	-	18.8	-	6.3	31.3	6.3	-
GaJj 12	139	41.2	8.6	18.8	0.7	1.4	10.8	8.6	8.6
FxJj 12	57	35.7	14.3	12.5	3.8	3.6	14.3	10.7	5.4
FxJj 12: Ex.	29	65.5	-	20.7	-	-	6.9	6.9	-
FxJj 12 North	146	24.7	21.9	16.4	-	1.4	22.6	13.0	-
Group 2									
GaJi 2: Ex.	5	40.0	-	60.0	-	-	-	-	-
GaJi 4: Ex.	155	54.8	4.5	36.8	-	-	-	2.6	1.3
FwJj 5: Sf.	143	40.0	28.7	15.4	-	5.6	-	9.8	2.8
FwJj 5: Ex.	80	42.5	16.3	31.3	-	2.5	-	5.0	2.5
Group 3									
GaJj 14	26	42.3	30.8	-	-	-	-	15.4	7.7
FxJj 102	211	10.9	40.3	23.2	-	9.0	0.5	9.0	7.1
Group 4									
GaJj 15	209	60.8	10.0	18.7	-	1.4	-	9.1	-
GaJk 2	80	78.8	5.0	3.8	-	6.3	-	6.3	-

Table 90

PERCENTAGE COMPOSITION OF MICROLITHIC TOOLS BY SITE

Site	(N)	Triangle	Trapezoid	Crescent	Curved Backed Flakes	Straight Backed Flakes	Double Backed Flakes	Truncations	Micro-Percoir	Misc.	Fragmentary
Group 1											
GaJj 1: Sf.	208	1.4	1.4	24.5	24.5	5.8	-	1.0	1.5	2.4	37.5
GaJj 1: Ex.	62	1.6	1.6	22.7	29.0	-	-	3.2	-	1.6	40.3
GaJj 2	54	1.9	1.9	44.3	13.0	-	-	3.8	3.7	11.1	20.3
GaJj 12	58	-	-	32.8	24.2	-	-	1.7	3.4	6.9	31.0
FxJj 12: Sf.	20	-	5.0	10.0	35.0	-	-	15.0	-	-	35.0
FxJj 12: Ex.	19	-	-	21.1	31.6	-	-	5.3	-	-	42.1
FxJj 12 North	36	-	-	31.0	31.0	2.5	-	-	5.0	-	31.0
Group 2											
GaJi 2: Ex.	2	-	-	50.0	-	-	-	-	-	-	50.0
GaJi 4: Ex.	85	-	-	-	5.0	-	14.0	34.0	7.0	5.0	35.0
FwJj 5: Sf.	54	-	-	7.4	7.4	1.9	5.6	1.9	22.2	9.3	40.7
FwJj 5: Ex.	34	-	-	12.6	5.9	-	-	5.9	-	8.8	61.8
Group 3											
GaJj 14	11	-	9.1	18.2	-	-	9.1	-	-	-	63.4
FxJj 102	23	-	-	39.1	-	-	-	13.0	-	17.4	30.4
Group 4											
GaJj 15	127	2.4	1.6	23.7	13.4	2.4	1.6	3.2	3.9	7.1	41.4
GaJk 2	63	1.6	1.6	7.9	14.3	9.5	7.9	3.2	1.6	5.0	47.6

Table 91

PERCENTAGE COMPOSITION OF CORES BY SITE

Site	(N)	Single Platform	Double	Multi	Disc	Casual	Fragmentary
Group 1							
GaJj 1: Sf.	392	17.9	22.0	11.1	0.3	30.6	18.1
GaJj 1: Ex.	69	10.1	17.3	17.6	2.9	24.6	27.5
GaJj 2	160	16.9	10.0	16.9	4.3	37.5	14.4
GaJj 11	18	33.3	11.1	44.5	-	11.1	-
GaJj 12	106	6.6	39.6	27.4	5.7	13.2	7.5
FxJj 12: Sf.	57	10.5	38.6	19.4	3.5	14.0	14.0
FxJj 12: Ex.	19	-	68.4	15.8	-	5.3	10.5
FxJj 12 North	209	18.2	31.7	21.1	1.4	20.5	8.1
Group 2							
GaJi 2: Ex.	-	-	-	-	-	-	-
GaJi 4: Ex.	18	22.0	44.0	22.0	-	6.0	6.0
FwJj 5: Sf.	52	13.5	26.9	19.2	3.8	11.5	25.0
FwJj 5: Ex.	33	12.1	15.1	21.3	-	21.2	30.3
Group 3							
GaJj 14	31	6.5	16.2	16.1	-	35.5	22.6
FxJj 102	399	11.5	27.8	23.6	7.8	21.3	8.0
Group 4							
GaJj 15	251	9.2	41.8	13.9	5.6	13.9	15.5
GaJk 2	92	12.0	48.9	2.2	1.1	9.8	26.1

Table 92

PERCENTAGE COMPOSITION OF STONE ARTIFACT ASSEMBLAGES AND RAW MATERIAL

Site	(N)	Shaped Tools	Cores	UTFF	Chert	Lava	Quartz	Obsidian
Group 1 sites								
GaJj 1	2938	9.2	13.4	77.4	93.3	4.8	1.6	0.3
GaJj 1: Ex.	917	7.7	7.3	85.0	96.4	2.1	1.4	0.1
GaJj 2	844	8.4	19.0	72.6	87.7	10.9	1.4	-
GaJj 11	458	3.5	3.9	92.6	71.2	27.9	0.7	0.2
GaJj 12	1292	10.8	8.2	81.0	48.1	47.4	3.8	0.7
FxJj 12	522	10.9	10.9	78.2	53.4	42.8	3.5	0.3
FxJj 12: Ex.	295	9.8	6.4	83.4	63.7	30.2	6.1	-
FxJj 12 North	2259	6.5	9.3	84.2	44.0	54.2	1.7	0.1
Group 2								
GaJi 2: Ex.	123	4.1	-	95.9	73.2	4.0	-	22.8
GaJi 4: Ex.	1848	8.4	1.0	90.6	4.3	0.3	0.2	95.2
FwJj 5	1596	9.0	3.0	88.0	85.0	3.0	6.4	5.6
FwJj 5: Ex.	1477	5.4	2.2	92.4	73.6	2.7	9.8	13.9
Group 3								
GaJj 14	314	8.3	9.9	81.8	75.5	13.4	6.1	5.1
FxJj 102	2970	7.1	13.4	79.5	78.3	6.7	14.9	0.1
Group 4								
GaJj 15	2597	8.0	9.7	82.3	91.8	2.0	5.6	0.6
GaJk 2	2307	3.5	4.0	92.5	96.0	1.8	1.7	0.5

Table 93

A COMPARISON OF PERCENTAGE COMPOSITIONS OF STONE ARTIFACT ASSEMBLAGES:
GROUP 1 SITES AND SELECTED FISHING SETTLEMENTS IN NORTHERN, EASTERN AND CENTRAL AFRICA

	Shaped Tools		Cores		UTFF		Total	Microliths		Scrapers		Core Tools		Outils Écaillés		Other	
	N	%	N	%	N	%	(N)	N	%	N	%	N	%	N	%	N	%
GaJj 1: Sf.	271	9.2	392	13.4	2,275	77.4	2,938	208	77.0	13	4.8	14	5.2	17	6.3	19	6.7
GaJj 1: Ex.	70	7.7	80	7.3	778	85.0	915	62	89.0	-	-	2	2.9	2	2.9	4	5.7
GaJj 2	71	8.4	160	19.0	613	72.6	844	54	76.1	7	13.8	5	7.0	-	-	5	7.1
GaJj 11	16	3.5	18	3.9	424	92.6	458	6	37.6	-	-	5	31.3	3	18.8	2	12.3
GaJj 12	139	10.8	106	8.2	1,047	81.0	1,292	58	41.2	12	8.6	15	10.8	26	18.8	28	19.3
FxJj 12: Sf.	57	10.9	57	10.9	408	78.2	522	20	35.7	8	14.3	8	14.3	7	12.5	14	23.5
FxJj 12: Ex.	29	9.8	19	6.4	247	83.4	295	19	65.5	-	-	2	6.9	6	20.7	2	6.9
FxJj 12 North	146	6.5	209	9.3	1,904	84.2	2,259	36	24.7	32	21.9	33	22.6	24	16.4	21	14.4
Lothagam (Middle Ex.)	479	2.3	160	0.7	20,050	97.0	20,654	175	36.5	234	48.8	24	5.0	-	-	46	9.7
Lowasera (Units 1, 2, 4)	79	3.4	71	3.1	2,136	93.5	2,281	58	73.4	18	22.8	2	2.5	1	1.3	-	-
Ishango	283	13.2	256	11.9	1,610	74.9	2,149	32	11.3	46	16.3	110	38.9	-	-	85	30.0
Khartoum ("Mesolithic")	-	-	-	-	-	-	-	952	67.9	198	14.1	-	-	-	-	253	18.0
Kiffian Industry (3 sites combined)	504	21.3	44	1.9	1,816	76.8	2,364	152	30.2	154	30.6	-	-	13	2.6	185	36.7

Table 94

A COMPARISON OF PERCENTAGE COMPOSITIONS OF STONE ARTIFACT ASSEMBLAGES: GROUP 2 SITES AND SELECTED EAST AFRICAN PASTORAL NEOLITHIC SETTLEMENTS

	Shaped Tools		Cores		UTFF		Total	Microliths		Scrapers		Outils Écaillés		Other		Burins	
	N	%	N	%	N	%	(N)	N	%	N	%	N	%	N	%	N	%
GaJi 4: Ex.	155	8.4	18	1.0	1,675	90.6	1,848	85	54.8	7	4.5	57	36.8	6	3.9	-	-
FwJj 5: Sf.	143	9.0	52	3.0	1,401	88.0	1,596	52	40.0	41	28.7	22	15.4	26	18.2	-	-
FwJj 5: Ex.	80	5.4	33	2.2	1,364	92.4	1,477	34	42.5	13	16.3	25	31.3	8	10.0	-	-
Kantsyore	511	-	-	-	-	-	-	113	22.1	254	49.7	142	27.8	-	-	-	-
Tunnel Rockshelter	523	-	-	-	-	-	-	145	27.7	121	23.1	257	49.1	-	-	-	-
Prolonged Drift	325	-	-	-	-	-	-	154	47.4	170	52.3	1	0.3	-	-	-	-
Narosura	867	8.4	307	3.0	9,088	88.6	10,262	415	47.9	165	19.0	125	14.4	42	4.8	120	13.8
Crescent Island (Ex. 21 - Onyango-Abuje)	1,395	17.0	46	0.5	6,744	82.5	8,185	219	15.7	558	40.0	14	0.1	551	39.5	53	3.8
Hyrax Hill (M.D. Leakey)	3,208	6.6	280	0.6	45,457	92.8	48,945	2,756	86.0	137	4.3	59	1.8	27	0.8	229	7.1
Naivasha Rock Shelter	2,649	-	-	-	-	-	-	2,233	84.0	392	14.7	24	1.0	-	-	-	-

Table 95

INVENTORY OF BARBED BONE HARPOONS BY SITE

Site	Complete	Broken	Total
GaJj 1: Sf.	1	48	49
GaJj 1: Ex.	1	7	8
GaJj 2	4	21	25
GaJj 11: Sf.	2	139	141
GaJj 11: Ex.	-	1	1
GaJj 12	1	7	8
FxJj 12: Sf.	1	5	6
FxJj 12: Ex.	-	1	1
FxJj 12 North	1	8	9
GaJi 3	3	3	6
Survey sites	1	6	7
Localities	4	14	18
Total (N)	19	252	271
(%)	7.0	93.0	100.0

Table 96

INVENTORY OF ALL BONE ARTIFACTS

	(N)	(%)
Uniserial	243	89.6
Biserial	9	3.3
Triserial	5	1.9
Barbless	6	2.2
NID	8	3.0
Total (N)	271	100.0

Table 97

INVENTORY OF COMPLETE BARBED HARPOONS WITH SELECTED ATTRIBUTES

Site	Length (mm)	# of Barbs	Base
GaJj 1: Sf.	63	5	Circular grooves
GaJj 1: Sf.	51	2	Circular grooves
GaJj 1: Ex.	43	1	Notch
GaJj 2	97	4	Notch
GaJj 2	58	4	Notch
GaJj 2	71	4	Notch
GaJj 11	49	3	Notch
GaJj 11	107	10	Notch
GaJj 12	93	6	Notched & circular grooves
FxJj 12: Sf.	70	5	Notch
FxJj 12 North	92	8	Notch
GaJi 3	116	17	Circular grooves
GaJi 3	93	7	Notch
GaJi 3	78	7	Circular grooves
GaJj 16	110	10	Notched & circular grooves
GaJj 0/23	66	1	Notch
GaJi 0/28	128	11	Circular grooves
FxJj 0/9	79	6	Notched & circular grooves

Table 98

ATTRIBUTES OF BARBED BONE HARPOON BASES

Site	Notch	Circular Grooves	Both	Total
GaJj 1	1	11	-	12
GaJj 2	5	5	-	10
GaJj 11	57	-	-	57
GaJj 12	1	2	1	4
FxJj 12	2	1	-	3
FxJj 12 North	3	-	-	3
GaJi 3	1	2	-	3
Survey sites/ localities	3	4	2	11
Total (N)	72	25	3	100

Table 99

A COMPARATIVE INVENTORY OF BONE HARPOONS FROM SELECTED FISHING SITES IN NORTHERN, EASTERN AND CENTRAL AFRICA

	Uniserial	Biserial	Triserial	Barbless	NID	Total (N)
East Turkana (all sites)	243	9	5	6	8	271
Lothagam	263	-	-	15	-	278
Lowasera	72	2	-	6	-	81
Khartoum*	167	3	-	-	-	170
Ishango	89	290	-	-	30	419

*an estimate (see Arkell 1949, p. 75)

Table 100 A

INVENTORY OF STONE BOWLS BY SITE AND LOCALITY DESIGNATION

Site	Catalogue Number	Area	Raw Material
Ileret Stone (sf) Bowl Site (FxJj 5)	391	10	Basalt
	392	-	Basalt
	393	-	Quartzite
	394	-	Welded tuff or lava
	395	-	Basalt
	388	-	Lava
	389	-	Welded tuff or lava
	390	-	Vesicular basalt
Dongodien (sf)	427	102	Soft green stone--not identified

Table 100 B

Locality (sf)	Area	Raw Material
GaJi 0/5	102	Basalt
GaJi 0/20	-	Basalt
GaJi 0/21	-	Soft green stone--not identified
GaJj 0/13	102	Indurated sandstone
GaJj 0/17	100	Basalt
FxJj 0/1	105	Basalt
FxJj 0/3	105	Basalt
FxJj 0/10	105	Basalt
FxJj 0/14	117	Basalt

CHAPTER VII
SITE DISTRIBUTION

The reconstruction of late Pleistocene-Holocene site distribution patterns constituted a major objective of the present study. Such distribution patterns are often of great importance to archaeologists as they help discern critical interrelationships between social, economic and ecological aspects of past human societies (Isaac 1981).

A recurrent theme in site distribution studies has been the emphasis on field identification of discrete clusters of artifacts. Recent investigations from the Plio-Pleistocene deposits at Lake Turkana have stressed however, that a more complete treatment of site distribution should also include analysis of the low density scatters of archaeological objects that lie between the concentrated patches (Isaac and Harris 1978). This approach seeks to reconstruct not only the distribution of sites across the paleo-landscape but also the behavorial and ecological meaning of scatters between sites as well as of the sites themselves (Isaac 1981). In such a perspective, emphasis is directed towards reconstruction of the total area of human land use patterns rather than just the spacial distribution of archaeological clusters (Foley 1981).

Within the East Turkana research area, the reconstruction of site distribution and land use patterns was conducted on two separate levels. First, a general reconnaissance of known Galana Boi outcrops was undertaken. The objective was to initiate the building of a cultural-historical succession by locating concentrations of artifacts (sites) for future excavation. Sites with a broad array of artifacts, associated faunal debris and datable raw materials were given particular attention. Second, a much more systematic site survey focused specific attention on the reconstruction of land use patterns. This survey utilized a series of transects that recorded the distribution of archaeological materials in relationship to specific paleo-topographic features. In this survey both the distribution of sites and low density scatters of artifacts (localities) were recorded. The results of the transect survey permitted a more comprehensive understanding of late Pleistocene-Holocene land use patterns as well as strengthening the framework of the cultural- historical succession. A description of the transect survey's methodology and findings is presented in the following section.

TRANSECT SURVEY

Methodology

A total of six survey transects were located on several different paleo-topographic features (Fig. 117). These features included, former Holocene beach ridges along the crests of Plio-Pleistocene escarpments (Transects A and B); a major river drainage and a partially disected ridgetop upstream from the 75-80 m Holocene beach (C and D); an inland erosional escarpment (E); and the countryside east

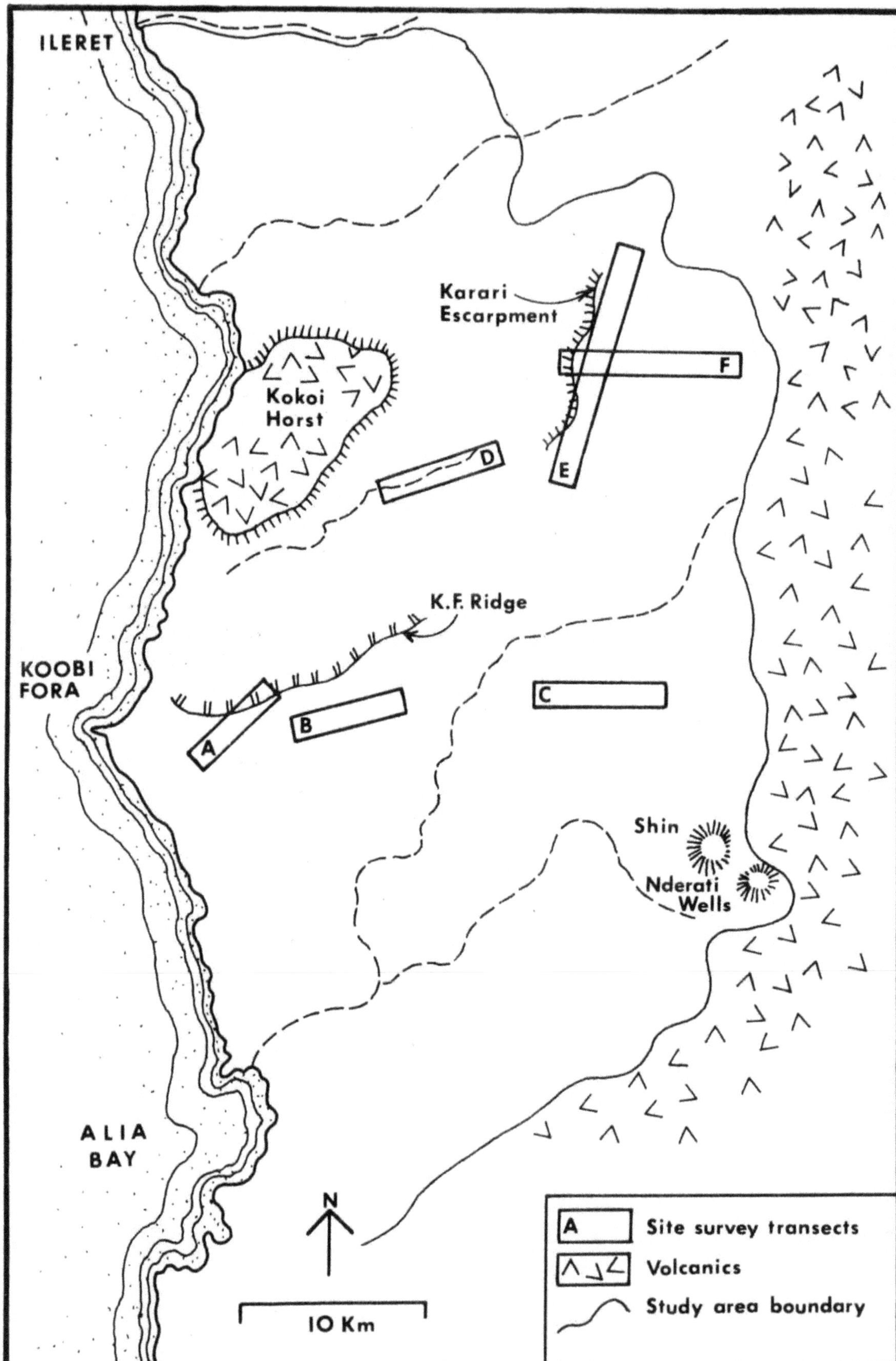

Figure 117. Location of site survey transects.

of the erosional escarpment to the margin of the eastern volcanic highlands (F).

Once a particular topographic feature was selected for survey, it was recorded on aerial photographs and traced on acetate overlays. Transects, each one kilometer in width, were then drawn on the overlays. Survey units, consisting of one kilometer squares, were subsequently drawn up for each transect. The length of each transect and number of survey units varied. For example:

Transect	Length	No. of Survey Units
A	6 km	3
B	6 km	2
C	5 km	2
D	7 km	4
E	13 km	3
F	10 km	2

During survey, each archaeological site and locality was marked on the transect overlays. Field notes, including sedimentary context, artifact description, land use reconstruction and future excavation potential were recorded. As well, a representative sample of artifacts was often collected from the newly discovered sites and localitites. It should be stressed that while the present methodology successfully identified both clusters and scatters of artifacts, the survey was essentially a qualitative approach to site distribution patterning. More rigorous quantitative studies, similar to Isaac (1981) and Foley (1981) are planned for future research.

Results

Transect A

Transect A was situated in Areas 100, 102 and 103 along the crest and eroded margins of the Koobi Fora Ridge (Fig. 118). Discontinuous outcrops of Galana Boi sediments had previously been surveyed during the 1975 field season and a rich diversity of artifacts had been identified. In survey unit 1, Galana Boi deposits mantled approximately 90% of the ground surface. Coarse beach sands, thick beds of fine grained diatomaceous silts and a widespread covering of reworked, poorly sorted, gravelly sand and shell were present. The majority of the artifacts were weathered and distributed as isolated finds either downslope from the beach and silt units or atop the reworked sands. No sites were identified. It appears that more recent fluvial-colluvial processes have substantially reworked earlier Holocene archaeological materials which resulted in dispersed scatters of reworked stone artifacts, pottery and barbed bone harpoons.

Within survey unit 3, excellent exposures of Galana Boi sediments occurred along the northern and southern erosional margins of the Koobi Fora Ridge. Previous excavations had yielded two Pastoral Neolithic sites (GaJi 2 and Dongodien) as well as numerous isolated artifacts. Holocene sediments in this area consisted of a thick suite of diatomaceous silts, coarse beach sands, silty clays and mantling reworked sandy gravels. The majority of the artifacts, as in unit 1, occurred as isolated pieces either on the slope or at the base of the eroded Galana Boi outcrops. Often, artifacts from two or more distinctive cultural traditions were found mixed together at the base of an outcrop. For example, stone artifacts, pottery (including Nderit Ware) and barbed bone harpoons were identified from at least five separate levels near Dongodien. The mixed and reworked provenience of many artifacts made it extremely difficult to assign surface pieces to a discrete horizon in the local stratigraphic section. However, the multiplicity of primary and reworked occupation levels did reveal numerous concentrated patches of cultural remains including the adult burials at GaJj 9.

Survey unit 5 was situated in Area 100 approximately 1/2 km behind the 75-80 m high lake stand. The area consisted primarily of an undisected ridge of the Koobi Fora Formation on which localized indurated sandstone beds cropped out. Few artifacts were recorded from this unit. A scattering of small heavily abraided chert flakes, including a weathered MSA unifacial point, occurred within the widespread ridgetop gravels. These artifacts may well represent the traces of former MSA occupations which were located on Koobi Fora Ridge. Several decorated, recent to sub-recent pot handles, a broken stone bowl and clusters of probably recent stone circles and cairns were also identified. Few traces of either fishing or early pastoralist sites were present in unit 5. Instead, the undisected ridgetops seem to have been a preferred location for Middle Stone Age as well as recent to sub-recent human encampments.

Transect B

Transect B was located parallel to three conspicuous Holocene beach terraces within Area 119 and to the south of Area 102 (Fig. 119). Reconstruction of the area's paleo-topography indicate the terraces were formed within a large embayment not unlike the modern configuration of Allia Bay. Cultural materials were widely distributed, probably due in part to the discontinuous nature and poor exposure of the beach terraces.

Within survey unit 2 all three Holocene beach ridges were exposed. The highest terrace (78-82 m) was composed of coarse clean sandy gravels with scattered broken Melanoides shells. Artifacts were widely spaced and consisted of pottery sherds and small pieces of chert debitage. No sites were identified.

The second beach terrace was composed of a fine to medium sand with abundant associated Melanoides, Corbicula and Pila ovata shells. Cultural remains were primarily distributed as small reworked clusters and isolated pieces. In one area, a modern gully had sliced through the broad Holocene beach platform and exposed a 50 cm section of sands, silts and a dense patch of pottery and chert debitage (GaJj 4). A second partially exposed site was discovered near the margins of a larger stream channel. The modern stream had also exposed large

Plio-Pleistocene sandstone blocks onto which freshwater oysters, _Etheria elliptica_, were anchored.

The third and lowermost terrace revealed a tremendous accumulation of _Corbicula_ shells, many of which were still articulated. Artifacts were widely dispersed along the terrace and clustered in one area (GaJj 5). Obsidian chippage and decorated pottery, some of which resembled Ileret type, were selectively sampled from the site.

Within survey unit 4, only discontinuous remnants of the upper and middle beach terraces were exposed. A large fishing site (GaJj 16) with barbed harpoons, pottery and considerable quantites of mammal and fish bone refuse was discovered on the upper terrace. Artifactual materials on the eroded margins of the second terrace were few in number and consisted of undiagnostic stone artifacts.

The distribution of cultural remains in Transect B indicates that large occupation sites were located on or near each of three paleo-lake terraces. The high 75-80 m terrace contained a single large fishing campsite as well as frequent reworked artifacts characteristic of this widespread lacustrine economic adaptation. Sites with decorated pottery and obsidian debitage that lacked bone harpoons were identified from the second and third beach terraces. These more recent lacustrine based sites have a number of similarites with the Pastoral Neolithic occupations in Area 102 and should be thoroughly investigated in the future. In contrast with Transect A, sites and localities along the three Holocene beach terraces were relatively uncommon. While such a difference may in large part be due to the discontinuous aspect and poor exposure of the beach sediments in Transect B, the relative dearth of artifacts may also reflect the deliberate preference for site location along the perimeter of the paleo-Koobi Fora penninsula.

Transect C

In order to gather site distribution data behind and to the east of the high 75-80 m shoreline, Transect C was located in Area 127 directly behind the large fishing encampment at GaJj 11 (Fig. 120). While surveying in unit 2 however, it soon become apparent that the high transgressive beach lay approximately one kilometer further east. These beach deposits consisted of a relatively thick sequence of coarse gravelly sands and fine sandy silts that were continuously distributed for nearly 3/4 of a kilometer. One very substantial fishing site (GaJj 12) was discovered and samples of bone harpoons and pottery sherds were collected. Cultural remains were dispersed in varying densities along the length of the outcrop but only the single large concentration of harpoons and other artifacts was identified. To the east of GaJj 12, survey revealed few cultural remains except for a broken grindstone.

Survey unit 4 was located on a broad plain of Plio-Pleistocene sediments that had been dissected by shallow ephemeral stream courses. Holocene aged lacustrine and channel/overbank sediments were not present. Four stone cairns, probably of recent origin, were the only cultural remains identified.

The site distribution evidence from Transect C indicated that large fishing settlements were located along the shoreline of the 75-80 m beach and, if the paleo-geography of GaJj 11 is interpreted correctly, also on offshore sand and gravel bars. Little evidence of other occupation was recorded behind the high beach stand.

Transect D

Transect D was located in Areas 106, 116 and 117 on both sides of Laga Aberegaya (Fig. 121). This placement, as in Transect C, permitted systematic survey of the laga's drainage upstream from the 75-80 m Holocene beach.

Two well defined outcrops of Galana Boi sediments were present in survey unit 1. Both outcrops occurred as relatively thin linear strips that mantled Plio-Pleistocene indurated sandstones and silty clays. Two large multi-component fishing settlements (FxJj 12 and FxJj 12 North) were excavated from the sequence of Holocene gravelly sands and fine silty sands. A widespread layer of dark brown silty clay was discovered lateral and to the east of the beach sediments. This layer is thought to represent a former lakeshore plain with localized deltaic mudflats. Very few artifacts were identified from this feature. Farther upstream a discontinuous series of low lying coarse overbank deposits occurred near the edge of the modern laga. A low density scatter of water-abraided, reworked stone artifacts was identified from these outcrops. As well, recent to sub-recent occupational debris was observed both adjacent to the overbank sands and on ridgetops overlooking the laga.

The site distribution evidence from survey unit 1 clearly indicates repeated occupation of the former high lake shoreline by early Holocene fishing groups. More recent occupants of the area apparently located their campsites along the laga's margins and on elevated remnants overlooking the sand river. With the exception of the recent campsites, few settlement traces were recorded behind the 75-80 m Holocene beach.

Within unit 3, well eroded Plio-Pleistocene deposits formed the largest portion of the landscape. When present Holocene sediments occurred as thin strips of coarse sands and gravels on both sides of the laga. Cultural materials were usually in secondary context and consisted primarily of isolated waterworn pottery sherds and microlithic debitage. Several sherds exhibited incised and impressed decorations.

Plio-Pleistocene units and coarse overbank Holocene sands, sometimes nearly one meter thick, also made up the majority of the sediments in unit 5. Artifacts once again showed evidence of water abrasion and were clearly in secondary context. Importantly, one weathered sherd of Nderit Ware and an associated group of water worn obsidian debitage were found. Although the pottery and stone artifacts may not necessarily come from the same occupation, the presence of Nderit Ware and obsidan microliths does argue for the existence of early pastoralist sites in the Laga Aberegaya area. Two recent to sub-recent campsites were also discovered near the laga's northern margin.

Survey unit 7 was located among the eroded Plio-Pleistocene deposits as well as alongside the upper drainage of the laga. Most of the identified cultural remains were large core tools and lava debitage that belonged to the Karari Industry. Several fragmentary pots from recent encampments were also recorded.

Transect E

Transect E was located in Areas 118, 129, 130 and 131 parallel to the erosional flanks of the Karari Escarpment (Fig 122). The transect was situated between the high Holcene lake shoreline and the rim of the eastern volcanic highlands.

Survey unit 1 was located on and behind the escarpment's western border. Middle Stone Age artifacts, including unifacial points and faunal debris, were exposed on the surface of a light brown sand that mantled the ridgetop. This sandy unit extended across large areas of the escarpment's western rim and, to judge from associated water worn cobbles, was probably deposited under fluvial conditions. Later Stone Age artifacts were not recorded from this unit.

Survey unit 7 was situated both on the eroded flanks and uneroded platform of the escarpment in Area 131. Widely scattered LSA artifacts were identified from lag gravels on the rolling erosional slopes. The majority of these pieces showed evidence of water abrasion. On the escarpment's surface, worn MSA artifacts, including levallois flakes, were present among a thin covering of lag cobbles and gravels. No discernible sites were recorded.

Survey unit 13 was located in Area 129 along the rim of the Karari Escarpment. Coarse, well sorted reddish brown sands extended over much of the area. The sands were of variable thickness but probably did not exceed one meter. MSA artifacts, including bifacial points, were exposed in an area coincident with a network of small erosional gullies. These pieces were invariably patinated and had smooth, rounded flake margins. Partially mixed with and stratigraphically above the MSA material a concentrated patch of LSA lithics, bone fragments and pottery sherds was identified (FxJj 102).

The distribution of cultural materials in Transect E suggests that the western rim of the Karari Escarpment was a favored settlement area for Middle Stone Age hunter-gatherers. Artifacts were widely dispersed along the escarpment and, in several localities, were concentrated in discrete patches. Evidence for Later Stone Age occupation was limited to reworked isolated stone artifacts along the upland's eroded flanks and to several sites along the northern border. Such sites suggest that LSA groups with pottery, and perhaps domestic animals, located their campsites along the highly visible escarpment midway between the high Holocene lake shoreline and the perimeter of the eastern highlands. No evidence for recent to sub-recent archaeological remains were recorded along the escarpment's border.

Transect F

Transect F was situated from the erosional lip of the Karari Escarpment eastward to the margins of the volcanic highlands (Fig. 122). Two kilometer units were surveyd.

Survey unit 2 was located on the eastern flanks of the escarpment along a flat lying ridgetop that had been repeatedly disrupted by small erosional gullies. A relatively thin lens of lag cobbles and gravels mantled the exposed Upper Member sediments. Artifacts were relatively uncommon and only several isolated MSA flakes with prepared platforms were identified from the rubble surface.

Unit 7 was situated to the east of the escarpment in Area 131. A thin veneer of lava cobbles and pebbles covered the ridgetop and the slopes of the numerous erosional gullies. No MSA or LSA artifacts were recorded. The only identified cultural remains consisted of a cluster of three stone cairns and a stone circle of recent to sub-recent origin. The research design for Transect F called for detailed survey of the landscape east of the Bura Hasuma drainage up to and including the eastern volcanic margin. However, the presence of a modern pastoralist encampment in the river's dry stream bed resulted in an abrupt halt to the survey.

SITE DISTRIBUTION: CONCLUSIONS

The combination of outcrop exploration and systematic transect survey yielded an important body of information on late Pleistocene-Holocene site distribution and land use patterns. Archaeologoical sites and localities were shown to be present on at least seven principal topographic features. A discussion of site distributions for each cultural phase is presented below.

Later Stone Age Sites Without Pottery (Group IV)

Clusters of Later Stone Age lithics without associated pottery sherds or barbed bone harpoons were recorded from two paleo-geographic features: (1) river drainages behind the 75-80 m Holocene lake stand, and (2) near permanent fresh water sources within the eastern volcanic highlands. Similar sites that may have originally been situated near Lake Turkana's shoreline before 10,000 years ago have since been destroyed by successive lake trangressions. The absence of suitable materials for radiocarbon dating made it impossible to determine whether the Group IV occupations preceded the widespread Holocene fishing and Pastoral Neolithic settlements or were, in part, contemporaneous with them. Comparable late Pleistocene-early Holocene LSA settlements without pottery are known throughout East Africa (Phillipson 1977b) and the East Turkana sites may well represent a period of occupation that was antecedent to the later lacustrine and pastoralist cultural traditions. Collection of dating samples and faunal remains from these sites should be given high priority in future fieldwork.

Fishing Settlements (Group I)

Sites and dispersed scatters of artifacts from this cultural phase were invariably located on or very near to early and mid Holocene lake shorelines. A single site (GaJj 11) was thought to have been situated on an offshore sandbar in proto-Allia Bay. A thorough exploratory and systematic transect survey of backbeach, estuarine and river drainage environments failed to yield any evidence of inland sites that were contemporary with the widespread lake shore encampments. It would appear that early to mid Holocene fishing villages were preferentially located very close to the lake's shoreline and not upstream along major river courses. Other Holocene fishing settlements in the Turkana Basin show similar distribution (Phillipson 1977a; Robbins 1974).

Pastoral Neolithic (Group II)

Archaeological remains attributed to the Pastoral Neolithic phase at East Turkana were recorded from a variety of paleo-geographic settings. Two large settlements (GaJi 2 and Dongodien) were located, like earlier fishing occupations, immediately adjacent to the lake shoreline. A third large settlement (FwJj 5) was situated several kilometers behind the shoreline, along the flanks of a major river drainage. Other sites and localities were located along river drainages and on the ridgetop of an inland erosional escarpment. No evidence of occupation was recorded within the volcanic highlands.

For the Pastoral Neolithic phase, it is possible to further refine site distribution patterns. It would appear that the earliest sites in the Basin were placed on or near lake shorelines while later PN occupations were more riverine based. The diversity of PN site locations may reflect, similar to modern Dassanetch pastoralists of the southern Omo delta (Almagor 1978) and the Koobi Fora-Ileret area of Lake Turkana (Gifford 1978), different settlement responses to seasonal grazing conditions, diet (especially fish), and for defensive purposes. Further distributional data are needed however, before these finer aspects of PN site location can be correctly recognized and interpreted.

Recent to Sub-Recent Sites and Localities

Cultural materials attributed to recent and sub-recent occupation of the East Turkana area were widely distributed across the landscape. Sites and localities were recorded adjacent to the modern lake shoreline, along major river courses east of the Karari Escarpment and within the volcanic highlands. The age and cultural affinity of the settlement traces lay outside the projects' main research objectives and consequently have been studied on only a preliminary basis. Nevertheless, several occupations can be associated with different contemporary Kenyan and Ethiopian cultural traditions. For example, along the modern lake shoreline the seemingly ubiquitous distribution of hearths with fire blackened stones and fishbones can be attributed to the Dassanetch gal dies hunting-fishing camps (Gifford 1978:85-86). Reconstructed pots collected from several sites along Laga Aberegaya strongly suggest the former presence of Samburu pastoralists. The cultural affinity of the makers of the widespread stone cairns and stone circles is not well understood but preliminary studies have suggested they are of possible Cushitic origin (Soper and Jacobs 1972).

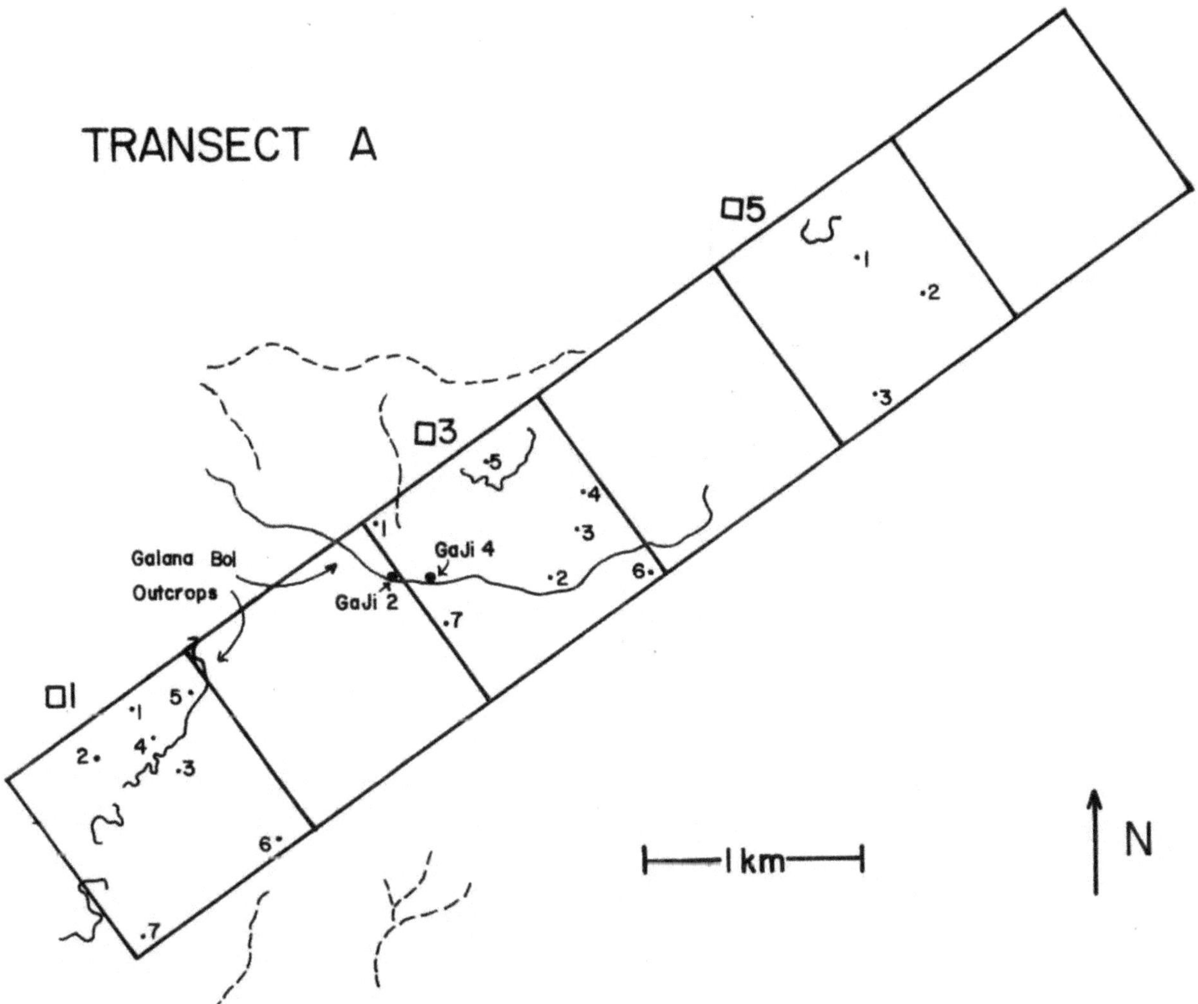

Figure 118. Transect A: distribution of surface finds.

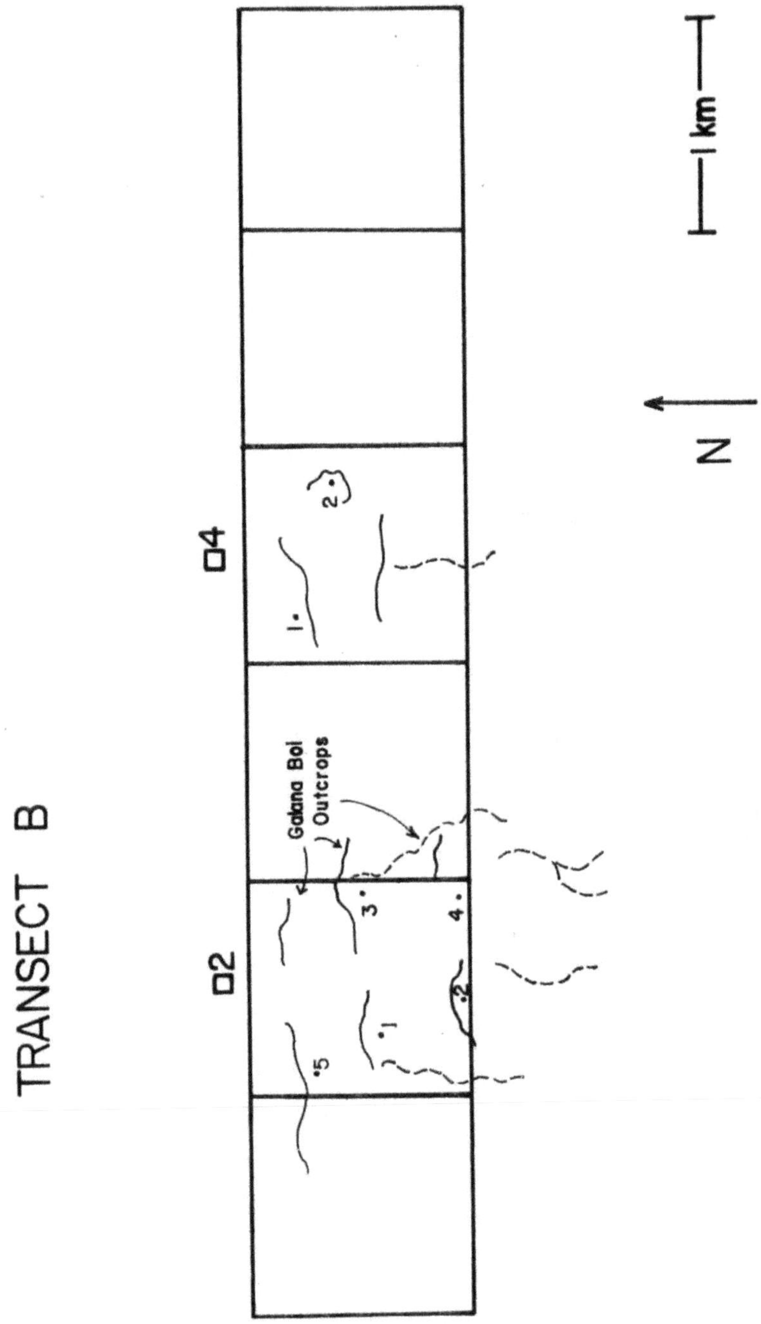

Figure 119. Transect B: distribution of surface finds.

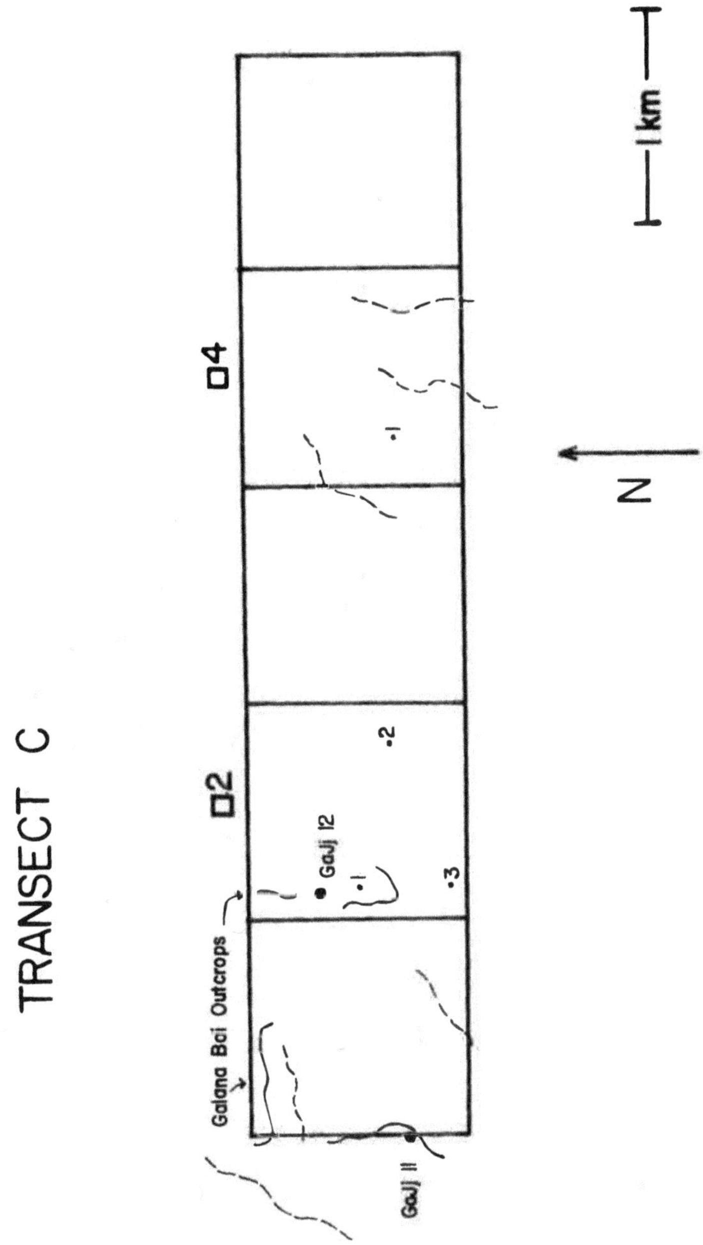

Figure 120. Transect C: distribution of surface finds.

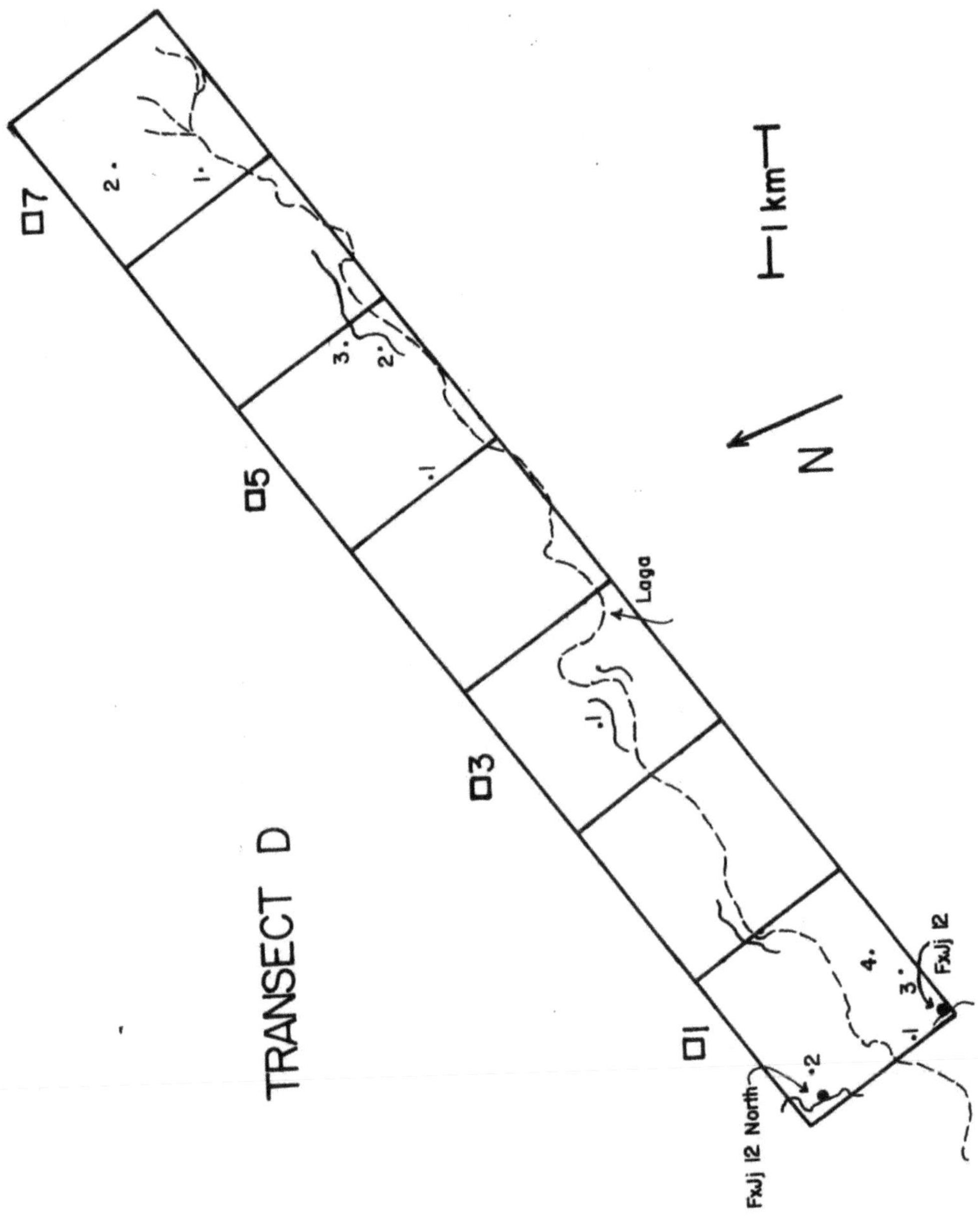

Figure 121. Transect D: distribution of surface finds.

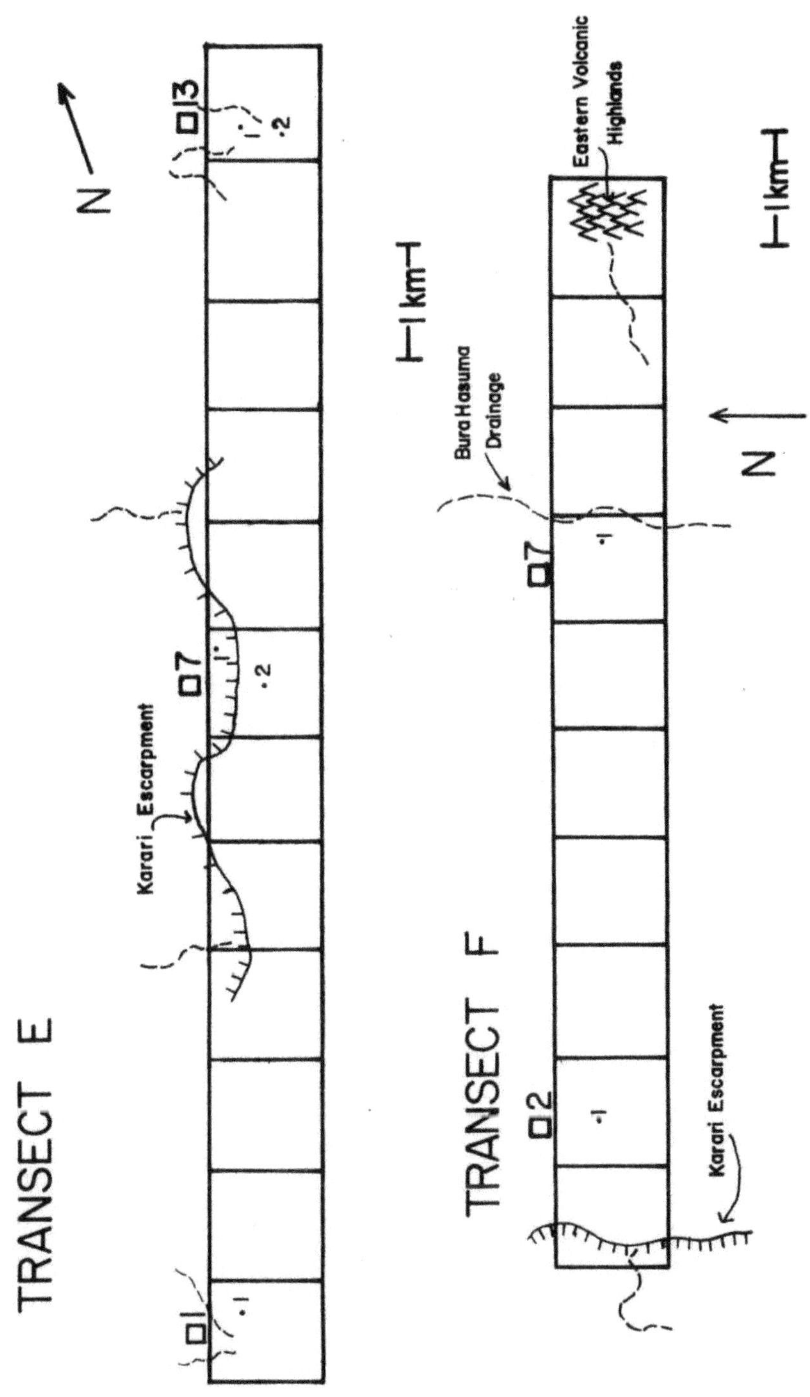

Figure 122. Transects E and F: distribution of surface finds.

CHAPTER VIII

SUMMARY AND CONCLUSIONS

This monograph has focused on the reconstruction of Holocene lake level changes, the establishment of a late Pleistocene-Holocene prehistoric chronology and the delineation of land use patterns for the area northeast of Lake Turkana. Studies of geomorphic features, altimeter readings, and a series of radiocarbon determinations identified three prominent lake transgressive-regressive cycles of Holocene age. Systematic location, sampling and excavation of archaeological sites yielded a cultural succession of gathering-hunting, fishing, Pastoral Neolithic and sub-recent to recent occupations within the research area. Site survey transects helped identify the distribution of archaeological sites and localities near specific paleo-topographic features.

LAKE LEVEL CHANGES

Three principal phases of high lake levels were identified from the Galana Boi beds. The first phase was reconstructed from outcrops of early Holocene lacustrine silts and littoral sands in Areas 102, 103, 117 and 127. These sediments and associated radiocarbon dates demonstrate that by 9500 BP, Lake Turkana had attained a maximum transgressive height of 75-80 m. The lake remained high, although elevation probably flucuated between 70-80 m, until approximately 7500-7000 BP. Pollen studies have indicated the presence of montane forest and thicket taxa which do not occur in the modern pollen rain. The pollen grains were probably transported by rivers that originated in the Ethiopian highlands and suggest both increased rainfall activity and the extension of regional highland forests (Owen et al. 1982).

Based on detailed regional geomorphologic and micro-stratigraphic studies on the Kibish Lake Plain, and on a single mixed shell date of 6600±150 BP, Butzer (1980) has suggested the level of Lake Turkana fell to modern levels during the middle Holocene. Although this interpretation may be correct, experience with mollusc shells in the Turkana Basin has shown that such shells are often reworked with little evidence of damage and single radiocarbon dates should therefore be interpreted with caution. If correct, the date from the Omo implies Lake Turkana experienced major dessication when the great majority of intertropical lakes were at high or intermediate status (Street and Grove 1979). Significantly, no evidence has been found for major erosion of the unconsolidated sediments in Areas 102 and 103 which span this time period. Documentation of a drop of such magnitude was also absent from investigations to the southeast of the lake (Phillipson 1977a). While it appears the lake may well have fallen to intermediate levels at this time, evidence for a fall to modern levels is lacking.

The second prominent transgressive-regressive phase occurred during the middle Holocene and was identified from stratigraphic and sedimentological features in Area 102. Lake elevations fluctuated between high and intermediate heights at this time. While the beginning of this phase has yet to be dated at East Turkana, complimentary studies to the north (Butzer 1980) and the southwest (Robbins 1974) have indicated a probable onset between 6500-6200 BP. Shell dates from Areas 102 and 103 yielded dates of 5060±245 and 4540±230 BP. Diatom and pollen studies indicate the lake remained fresh and the local vegetation was dominated by Gramineae (Owen et al. 1982). By 4500-4000 BP, when the lake stood 45-50 m, evidence from the Area 102 geological transect clearly demonstrates the waning stages of this second high lake phase.

During the late Holocene, Lake Turkana underwent a third major lake transgressive between 3500-3250 BP (Phillipson 1977a; Butzer 1980). Evidence exists for raised lake levels of roughly 50m south of Koobi Fora at this time, but data are lacking to confirm a rise of 75m as reported from the Omo area. During the last 2500 years, the lake has become increasingly alkaline and shoreline levels have fluctuated around those of the modern lake.

Today, fluctuations in the level of Lake Turkana are primarily controlled by the evaporation-precipitation balance. During the early to late Holocene, however, these fluctuations were probably the result of significant variations in the water balance caused by climatic change as well as stable outlet heights. Several researchers have argued that a series of overflows linked Lake Turkana to the White Nile drainage (Butzer et al. 1972, 1980) and that overflows from Lake Chew Bahir in Ethiopia (Grove et al. 1975) and Lakes Alablab and Logipi in the Suguta Basin (Owen and Renaut 1980) similarly overflowed and joined with Lake Turkana. Butzer (1980) has suggested that, while field demonstration of Holocene connections with the White Nile drainage are still tenuous, altimeter data indicate elevations of 75-80 m stood close to 5 m above the critical watershed level during brief periods, perhaps seasonal, of the early and middle Holocene. Other sources of evidence, especially molluscan (Van Damme and Gautier 1972), strongly indicate a Turkana-White Nile connection. The altimeter recordings presented in this monograph support Butzer's findings. Confirmation will come with careful topographic mapping of the shallow basins and low watersheds between Sanderson's Gulf in Lake Turkana and the Pibor-Sobat river drainages that empty into the Nile.

A second overflow hypothesis suggests that when Lakes Abaya, Chamo and Chew Bahir (Stephanie) in southern Ethiopia were periodically connected during the Holocene by a series of overflow channels and rivers, Lake Chew Bahir periodically reached its overflow level and emptied into Lake Turkana (Grove et al. 1975). Such a reconstruction, connecting rainfall south of Addis Ababa with Lake Turkana, the White Nile and ultimately with the Mediterranean Sea has however, yet to be satisfactorily established. Primary evidence linking Lakes Chew Bahir and Turkana rests on a well defined upper limit of an algal limestone that is 20 m above the current level of Lake Chew Bahir and that has been radiocarbon dated to 5660±110 BP (Grove et al. 1975:183). While clear wave-cut notches or beach strandlines were not identified, the height of the limestone is considered a possible overflow level. Secondary evidence includes the presence of Nile

Perch in Lakes Abaya and Chamo and the Bakate Gap structure, the probable overflow channel into the Turkana Basin.

In order to test this overflow hypothesis with additional field date, excursions were taken into the southeastern sector of the Chew Bahir Basin as well as into the upper regions of the Bakate Gap. Both areas were prospected for possible Holocene wave-cut notches, lacustrine, littoral and fluvial sediments. Although field survey within each area did not yield supporting information, the overflow hypothesis may well be correct. It is suggested that careful topographic mapping of watershed elevations between both basins should be undertaken in future field work.

An increasing number of studies on the geology, chronology, altimetry, diatoms and pollen indicate that Lake Turkana was a much larger freshwater lake during the early and middle Holocene. At least three major transgressive-regressive cycles have been identified and connection with the White Nile drainage appears well documented during the first and probably the second transgressive phase. The stratigraphic, geomorphic and geo-archaeological studies of Holocene lake levels at Lake Turkana are in general agreement with similar studies in other East African and Ethiopian lake basins and provide supporting evidence for widespread major climatic changes during the Holocene.

CULTURAL-HISTORICAL SUCCESSION

The area to the northeast of Lake Turkana proved to be exceptionally rich in late Pleistocene-Holocene archaeological sites. A total of 30 sites and 74 localities were discovered. Fourteen sites were either sampled and/or excavated and the abundance of cultural remains permitted the formulation of a preliminary prehistoric chronology for the research area.

Situated near the basin's perimeter and within the eastern volcanic highlands was a group of open air Later Stone Age sites (Group IV) that lacked pottery and contained a distinctive lithic industry dominated by chert microliths and microblades. The occupations, like GaJk 2 near Nderati Wells, were located on high ground, usually overlooking permanent sources of fresh water. The absence of faunal remains and radiocarbon dating materials made it impossible to identify subsistence patterns or to date the occupations. Nevertheless, the distinctive stone tool assemblages and lack of associated pottery may well represent an inland late Pleistocene-early Holocene hunting and gathering tradition that preceded, or coexisted with, the lacustrine based early Holocene fishing settlements. Comparable late Pleistocene LSA sites are known from the western side of Lake Turkana (Robbins 1972, 1979). Future research should include the location of inland rockshelters or caves with a stratified occupation sequence and the recovery of radiocarbon dating samples.

The most numerous archaeological sites at East Turkana were the early to middle Holocene fishing settlements (Group I). Seven large occupations were discovered along the 75-80 m lake shoreline. At least two of these fishing settlements were associated with the first high lake transgression. The remaining five sites have not been dated and could be affiliated with either high levels during phase 1 or phase 2. A single site was associated with the second major Holocene regression.

The fishing settlements as a whole exhibited broadly similar technological and economic patterns. The lithic assemblages were primarily composed of microliths, unstandardized scrapers, outils écaillés, and core tools. Crescents and curved backed blades were the most common microlithic types but straight and double backed forms also were collected. Scrapers were comparatively rare and burins, except for a few dubious examples, were absent. Outils écaillés and core choppers were well represented at all fishing settlements. Cores were represented by a number of forms including bipolar, multi-platform and casually flaked examples. Different raw materials were consistently selected for specific shaped tools. For instance, cryto-crystalline silicas provided the raw material for microliths, outils écaillés, and some scraper elements. Quartz was preferentially selected for outils écaillés and hammerstones. Lavas were primarily used in the manufacture of scrapers and heavy-duty tools. Obsidian was extremely scarce. No microlithic forms were manufactured from either lava or obsidian. Grinding stones and fire cracked rocks were consistent features of the occupation surfaces.

Uniserially barbed bone harpoons were sampled or excavated from each fishing settlement. Several biserial, triserial and barbless bone artifacts were also recovered. Unbroken uniserial harpoons had a mean length of nearly 80 mm. Both notched and circular grooved bases were identified from the harpoon collection.

Decorated and undecorated pottery was associated with all but two of the fishing settlements. One of the sites, FxJj 12, is thought to represent an early Holocene aceramic phase of the lacustrine tradition in the Turkana Basin. A second site, FxJj 12 North, contained abundant evidence of pottery similar to the well known wavy-line pottery of the Sudan and Sahel. No examples of dotted wavy-line decoration were identified. Ostrich eggshell beads and several examples of stone beads were recovered from most of the sites. Archaeological features, such as hearths, pottery firing areas, or windbreak foundations were regrettably not encountered.

Fish, primarily several species of catfish and Nile Perch, constituted a substantial proportion of the faunal refuse at each site. Hippopotamus and crocodile remains, along with a wide variety of land mammals, were well represented. Preliminary identification of the faunal remains suggested that there were major intersite differences in the proportion of fish and land mammals. At GaJj 2, a terrestrial fauna, ranging in size from dik-dik to elephant, was identified. By contrast, the faunal remains from GaJj 11 consisted entirely of fish bone and a few elements of hippopotamus and crocodile. Such dissimilar dietary patterns may be provisionally attributed to different subsistence preferences, or, as with GaJj 11, be primarily the consequence of specific site locations. A more detailed economic comparison of the shoreline fishing settlements must await a complete analysis of the land mammal and fish remains. While freshwater mollusc shells, including such edible species as Pila ovata and Etheria elliptica, were commonly associated with the occupation surfaces, direct evidence for their consumption was not obtained. No evidence for the consumption of plant foods was identified, though no doubt they contributed a substantial part of the prehistoric diet.

Studies of the human remains associated with the early to middle Holocene beach deposits, indicate the individuals were robustly built and exhibited a marked degree of tooth wear. Several clavicles show pronounced muscle attachment areas on the sternal end which, it has been suggested, may be the result of prolonged pressure on the bone due to canoe paddling (Mbua 1984). Artificial cranial deformation has been tentatively identified for one individual.

Current evidence indicates that very early occurrences of barbed bone harpoons in Africa come from the Turkana and Lower Omo Basins. At Harpoon Hill in the Omo Basin, a fishing station with uniserial and biserial harpoons has been radiocarbon dated to at least 9100 BP (Butzer 1980). At Lowasera, uniserial harpoons have yielded an age of roughly 9000 BP (Phillipson 1977a). Robbins' (1974) extensive work at Lothagam has documented a rich assemblage of barbed bone harpoons recovered directly above beach deposits dated to 7160 BP. In Northern Africa, early dates associated with bone harpoons come from sites located in the Sahara and along the White Nile. Near Hassi el Abiod, in northern Mali, a date of 8450 BP is associated with early Holocene beach sediments and numerous bone harpoons (Petit-Maire et al. 1983). At Tagra, along the Sudanese Nile, an isolated broken uniserial harpoon is associated with a radiocarbon date of 8130 BP (Adamson et al. 1974). Other early dates come from the site of Tamaya Mellet in the southern Sahara (Delibras et al. 1974). Bone harpoons have also been excavated in Egypt at Catfish Cave, and have a minimum date of 7060 BP (Wendt 1966). The great majority of the Saharan fishing settlements with bone harpoons appear to date between 8000-6000 BP (Sutton 1977). Although no radiocarbon dates are available from the fishing or Mesolithic occupation at Khartoum, comparisons of the barbed bone harpoon collection with similar bone artifacts from Shabona and Tagra suggest the site may date to between 7000-8000 BP (Adamson et al. 1974). The age of the fishing settlement at Ishango remains uncertain, although general estimates of 8500-11,000 BP have been made (de Heinzelin 1957, 1962; Sutton 1977). Recent shell dates from Ishango have unfortunately proved inconclusive (de Maret et al. 1977).

At East Turkana radiocarbon dates of 8710±130 BP from GaJj 11 and 8394±270 BP from FxJj 12 were directly associated with barbed bone harpoons. A low density scatter of fish bone and stone artifacts, including microliths, was identified from a Melanoides coquina near the base of Trench B at FxJj 12. This horizon yielded dates of 9540±260 and 9260±235 BP. While the small excavation did not produce bone harpoons, the stone artifacts indicated that early Holocene human groups, possessing a typical Later Stone Age tool kit, were camped along the 75-80 meter transgressive shoreline over 9500 years ago. It is not unreasonable to suggest that by this early date these groups were manufacturing barbed bone harpoons especially since the horizon contained abundant fish bone. Radiocarbon dates are currently available from only two of the large fishing settlements, although additional dating samples have been submitted. The remaining six occupations on the 75-80 meter shoreline are thought to be associated with either the first or second Holocene high transgressions and to date between 10,000 and 5500 BP.

It is tempting to suggest, as other workers have previously done (de Heinzelin 1962; Robbins 1974; Sutton 1974, 1977; and Butzer 1980), that the use of barbed bone harpoons in the early Holocene fishing cultures of Africa originated in the East African lake basins and later diffused northward, along the Nile and into the Sahara. The early dates for harpoons at East Turkana suggest that the manufacture and use of barbed bone harpoons in Africa may have started over 9500 years ago.

The origin and development of pottery in Africa has not been clearly reconstructed. Early dates for pottery in North Africa come from Amekni in the Sahara (Camps 1969) and along the Nile in Egypt (Wendorf et al. 1976). These sites are dated to 8670 BP and 8200 BP respectively. Wavy-line pottery, comparable to that from Amekni, is widespread throughout the Sahara (Hays 1975) and in the Sudan (Arkell 1949).

In East Africa, 'stamped ware' pottery has been dated to 8420 BP at Lothagam, while single wavy-line and wavy-line decorated sherds are associated with dates of 7960 BP and 6200 BP (Robbins 1972, 1974). A single sherd of putative dotted wavy-line pottery has also been collected at Gamble's Cave and is estimated to date between 8500-8000 BP (L.S.B. Leakey 1931; Sutton 1977; Bower et al. 1978). At Lowasera, one weathered sherd with indistinct decoration, has been attributed to wavy-line pottery (Phillipson 1977a). The sherd was collected on the surface and may be associated with either the early or middle Holocene fishing settlements.

At East Turkana, wavy-line decorated pottery was collected from two of the large fishing settlements on the 75-80 m beach. Four other large sites along the same shoreline contained substantial amounts of undecorated pottery. Radiocarbon dates have not yet been received for these sites. Two fishing occupations that lacked pottery have been dated to between 8700-8400 BP.

The evidence from East Turkana and the Lower Omo Basin suggests then that prior to 8500 BP, early Holocene fishing settlements contained barbed bone harpoons but lacked pottery. At some point afterwards, decorated pottery, including vessels decorated with wavy lines, and undecorated pottery were manufactured in the Basin. It would appear, based on the tentative chronology of early African ceramics, that the production of pottery may well have originated in the Sahara during the early Holocene and spread into East Africa via the Nile and the expanded Rift Valley lakes. However, the early dates from the Sahara also indicate that pottery may have been independently invented in eastern Africa (L.S.B. Leakey 1931; Sutton 1974).

The Pastoral Neolithic, the third major cultural tradition identified, was present at East Turkana by 4500-4000 BP. This date represents, along with the radiocarbon determinations from North Horr (Phillipson 1977b; Stiles 1982) the earliest well dated evidence for food production and domestic stock in Eastern Africa.

Domestic animals were excavated from three sites. At Dongodien, Nderit Ware and burnished pottery were recovered in direct association with domestic caprines and domestic cattle. Importantly, the site was situated adjacent to the lake shoreline when the lake stood 45 m higher than present. While fishing continued to play an important subsistence role, the absence of barbed bone harpoons strongly suggests the early pastoralists had devised an alternative, as yet unidentified, procurement strategy. In marked contrast to the fishing settlements, early pastoralists also relied extensively on the use of obsidian as a raw material for stone implements.

At the Ileret Stone Bowl site (FwJj 5), stone bowls were identified in association with domestic cattle and domestic caprines. Unlike the shoreline encampments at GaJi 2 and Dongodien, this site was situated substantially east of the lake shoreline, along the margins of a major river valley. Fishing apparently continued to play an important role in the diet, but analogous with Dongodien, barbed bone harpoons were not recovered. The site contained abundant pottery sherds and numerous execution techniques and decorative motifs were identified, including the possible antecedents of Narosura Ware (Ambrose 1984). Ileret type, a new pottery taxon, was described. Its early distribution was apparently centered in northwestern Kenya and the Turkana Basin. The site has been dated to 4000±140 BP. However, single dates should be viewed with extreme caution, especially when run on bone apatite (Collett and Robertshaw 1983). The occupation will be redated but, until then, a conservative age of 3000-4000 BP is best observed for the age of the site.

The archaeological evidence for pastoralism at East Turkana provides both a geographical and chronological link between the early food producing traditions in the Sudan (Arkell 1952; Krzyzaniak 1978) and the later pastoralist settlements in the Central Rift of Kenya (Bower and Nelson 1978; Ambrose 1984). The appearance of early pastoralism in the Turkana Basin may have been the result of small scale population movements, set initially into motion by middle Holocene climatic deterioration in the Sahel and Savannah zones, which resulted in the introduction of new cultural traditions. Foremost among the new introductions was animal domestication. Subsequently, due to continued small scale population movements or to the diffusion of pastoralist techniques to pre-existing local non-pastoralist groups (Phillipson 1977b:83) domestic animals were introduced into central Kenya and eventually into northern Tanzania. Ehret (1974) has proposed that linguistic evidence supports an ancestral Southern Cushitic identity for these early pastoralists. His hypothesis suggests Southern Cushitic speaking populations entered northern Kenya from southern Ethiopia some 5000 years ago. While careful testing of this hypothesis is required, studies of human skeletal remains (Rightmire 1975) as well as archaeological evidence (Phillipson 1977b; Ambrose 1982) lend support to a middle Holocene introduction of early Southern Cushitic speaking pastoralists into the Turkana Basin.

Recently, reports of early pastoralism from central Kenya that date to 15,000-5,000 BP have been made (Bower and Nelson 1978; Nelson 1979). The reports are primarily based on inferred sedimentary rates, unclear association of pastoralism with particular pottery wares, preliminary identifications of faunal elements and the widespread use of bone apatite dates. Such claims have been criticized (Owen et al. 1982; Collett and Robertshaw 1983) and require considerably more substantive data before they can be adequately evaluated.

A fourth cultural tradition identified at East Turkana was represented by numerous recent to sub-recent campsites, several of which contained pottery comparable to historic Sambury pottery. Other sites were represented by stone fortifications and burial cairns that may have been constructed by Eastern Cushitic speaking peoples (Soper and Jacobs 1972). Little systematic study of the more recent time periods in the research area was conducted. The material is of sufficient interest however, that a separate research project should be undertaken.

LAND USE

The reconstruction of land use patterns for the last 10,000-15,000 years at East Turkana is still in the preliminary stages. Nevertheless, it is possible to interpret shifts in land use as both adaptive responses to changing environments by pre-existing late Pleistocene populations and as the result of probable migration of populations with new economic and cultural traditions into the Basin. By 9500 years ago, and probably earlier, human groups had settled along the shorelines of a greatly expanded Lake Turkana. Fishing had become an important element of the diet although the hunting of wild land mammals continued. Barbed bone harpoons were manufactured to spear fish and possible aquatic mammals. Pottery, utilizing locally available silty clays and beach sands, appeared by at least 7000 BP. Site survey results indicate that the settlements of this time period were invariably located either very close to the lake's shoreline or on offshore sandbars. No evidence of settlements upstream from the lake shoreline, either within deltaic or riverine environments, was identified. It would appear, on the basis of early radiocarbon dates, that early Holocene hunter- gatherer groups responded to the steadily expanding lake by locating their campsites along the lake's shoreline and increasing their reliance on fishing. Thereafter, even though the lake experienced height fluctuations of varying magnitude, settlements were consistently placed adjacent to the water's edge. Smaller scale encampments, perhaps inland stations to hunt land mammals, if originally present, have yet to be identified.

The second significant shift in land use took place between 4500-4000 BP and was marked by the introduction of domestic animals. Occupation areas were now located on a variety of geographic features including beach shorelines, river drainages and possibly on the crest of inland escarpments. Fishing continued to play an important role in the diet of both the beach shoreline and riverine situated sites. The hunting of land mammals however, had decreased significantly. In addition to the well defined settlement locations, site survey revealed a low density scatter of pastoralist artifacts across most sectors of the research area. Such evidence suggests increased mobility and may be related, as among the nearby Dassanetch today (Almagor 1978:47), to seasonal movements in search of fresh pasture for domestic stock. The chronology and land use patterns of these food producing groups after circa 3500 BP is poorly understood and should be actively pursued in future fieldwork.

The geo-archaeological research presented in this monograph represents the initial effort towards reconstructing major questions of stratigraphy, prehistoric chronology and land use for the late Pleistocene-Holocene cultural sequence at East Turkana. Future fieldwork will need to become increasingly problem oriented. As well, a number of existing projects will require further refinement including continued field mapping and increased chronometric control of the Holocene high lake cycles; additional radiocarbon dates from fishing sites; larger faunal samples from each Pastoral Neolithic settlement and the exploration of new geographic areas especially within the eastern volcanic highlands. Evidence from these and other, as yet unformulated, studies will enhance our understanding of the interplay between climatic change and human adaptation at East Turkana.

ACKNOWLEDGEMENTS

Ten years ago I climbed into a well provisioned Landrover and began an investigation into the late Pleistocene-Holocene deposits along the northeastern margin of Lake Tukana. With the publication of this monograph, that journey is now completed. I want to acknowledge the many individuals who professionally assisted and emotionally sustained my efforts along the way. Indeed, without their constant support and encouragement, I realize this monograph would never has been finished.

Professor Glynn Isaac first introduced me to the study of prehistoric human behavior in Eastern Africa. He shepherded my professional development both in the field and around the seminar table. While he often commented on the direction and progress of the research project, he gave me the invaluable freedom to realize my own accomplishments and errors. No true teacher could do otherwise. For these opportunities and personal guidance I am most grateful. Professors Desmond Clark and David Phillipson also provided constant support and critical evaluation.

A number of colleagues on the Koobi Fora Research Project have generously given of their time and expertise. Richard Leakey, Director of the National Museums of Kenya, offered continuous use of the camp facilities at Koobi Fora and provided much need administrative counsel. Ian Findlater and Kay Behrensmeyer assisted in the completion of the Area 102 geological transect. They patiently answered my questions on stratigraphic nomenclature and facies relationships. Ron Watkins kindly provided raw material identifications of several stone bowl fragments. John Harris generously worked on the preliminary identifications of faunal remains from several fishing settlements and Peter Williamson introduced me to the bewildering field of molluscan taxonomy.

During field excavations I was assisted by a conscientious and hard working crew of skilled archaeological excavators from the KFRP. They included Muteti Nume, Mukilya Mangoka, Musau Mbova, Kimolo Mulwa, Bernard Kanunga, N'ungalu and Mutuse Muoki. To Muteti Nume, I am especially indebted. In times of mental exhaustion and bureaucratic frustration, his hospitality and friendship were most appreciated. To all the Kenyan staff who assisted in this research project, I gratefully say asante sana.

While in Nairobi, Andrew Hill offered his spare bedroom in a spirit of resigned good humor. I made repeated use of it and am most grateful. Charles Nelson generously let me use the living and research facilities at Lukenya Hill. I cannot think of a more lovely setting in which to measure stone artifacts.

I must give special thanks to John Ochieng of the British Institute in East Africa, Jeremy Ketchem of the National Museum and Corrine Glesne for their excellent illustrations of the pottery, stone and bone artifacts. John K. Mengitch of the National Museum broke away from his busy schedule to identify portions of the faunal remains from GaJi 2 and the Ileret Stone Bowl site. Simiyu Wandibba attempted, with intermittent success, to teach me the confusing pottery wares of Kenya. I gratefully acknowledge the assistance of these individuals.

In Berkeley, I was greatly supported by the friendship and warm optimism of Christine Schindler. To my student colleagues at Berkeley I extend my deepest gratitude for intellectual as well as personal encouragement. Ruth Dual, Computer Consultant for the Department of Anthropology at U.C. Berkeley, and Jean Sept worked long hours helping to transcribe and process mounds of field data. The excellent quality of the site plans and section diagrams is the work of Judith Ogden. Fiona Marshall undertook a study of the mammal remains from Dongodien. I am grateful for her meticulous efforts.

While at St. Lawrence University, a number of individuals aided in the completion of this monograph. Dr. George Gibson, Vice President and Dean of the University, released me from teaching duties during the 1983 Interterm session and has been a constant source of encouragement. Gail Colvin typed the final manuscript, including the seemingly endless figure captions and tables, on her word processor. I am indebted to her disciplined sense of consistency, good cheer and excellent work.

Fieldwork and living expenses in Kenya were provided by the National Science Foundation through a grant awarded to Professor Glynn Isaac. Other financial backing was received from a National Science Foundation Doctoral Dissertation Improvement Grant that helped provide crucial funds for radiocarbon dates. Other funds for radiocarbon dates were provided by the Archaeological Research Facility at Berkeley, the British Institute in Eastern Africa, through Dr. Richard Gillespie of the Sydney University Radiocarbon Laboratory, and Drs. Charles Nelson and Ari Siiriäinen. I am grateful for their financial support.

Several close friends, Todd Olson and Michael Martin, helped sustain and strengthen my sometimes wavering resolve to finish this monograph. My wife, Susan Cypert, remained optimistic during periods of mutual ill health. To each, I owe a tremendous debt.

Finally, I acknowledge the crucial influence of my father, Peter Andrew Barthelme, and my mother, Muriel Joyce Barthelme. Their spirit of free inquiry, tolerance, love and support encouraged me at an early age to choose my own direction in the world. To them, I am humbly grateful and dedicate this monograph.

LIST OF FIGURES

FIGURE

Site Distribution

LIST OF TABLES

Artifact Analysis

LIST OF PLATES

BIBLIOGRAPHY

Adamson, D., J.D. Clark, and M.A.J. Williams (1974) 'Barbed bone points from Central Sudan and the age of the "Early Khartoum" tradition', Nature, 249: 120-123.

Almagor, V. (1978) Pastoral Partners: Affinity and Bond Partnership among the Dassanetch of South-West Ethiopia. Manchester: University Press.

Ambrose, S.H. (1982) 'Archaeology and linguistic reconstructions of history in East Africa', pp. 104-157 in C. Ehret and M. Posnansky (eds.), The Archaeological and Linguistic Reconstruction of African History. Berkeley: University of California Press.

Ambrose, S.H. (1984) 'The introduction of pastoral adaptations, to the highlands of East Africa, pp. 212-239 in J.D. Clark and S. Brandt (eds.), From Hunters to Farmers: Considerations of the Causes and Consequences of Food Production in Africa. Berkeley: University of California Press.

Arambourg, C., P. Lester and J. Roger (1943) 'Mission Scientifique de l'Omo 1932-1933: géologie et archéologie', Bulletin du Musée d'Histoire Naturelle, 1: 61-230.

Arkell, A.J. (1949) Early Khartoum. London: Oxford University Press.

Arkell, A.J. (1953) Shaheinab. London: Oxford University Press.

Arkell, A.J. (1962) 'The distribution in central Africa of one early Neolithic ware (dotted wavy-line pottery) and its possible connection with the beginning of pottery', Pan-African Congress, 4: 283-287.

Baker, B.H. (1958) 'Geology of the Magadi area', Geological Survey of Kenya, report 42. Nairobi: Government Printer.

Baker, B.H., P.A. Mohr and L.A. Williams (1972) 'Geology of the Eastern Rift System', Geological Society of America Special Paper, 136.

Bailloud, G. (1969) 'L'Evolution des styles céramiques en Ennedi (République du Tchad)', Actes le Collogue Int. d'Archéologie africaine, 31-45.

Barthelme, J.W. (1977) 'Holocene sites north-east of Lake Turkana: a preliminary report', Azania, 12: 33-41.

Barthelme, J.W. (1981) Late Pleistocene-Holocene Prehistory to the North East of Lake Turkana, Kenya. Ph.D. Dissertation, University, of California, Berkeley. Ann Arbor: University Microfilms.

Barthelme, J.W. (1984) 'Early evidence for animal domestication in eastern Africa', in J.D. Clark and S. Brandt (eds.), From Hunters to Farmers: Considerations of the Causes and Consequences of Food Production in Africa. Berkeley: University of California Press.

Behrensmeyer, A.K. (1970) "preliminary geological interpretation of a new hominid site in the Lake Rudolf Basin', Nature, 226: 225-228.

Binford, L.R. and S.R. Binford (1966) 'A preliminary analysis of functional variability in the Mousterian of Levallois facies', American Anthropologist, 68, 2: 238-295.

Binford, L.R. (1979) Nunamiut Ethnoarchaeology. New York: Academic Press.

Bishop, W.W. (1976) in R.E.G. Williams and A.S. Johnson, 'Birmingham University radiocarbon dates' Radiocarbon, 18: 248-267.

Bishop, W.W., A Hill and M. Pickford (1978) 'Chesowanga: a revised geological interpretation', pp. 303-328 in W.W. Bishop (ed.), Geological Background to Fossil Man. Edinburgh: Scottish Academic Press.

Bowen, B.E. (1974) The geology of Upper Cenozoic sediments in the East Rudolf embayment of the Lake Rudolf Basin, Kenya. Unpublished Ph.D. Dissertation, Iowa State University.

Bowen, B.E. and C.F. Vondra (1973) 'Stratigraphical relationships of the Plio-Pleistocene deposits, East Rudolf, Kenya', Nature, 242: 391-393.

Bower, J.R.F. (1973) 'Seronera: excavations at a stone bowl site in the Serengeti National Park', Azania, 8: 71-104.

Bower, J.R.F. (1977) 'Preliminary report on a study of prehistoric cultures of the Serengeti National Park', Nyame Akuma, 11: 20-27.

Bower, J.R.F., C.M. Nelson, A.F. Waibel and S. Wandibba (1977) 'The University of Massachusett's Late Stone Age/Pastoral "Neolithic" comparative study in Central Kenya: an overview', Azania, 12: 119-146.

Bower, J.R.F. and C.M. Nelson (1978) 'Early pottery and pastoral cultures of the Central Rift Valley, Kenya', Man, 13: 554-566.

Brown, F.H. (1975) 'Barbed bone points from the Lower Omo Valley, Ethiopia', Azania, 10: 144-149.

Brown, J. (1966) 'The excavations of a group of burial mounds in Ilkek near Gilgil, Kenya', Azania, 1: 59-77.

Butzer, K.W. (1971) Recent history of an Ethiopian delta. University of Chicago, Department of Geography Research Paper, 136: 1-184.

Butzer, K.W. (1976) 'The Mursi, Nkalabong, and Kibish formations, Lower Omo Basin, Ethiopia', pp. 12-23 in Y. Coppens, et al. (eds.), Earliest Man and Environments in the Lake Rudolf Basin. Chicago: University of Chicago Press.

Butzer, K.W. (1980) 'The Holocene lake plain of north Rudolf, East Africa', Physical Geography, 1: 42-58.

Butzer, K.W., F.H. Brown, and D.L. Thurber (1969) 'Horizontal sediments of the lower Omo valley: the Kibish Formation', Quarternaria, 11: 15-30.

Butzer, K.W., G. Ll. Isaac, J.L. Richardson and C. Washbourn-Kamau (1972) 'Radiocarbon dating of the East African lake levels', Science, 175: 1069-1076.

Camps-Fabrer, H. (1966) Matière et art mobilier dans la préhistoire nord-Africaine et Saharienne. Paris: Mémoirs du C.R.A.P.E., No. V.

Camps, G. (1974) Les Civilisations Préhistoriques de l'Afrique du Nord et du Sahara. Paris.

Caton-Thompson, G. and E.W. Gardner (1934) The Desert Fayum. London: Royal Anthropological Institute of Great Britain and Ireland.

Chaplin, R.E. (1971) The Study of Animal Bones from Archaeological Sites. New York: Seminar Press.

Chapman, S. (1967) 'Kantsyore Island', Azania, 2: 165-191.

Clark, J.D. (1971a) 'A re-examination of the evidence for agricultural origins in the Nile Valley', Proceedings of the Prehistoric Society, 38: 34-79.

Clark, J.D. (1971b) 'An archaeological survey of the northern Aïr and Tenère. The British Expedition to the Aïr Mountains', The Geographical Journal, 37, 4: 455-457.

Clark, J.D. (1973) Preliminary report of an archaeological and geomorphological survey in the Central Sudan, January to March, 1973. Unpublished manuscript.

Clark, J.D. (1976) 'The domestication process in sub-Saharan Africa with special reference to Ethiopia', pp. 56-115 in E. Higgs (ed.), Colloque XX: Originé de l'élevage et de la domestication. Nice: Ninth Congress of Prehistory and Protohistory.

Clark, J.D. and S.A. Brandt (eds.) (1984) From Hunters to Farmers: Considerations of the Causes and Consequences of Food Production in Africa. Berkeley: University of California Press.

Clark, J.G.D. and M.W. Thompson (1953) 'The groove and splinter technique of working antler in Upper Paleolithic and Mesolithic Europe', Proceedings of the Prehistoric Society, 19: 148-160.

Cole, S. (1963) The Prehistory of East Africa. New York: Macmillan.

Collett, D. and P. Robertshaw (1983) 'Problems in the interpretation of radiocarbon dates: the Pastoral Neolithic of East Africa', The African Archaeological Review, 1: 57-74.

Connah, G. (1976) 'The Daima sequence and the prehistoric chronology of the Lake Chad region of Nigeria', Journal of African History, 27: 321-352.

Cooke, H.B.S. (1958) 'Observations relating to Quaternary environments in East and Southern Africa', Trans. Geol. Soc. S. Africa, annexure to Vol. LX: 1-73.

Damon, P.E., C.W. Ferguson, A. Long and E.I. Wallick (1974) 'Dendrochronologic calibration of the radiocarbon time scale', American Antiquity, 39: 350-366.

Delibrias, G., M.T. Gullier and J. Labeyrie (1974) 'Gif natural radiocarbon measurements VIII, Radiocarbon, 16: 15-94.

Dodson, R.G. (1963) 'Geology of the South Horr area' Geological Survey of Kenya, report 60. Nairobi: Government Printer.

Ehret, C. (1971) Southern Nilotic History: Linguistic Approaches to the Study of the Past. Evanston: Northwestern University Press.

Ehret, C. (1974) Ethiopians and East Africans: The Problem of Contacts. Nairobi: East African Printing House.

Ehret, C. (1976) 'Linguistic evidence and its correlation with archaeology', World Archaeology, 8: 5-18.

Ehret, C. and M. Posnansky (1982) The Archaeological and Linguistic Reconstruction of African History. Berkeley: University of California Press.

Epstein, H. (1969) The Origin of the Domestic Animals of Africa. New York: Holmes and Meier.

Faure, H. (1966) 'Evolution des grands lacs sahariens à l'Holocene', Quaternaria, 8: 167-175.

Findlater, I.C. (1978) 'Isochronous surfaces within the Plio-Pleistocene sediments east of Lake Turkana, Kenya', pp. 415-420 in W.W. Bishop (ed.), Geological Background to Fossil Man. Edinburg: Scottish Academic Press.

Fitch, F.J. and C.F. Vondra (1976) 'Tectonic and Stratigraphic Framework', pp. 8-11 in Y. Coppens, et al. (eds.), Earliest Man and Environments in the Lake Rudolf Basin. Chicago: University of Chicago Press.

Flint, R.F. (1959) 'On the basis of Pleistocene correlation in East Africa', Geol. Mag., 96: 265-284.

Fuchs, V.E. (1936) 'The geological history of the Lake Rudolf Basin, Kenya Colony', Phil. Trans. Roy. Soc. of London, 229: 219-274.

Gasse, F. and P. Rognon (1973) 'Le Quaternaire des bassins lacustres de l'Afar', Rev. Geogr. Phys. Geol. Dyn., (2) 15, 4: 405-414.

Gifford, D.P. (1978) 'Ethnoarchaeological observation of material processes affecting cultural materials', pp. 77-101 in R.A. Gould (ed.), Explorations in Ethnoarchaeology. Albuquerque: University of New Mexico Press.

Gifford, D.P., G. Ll. Isaac and C.M. Nelson (1980) 'Evidence for predation and pastoralism at Prolonged Drift: a Pastoral Neolithic site in Kenya, Azania, 15: 57-108.

Gramly, R.M. (1976) 'Upper Pleistocene archaeological occurrences at site GvJm 22, Lukenya Hill, Kenya', Man, 11: 468-474.

Gregory, J.W. (1921) The Rift Valleys and Geology of East Africa. London: Seeley Service.

Grove, A.T., F. Alayne Street and A.S. Goudie (1975) 'Former lake levels and climatic change in the rift valley of southern Ethiopia', Geographical Journal, 141: 177-194.

Hansen, C.L. and C.M. Keller (1971) 'Environment and activity patterning at Isimila karongo, Iringa District, Tanzania: a preliminary report', American Anthropologist, 73: 1201-1211.

Harlan, J.R. (1971) 'Agricultural origins: centers and noncenters', Science, 174: 468-474.

Harlan, J.R., J.M.J. deWet and A.B.L. Stemler (eds.) (1976) Origins of African Plant Domestication. The Hague: Mouton.

Harris, J.W.K. (1978) The Karari Industry: Its Place in East African Prehistory. Ph.D. Dissertation, University of California, Berkeley. Ann Arbor: University Microfilms.

Hays, T.R. (1975) 'An examination of the Sudanese Neolithic', in A. Berhanou, J. Chavaillon and J.E.G. Sutton (eds.), Proceedings of the Seventh Pan-African Congress on Prehistory. Addis Ababa.

Heinzelin, Jean de (1957) Les Fouilles d'Ishango. Brussels: Institut des Parcs, Nationaux du Congo Belge.

Heinzelin, Jean de (1962) 'Ishango', Scientific American, 206, 6: 105-116.

Hill, A. (1980) 'Early postmortem damage to the remains of some contemporary East African mammals', pp. 131-155 in A.K. Behrensmeyer and A. Hill (eds.), Fossils in the Making. Chicago: University of Chicago Press.

Höhnel, L. von (1894) Discovery of Lakes Rudolf and Stefanie. London: Longmans, Green. (Frank Cass Press Reprint, 1968).

Huard, P. and J.M. Massip (1964) 'Harpons en os et ceramique à décor en vague (wavy-line) au Sahara Tchadien', BSPF, LXI: 105-123.

Isaac, G. Ll. (1972a) 'Comparative studies of Pleistocene site locations in East Africa', in P.J. Ucko, R. Tringham and G.W. Dimbleby (eds.), Man, Settlement and Urbanism. London: Duckworth.

Isaac, G. Ll. (1972b) 'Early phases of human behavior: models in Lower Paleolithic archaeology', in D.L. Clarke (ed.), Models in Archaeology. London: Methuen and Co.

Isaac, G. Ll. (1981) 'Stone age visiting cards: approaches to the study of early land use patterns', pp. 131-155 in I. Hoddler, G. Isaac and N. Hammond (eds.), Pattern of the Past: Studies in Honour of David Clarke. Cambridge: Cambridge University Press.

Isaac, G. Ll., H.V. Merrick and C.M. Nelson (1972) 'Stratigraphic and archaeological studies in the Lake Nakuru Basin, Kenya', Paleoecology of Africa, 6: 225-232.

Isaac, G. Ll. and J.W.K. Harris (1978) 'Archaeology. Koobi Fora Research Project, Vol. 1', in M.G. Leakey and R.E.F. Leakey (eds.), The Fossil Hominids and an Introduction to their Context, 1968-1974. Oxford: Clarendon Press.

Jacobs, A.H. and R.C. Soper (1972) 'A report on surface pottery from Eliye Springs, Turkana District, northern Kenya'. Unpublished manuscript, Institute of African Studies, University of Nairobi.

Jochim, M.A. (1976) Hunter-Gatherer Subsistence and Settlement: A Predictive Model. New York Academic Press.

Kendall, R.L. (1969) 'An ecological history of the Lake Victoria basin', Ecological Monographs, 39: 121-176.

King, B.C. (1978) 'Structural and volcanic evolution of the Gregory Rift Valley', pp. 29-54 in W.W. Bishop (ed.), Background to Fossil Man. Edinburgh: Scottish Academic Press.

Klein, R.G. (1980) 'The interpretation of mammalian faunas from stone-age archaeological sites, with special reference to sites in the southern Cape Province, South Africa', pp. 223-246 in A.K.

Behrensmeyer and A. Hill (eds)., Fossils in the Making. Chicago: University of Chicago Press.

Krzyaniak, L. (1975) 'Kadero (first season, 1972)', Etudes et Travaux (Warsaw), VIII: 361-366.

Krzyaniak, L. (1978) 'New light on early food production in the Central Sudan', Journal of African History, 19, 2: 159-172.

Krzyzaniak, L. (1982) 'Radiocarbon measurements for the Neolithic settlements at Kadero', Nyame Akuma, 21: 38.

Leakey, L.S.B. (1931) The Stone Age Cultures of Kenya Colony. Cambridge: Cambridge University Press.

Leakey, L.S.B. (1935) The Stone Ages Races of Kenya. London: Oxford University Press.

Leakey, M.D. (1945) 'Report on the excavations at Hyrax Hill, Nakuru, Kenya Colony', Trans. Roy. Soc. S. Af., 30: 271-409.

Leakey, M.D. (1971) Olduvai Gorge, Vol. 3: Excavations in Beds I and II, 1960-1963. Cambridge: Cambridge University Press.

Leakey, M.D. and L.S.B. Leakey (1950) Excavations at the Njoro River Cave. London: Oxford University Press.

Leakey, M.G. and R.E. Leakey (1978) Koobi Fora Research Project Vol. 1. The Fossil Hominids and an Introduction to their Context, 1968-1974. London: Oxford University Press.

Livingstone, D.A. (1975) 'Late Quaternary climatic change in Africa', Annual Review of Ecology and Systematics, 6: 249-280.

Maley, J. (1977) 'Paleoclimates of the Central Sahara during the early Holocene', Nature, 169: 573-577.

Mandahl-Barth, G. (1954) 'The fresh water mollusks of Uganda and adjacent territories', Annales Du Musée Royal Du Congo Belge, Tervuren, 32.

de Maret, P., F. van Noten and D. Cahen (1977) 'Radiocarbon dates from West Central Africa: a synthesis', Journal of African History, 28: 481-505.

Marshall, F.B. (1977) The analysis of faunal material from site GaJi 4; N.E. Lake Turkana; with a discussion on the methodology of faunal analysis. Unpublished B.A. Honors Thesis, Reading University, Bristol.

Marshall, F., K. Stewart, D. Gifford-Gonzalez and J. Barthelme (1985) 'Evidence for early domestic stock at Dongodien (GaJi 4), in northern Kenya', in press.

Mbua, E. (1984) 'Human remains from the most recent deposits of Lake Turkana, northern Kenya', Kenya Past and Present, 16: 38-41.

Merrick, H.V. (1973) 'Aspects of size and shape variation of the East African stone bowls', Azania, 8: 115-130.

Merrick, H. V. and F. H. Brown (1984) 'Obsidian sources and patterns of source utilization in Kenya and northern Tanzania: some initial findings', The African Archaeological Review, 2: 129-152.

Monod, T. (ed.) (1975) Pastoralism in Africa. London: Oxford University Press.

Monod, T. and R. Mauny (1957) 'Découverte de nouveau instruments en os dans l'Ouest Africain', in J.D. Clark and S. Cole (eds.), Third Pan-African Congress on Prehistory. Livingstone.

Mori, F. (1964) 'Some aspects of the rock art of the Acacus (Fezzan Sahara) and dating regarding it', pp. 247-259 in L. Pericot Garcia and E. Rippol Perello (eds.), Prehistoric Art of the Western Mediterranean and the Sahara. New York.

Mueller, J.W. (ed.) (1975) Sampling in Archaeology. Tucson: University of Arizona Press.

Muriuki, G. (1974) A History of the Kikuyu 1500-1900. Nairobi: Oxford University Press.

Nelson, C.M. (1971) 'A standard site enumeration system for the continent of Africa', Bull. Comm. Nomen. Term. Panafr. Congr. Prehist. Quat. Stud., 4: 6-12.

Nelson, C.M. (1973) A Comparative Analysis of Thirteen Later Stone Age Sites in East Africa. Unpublished Ph.D. dissertation, University of California, Berkeley.

Nelson, C.M. (1979) 'Continuing work at Lukenya Hill, Kenya', Nyame Akuma, 14: 28-30.

Nilsson, E. (1931) 'Quaternary glaciations and pluvial lakes in British East Africa', Geografiska Annaler, 13: 249-349.

Nilsson, E. (1940) 'Ancient changes of climate in British East Africa and Abyssinia', Geografiska Annaler, 22: 1-79.

Oakley, K.P. (1961) 'A bone harpoon from Gamble's Cave, Kenya', Antiquaries Journal, 41: 86-87.

Odner, K. (1972) 'Excavations at Narosura, a stone bowl site in the southern Kenyan Highlands', Azania, 7: 25-92.

Onyango-Abuje, J.C. (1977) 'Crescent Island: an East African Neolithic site; a preliminary report', Azania, 12: 147-160.

Onyango-Abuje, J.C. (1980) 'Temporal and spatial distribution of Neolithic cultures in East Africa', pp. 288-292 in R.E. Leakey and B.A. Ogot (eds.), VIII Panafrican Congress. Nairobi: TILLMIAP, Kenya.

Owen, R.B. and R.W. Renaut (1980) 'The early Holocene paleogeography of the northern Kenya Rift Valley', Recherche Géologie Africaine, 5: 143-146.

Owen, R.B., J.W. Barthelme, R.W. Renaut and A. Vincens (1982) 'Paleolimnology and archaeology of Holocene deposits north-east of Lake Turkana, Kenya', Nature, 298: 523-529.

Petit-Maire, N., J.C. Celles, D. Commelin, G. Delibrias and M. Raimbault (1983) 'The Sahara in northern Mali: man and his environment between 10,000 and 3,500 years bp.' The African Archaeological Review, 1: 105-126.

Pilsbry, H. and J. Bequaert (1927) 'The aquatic mollusks of the Belgian Congo, with a geographical and ecological account of Congo malecology', Bull. Amer. Mus. Nat. His., 53, 2: 69-602.

Raynolds, R.G.H. (1973) 'Summary of Mapping-Area 103, Galana Boi beds'. Unpublished manuscript, Koobi Fora Research Project.

Renaut, R.W. and R.B. Owen (1980) 'Late Quaternary fluvio-lacustrine sedimentation and lake levels in the Baringo basin, northern Kenya Rift Valley, Recherche Géologie Africaine, 5: 130-133.

Richardson, J.L. and A.E. Richardson (1972) 'History of an African Rift lake and its climatic implications', Ecological Monographs, 42: 499-534.

Rightmire, G.P. (1975) 'New studies of post- Pleistocene human skeletal remains from the Rift Valley, Kenya', American Journal of Physical Anthropology, 42: 351-370.

Robbins, L.H. (1972) 'Archaeology in Turkana District, Kenya', Science, 176- 359-366.

Robbins, L.H. (1974) The Lothagam Site: A Late Stone Age fishing settlement in the Lake Rudolf Basin, Kenya. Michigan State University Museum Anthropology Series, 1 (2).

Robbins, L.H. (1980) Lopoy: A Late Stone Age fishing and pastoralist settlement west of Lake Turkana, Kenya. Michigan State University Museum Anthropology Series, 2 (1).

Savage, R.J.G. and P. Williamson (1978) 'The early history of the Turkana depression', pp. 375-394 in W.W. Bishop (ed.), Geological Background to Fossil Man. Edinburgh: Scottish Academic Press.

Servant, M., S. Servant and G. Delibrias (1969) 'Geologie du Quaternaire. Chronologie du Quaternaire récent des basses regions du Tchad', C. R. Acad. Sci., 269: 1603-1606.

Shepard, Anna O. (1956) Ceramics for the archaeologist. Washington, D.C.: Carnegie Institution of Washington. Publication 609.

Siiriainen, A. (1977) 'Later Stone Age investigation in the Laikipia Highlands, Kenya: a preliminary report', Azania, 12: 161-186.

Smith, A.B. (1974) 'Preliminary report of excavations at Karkarichinkat nord and Karkarichinkat sud, Tilemsi valley, Republic of Mali, spring 1972', West African Journal of Archaeology, 4: 33-55.

Soper, R.C. and A.H. Jacobs (1972) 'Report on stone structures at Koobi Fora, East Rudolf'. Unpublished manuscript, Koobi Fora Research Project.

Stewart, K. (n.d.) 'A brief report on the faunal fish remains of FxJj 12 and GaJi 4.' Unpublished manuscript.

Stiles, D. (1982) 'Preliminary results of archaeological and paleoenvironmental research in northern Kenya', Nyame Akuma, 20: 19-25.

Street, F.A. and A.T. Grove (1979) 'Global maps of lake level fluctuations since 30,000 yr. B.P. Quaternary Research, 12: 83-118.

Struever, S. (1968) 'Woodland subsistence-settlement systems in the lower Illinois Valley', in S. Binford and L. Binford (eds.), New Perspectives in Archaeology. Chicago: Aldine Publishing Co.

Sutton, J.E.G. (1973) The Archaeology of the Western Highlands of Kenya. Nairobi: Memoirs of the British Institute in East Africa, No. 3.

Sutton, J.E.G. (1974) 'The aqualithic civilization of middle Africa', Journal of African History, 15: 527-546.

Sutton, J.E.G. (1977) 'The African aqualithic', Antiquity, 51: 25-34.

Trigger, B.G. (1968) Beyond History: The Methods of Prehistory. New York: Holt, Rinehart and Winston.

Truckle, P.H. (1976) 'Geology and Cainozoic lake sediments of the Sugata Trough, Kenya', Nature, 263: 380-383.

Van Damme, D. and A. Gautier (1972) 'Molluscan assemblages from the late Cenozoic of the Lower Omo Basin, Ethiopia', Quaternary Research, 2: 25-37.

Vavilov, N.I. (1926) 'Studies on the origin of cultivated plants', Bull. Appl. Bot. and Plant Breeding, 16 (2): 1-243. Leningrad.

Vita-Finzi, C. and E.S. Higgs (1970) 'Prehistoric economy in the Mount Carmel area of Palestine: site catchment analysis', Proc. Prehis. Soc., 36: 1-37.

Vondra, C.F., G.D. Johnson, B.E. Bowen and A.K. Behrensmeyer (1971) 'Preliminary stratigraphical studies of the East Rudolf Basin, Kenya', Nature, 231: 245-248.

Vondra, C. and B. Bowen (1978) 'Stratigraphy, sedimentary facies and paleoenvironments, East Lake Turkana, Kenya', pp. 395-414 in W.W. Bishop (ed.), Geological Background to Fossil Man. Edinburgh: Scottish Academic Press.

Walsh, J. and R. Dodson (1969) Geology of Northern Turkana. Geological Survey of Kenya, report 82. Nairobi: Government Printer.

Washbourn-Kamau, C.K. (1971) 'Late Quaternary lakes in the Nakuru-Elmenteita Basin, Kenya', The Geographical Journal, 137: 522-535.

Washbourn-Kamau, C.K. (1975) 'Late Quaternary shorelines of Lake Naivasha, Kenya', Azania, 10: 77-92.

Watkins, R.T. (1981) The Geochemistry of Rhyolitic and Tholeiitic Rocks from East of Lake Turkana, Northern Kenya. M. Sc. thesis, University of Leeds.

Wayland, E.J. (1923) 'Paleolithic types of implements in relation to the Pleistocene deposits of Uganda', Proc. Prehist. Soc. E. Anglia, 4: 96-112.

Wayland, E.J. (1929) 'African pluvial periods', Nature 123: 607.

Wendorf, F., R. Said and R. Schild (1970) 'Egyptian prehistory: some new concepts', Science, 169: 1161-1171.

Wendorf, F. and R. Schild (1976) Prehistory of the Nile Valley. New York: Academic Press.

Wendorf, F., R. Schild, N. el Hadidi, A.E. Close, M. Kobusiewicz, H. Wieckowska, B. Issawi, and M. Haas (1979) 'Use of barley in the Egyptian Late Paleolithic', Science, 205: 1341-1347.

Wendt, W.E. (1966) 'Two prehistoric archaeological sites in Egyptian Nubia', Postilla, 102: 1-46.

Whitworth, T. (1965a) 'The Pleistocene Beds of Kabua, Northern Kenya', Durham University Journal, 5712: 88-100.

Whitworth, T. (1965b) 'Artifacts from Turkana, northern Kenya', South African Archaeological Bulletin, 20: 75-78.

Williams, M.A.J. and D.A. Adamson (1974) 'Late Pleistocene desiccation along the White Nile', Nature, 248: 584-585.

Young, J.A.T. and R.W. Renaut (1979) 'A radiocarbon date from Lake Bogoria, Kenya Rift Valley', Nature, 278: 243-245.

Yuretich, R.F. (1979) 'Modern sediments and sedimentary processes in Lake Rudolf (Lake Turkana) eastern Rift Valley, Kenya', Sedimentology, 26: 313-331.

www.ingramcontent.com/pod-product-compliance
Ingram Content Group UK Ltd.
Pitfield, Milton Keynes, MK11 3LW, UK
UKHW061844041026
15591UKWH00015B/905

9 780860 543251